Tropical Forages: Their role in sustainable agriculture

Tropical Forages: Their role in sustainable agriculture

L.R. Humphreys

Emeritus Professor of Agriculture,
The University of Queensland

Longman Scientific & Technical
Longman Group Limited
Longman House, Burnt Mill, Harlow
Essex CM20 2JE, England
and Associated Companies throughout the world

*Copublished in the United States with
John Wiley & Sons, Inc., 605 Third Avenue, New York
NY 10158*

© Longman Group Limited 1994

All rights reserved; no part of this publication may be
reproduced, stored in any retrieval system, or transmitted in
any form or by any means, electronic, mechanical,
photocopying, recording, or otherwise without either the prior
written permission of the Publishers or a licence permitting
restricted copying in the United Kingdom issued by the
Copyright Licensing Agency Ltd., 90 Tottenham Court Road,
London W1P 9HE.

First published 1994

British Library Cataloguing in Publication Data
A catalogue entry for this title is available from the British Library.

ISBN 0-582-07868-7

Library of Congress Cataloging-in-Publication data
Humphreys, L. R.
 Tropical forages : their role in sustainable agriculture / L. R. Humphreys.
  p. cm.
 Includes bibliographical references and index.
 ISBN 0-470-23433-4
 1. Forage plants—Tropics. 2. Pastures—Tropics. 3. Tropical crops.
4. Cropping systems—Tropics. 5. Sustainable agriculture—Tropics. I. Title.
SB193.3.T76H85 1995
633.2'0913—dc20   94-25797
                         CIP

Printed and bound in Great Britain by Bookcraft (Bath) Ltd.

To ILH

CONTENTS

Preface xi
Acknowledgements xiii
List of abbreviations xv

1. Introduction 1
   1.1 The new technology of pasture improvement in the tropics and subtropics 1
   1.2 Processes of land degradation in cropping systems 6
       Trends in crop yields 7
       Soil erosion 8
   1.3 Deterioration of pastures 16
       Decreasing productivity as pastures age 17
       Deterioration of botanical composition 19
   1.4 Concepts of sustainability 20

Part 1: Objectives of Crop–Pasture–Animal Interactions 23

2. The maintenance of soil fertility I. Nitrogen and organic matter 23
   2.1 Introduction 23
       The function of soil nitrogen and organic matter in sustaining crop production 23
   2.2 Organic matter and plant production 24
       Organic matter as a source of nutrients 24
       Organic matter as an energy source for biological activity 31
   2.3 Organic matter accretion and decomposition 34
       Organic matter accumulation 35
       Organic matter depletion 38

	2.4	Soil nitrogen	47
		The availability of nitrogen for plant growth	48
		Gains and losses of nitrogen in the agricultural system	53
	2.5	Biological nitrogen fixation	55
		Sources of nitrogen fixation	56
		Management of biological nitrogen fixation	60
	2.6	Conclusions	71
3.	The maintenance of soil fertility II. Soil structure and erosion		73
	3.1	Introduction	73
	3.2	The nature of soil structure	74
		Soil architecture	74
		Formation of surface crusts	77
		Physical properties of soils	81
	3.3	Soil erosion	92
		Processes of soil erosion	92
		The Universal Soil Loss Equation	100
	3.4	Approaches to soil conservation	106
	3.5	Conclusions	110
4.	Efficiency of resource use		111
	4.1	Introduction	111
	4.2	Integration of land resources	112
	4.3	Energy transfer and the use of crop residues	114
	4.4	Utilization of moisture and nutrients	117
		Moisture use	117
		Nutrient flow and nutrient cycling	119
	4.5	The modification of interference between components of the system	127
	4.6	Control of environmental pollution	129
5.	Crop protection systems		133
	5.1	Introduction	133
	5.2	Weeds	134
	5.3	Pests	135
	5.4	Disease	138
6.	Outputs from animal production in farming systems		141
	6.1	Introduction	141
	6.2	Entrepreneurial objectives for animal production	144
	6.3	Other farmer objectives for animal production	150

7.	The diversification of income and flexibility of production	158
	7.1 Introduction	158
	7.2 The minimization of risk	159
	7.3 Flexibility of production	161
	7.4 Positive indicators for diversification	163

Part 2: Management of crop–pasture systems   166

8.	Tree crops with pastures	166
	8.1 Introduction	166
	Farming systems of the tropics	166
	The rationale for agroforestry	169
	8.2 Tree–pasture relations	171
	The shaded environment of the plantation	171
	Interference between trees and pastures	195
	8.3 Management of tree–pasture associations	205
	Pasture management	205
	Management of tree crops	213
	8.4 Management of particular crops	216
	Coconut	216
	Rubber	218
	Oil palm	219
	Other tree crops	221
	Timber production	222
	8.5 Conclusions	227
9.	Shrub legumes with annual crops	229
	9.1 Introduction	229
	The rationale for planting shrub legumes	229
	Sustained yield in annual cropping systems	232
	9.2 The objectives of alley farming	233
	Experience with alley farming	233
	The balance of different outputs	241
	Maintenance of the soil resource and its fertility	245
	Effective use of climatic growth factors	252
	Nutritive value of shrub legumes	258
	9.3 Alley farming practice	265
	Hedgerow establishment and density	265
	Cutting and grazing management	267
	Choice of shrub legume species	271
	9.4 Conclusions	275
10.	Pastures with annual crops	277
	10.1 Introduction	277
	The philosophy of the ley	277

	The failure of past attempts at ley farming in the tropics	278
	New prospects for ley farming	280
10.2	Duration of the pasture and crop phases	283
	Soil nitrogen	283
	Soil organic matter	288
	Soil structure and soil loss	292
	Crop protection	293
	Economic considerations	294
10.3	Pasture and crop establishment	296
	Management of the soil surface	296
	Choice of pasture species	308
	Techniques of establishment	310
10.4	Fertilizer practice	315
	Fertilizer use on the legume ley	316
	Crop fertilizer needs following the legume ley	317
	Policy for N fertilizer use on grass leys	318
10.5	Utilization of the ley	319
10.6	Conclusions	321

11. Annual crops with forage crops 323
    11.1 Introduction 323
        The rationale for intercropping annuals with forage 323
    11.2 Relations between crops 330
        Light 330
        Moisture 334
        Nutrients and the nitrogen economy 337
    11.3 Cropping practice 340
        Intercropping 340
        Relay cropping 345
        Undersowing crops for pasture establishment 347
    11.4 Conclusions 348

12. Conclusions 350
    12.1 Central issues for the development of appropriate technology 350
    12.2 Facilitating farm practice which sustains tropical cropping 353

References 358
Index of scientific names 407
Subject index 411

PREFACE

Pressure on the land resources of the tropics and subtropics has increased as the numbers of people supported by the land have risen. The cultivation of uplands which are marginally suited to annual cropping has led to decreasing soil fertility and declining crop yields. Increased run-off from these lands has altered the hydrological characteristics of the watersheds, and the soil eroded represents the loss of irreplaceable resources. Continuous monocropping leads to the development of intransigent problems of disease, pest or weed infestation.

In the humid tropics plantation agriculture is vulnerable to world price fluctuations, and farmers dependent upon a single export commodity seek to diversify their operations.

Tropical forages are playing an increasing role in the development of sustainable cropping systems, and a new technology of forage improvement has arisen. This book begins by outlining the objectives of the interaction between crops, pastures and animals. These may be directed primarily to the maintenance or repair of soil fertility, as reflected by the nitrogen accretion due to forage legumes, the organic matter status, attributes of soil structure, the control of erosion and the spatial transfer of nutrients. The integration of forages in cropping systems may also lead to more efficient utilization of environmental resources, as represented by energy flows, moisture conservation and crop protection, whilst the added flexibility of production can lead to diversification and stabilization of farm income.

Principles for the management of various types of crop/forage systems are described. Pasture with tree crops may involve coconut, rubber, oil palm and fruit crops, representing developed land whose use may be intensified; or agroforestry in which timber and animal products are the outputs. Alley farming, in which shrub legumes are grown in conjunction with annual crops and are fed to livestock, has gained favour in some tropical regions as an innovative technology. The concept of ley pastures in rotation with annual crops requires revaluation, whilst the planting or regeneration of pastures after an annual cropping phase is widely practised. The integration of animals in annual cropping systems is favoured by companion or relay cropping with forage crop legumes, whilst the complementary use of crop and pasture lands is a feature of some systems where ruminants are used for draught as well as meat or milk.

Environmental protection and the maintenance of agriculture require the wide adoption of forages in tropical cropping systems. This book outlines the science upon which this adoption may be based.

Department of Agriculture L.R. HUMPHREYS
The University of Queensland
January 1994

ACKNOWLEDGEMENTS

I am indebted to Sir Raymond Hoffenberg KBE and Wolfson College, Oxford and to Professor C.J. Leaver FRS and the Department of Plant Sciences, University of Oxford for support. Special assistance in the preparation of this book was given by L.C. Bell, R.A. Cramb, H.M. Shelton, H.B. So and S.A. Waring of the Department of Agriculture, University of Queensland and by R.L. Ison of the Open University, Milton Keynes, UK. The manuscript was typed by Mrs Ann Hansen and Mrs Carolyn Smith.

I am also grateful for advice or assistance from: C.M. Coughenour, T.M. Davison, J.P. Dimer, I. Darnhofer, D.G. Edwards, R.C. Gutteridge, M.J. Fisher, R.L. Hall, A.M. Heineman, C.E. Hughes, P.J. Kanowski, C. Lascano, V. Lawrence, D.L. Lloyd, C. Manidool, M.A. Mohammed Saleem, E.A. Mortiss, B.W. Norton, W.J.A. Payne, M.E. Probert, G.B. Robbins, A.J. Simons, A. Seabrook, J.M. Spain, A.W. Speedy, G.R. Stewart, J.L. Stewart, W.W. Stür, P.J.M. Thompson, P. Turner, R.R. Vera, I. Vallis, M.K. Wegener, J. Whelan and N.D. Young.

We are indebted to the following for permission to reproduce copyright material:

American Society of Agronomy: Figures 2.4 and 11.1, Tables 8.8 and 11.1; Australian Centre for International Agricultural Research: Figures 4.1, 8.1, 8.2, 8.3, 8.11 and 8.15, Tables 4.1, 8.4, 8.5, 8.10, and 8.12; Australian Rangeland Society: Table 1.5; Blackwell Scientific Publications: Figures 3.1 and 8.6; CAB International: Figures 9.8 and 9.9, Tables 9.8, 9.9, 9.10, 9.12, 9.13 and 9.14; Cambridge University Press: Figures 1.3, 9.2 and 10.2, Tables 8.2, 8.6 and 11.7; Centro International Agricultura Tropical: Figures 2.10 and 10.1, Tables 1.1,

1.4, 2.8, 6.1 and 11.3; CSIRO Journals, Australia: Figures 2.5, 2.6, 2.8, 2.11, 3.2, 3.5, 3.8, 3.9, 4.6, 8.5, 8.12 and 8.13, Tables 2.7, 2.9, 2.12, 3.2 8.11, 10.4 and 10.5; J.P. Dimes: Figure 2.7; Elsevier Science Publishers: Figures 3.4, 3.6, 4.3, 9.5, 9.6, 11.2, 11.3, 11.5 and 12.1, Tables 2.1, 6.2, 6.3, 9.8, 10.9, 11.4, 11.5, 11.6 and 11.8; Food and Agriculture Organization of the United Nations: Figure 8.14, Table 10.10; L.R. Humphreys: Figures 2.9, 8.4 and 10.11, Tables 2.13 and 8.3; International Board for Soil Research and Management: Table 9.3; International Centre for Research in Agroforestry: Figure 9.4; International Development Research Council, Ottawa: Tables 5.1 and 5.3; International Grassland Congress, A.J. Kruger and J.C. Scanlan: Figures 8.9 and 11.2; International Livestock Centre for Africa: Figure 4.1, Table 10.6; International Society of Soil Science; Figure 1.1; Instututo Interamericano de Cooperacion para la Agricultura: Table 1.4; James Cook University of North Queensland: Tables 6.5, 6.6 and 6.7; Kenya Agricultural Research Institute: Figure 4.5; Kluwer Academic Publishers; Figures 2.1, 2.2, 2.3, 2.12, 3.3, 4.4, 7.1, 8.7, 8.8, 8.10, 9.1, 9.3 and 9.7, Tables 2.4, 2.10, 2.11, 2.14, 3.3, 8.1, 9.1, 9.2, 9.4, 9.5, 9.6, 9.7 and 9.11; R.C. Muchow, CSIRO Division of Tropical Crops and Pastures, Australia: Figures 10.6, 10.7 and 10.8; Pergamon Press: Tables 2.5 and 2.6; A.J. Pressland: Figure 3.10; Queensland Department of Primary Industries: Tables 3.1, 5.2, 10.1 and 10.7; Soil and Water Conservation Society, Iowa: Figures 3.11 and 3.12, Table 1.3; Metha Wanapat, Khon Kaen: Table 6.4; Waverly, Baltimore: Table 10.8; Tropical Grassland Society of Australia: Figures 1.5, 4.2, 9.10, 10.3, 10.10 and 10.11, Tables 1.2, 8.7 and 9.9; Department of Research and Specialist Services, Zimbabwe: Figures 1.2 and 10.4, Tables 10.1 and 10.11.

Whilst every effort has been made to trace the owners of copyright material, in a few cases this has proved impossible and we take this opportunity to offer our apologies to any copyright holders whose rights we may unwittingly have infringed.

LIST OF ABBREVIATIONS

ACIAR	Australian Centre for International Agricultural Research
ai	active ingredient
Al	aluminium
AU	animal unit
b	beasts
c.	approximately
C	carbon
Ca	calcium
CEC	cation exchange capacity
CIAT	Centro Internacional de Agricultura Tropical
DE	digestible energy
DM	dry matter
EMBRAPA	Empresa Brasiliera de Pesquisa Agropecuária
Fe	iron
FAO	Food and Agriculture Organization of the United Nations
GM	green matter
h	hour
ha	hectare
hd	head of livestock
IDRC	International Development Research Council
IITA	International Institute of Tropical Agriculture
ILCA	International Livestock Centre for Africa
IRRI	International Rice Research Institute
IVDOM	*in vitro* digestibility of OM
K	potassium
LAI	leaf area index or leaf area per unit ground surface
LER	land equivalent ratio
LSD	least significant difference
LW	liveweight

LWG	liveweight gain
m	metre
M	million
ME	metabolizable energy
mEq	milli-equivalents
Mg	magnesium
min	minute
Mn	manganese
Mo	molybdenum
N	nitrogen
Na	sodium
NA	not available
nm	nanometre, 10^{-9} m
NS	not significant
OM	organic matter
P	phosphorus
P	probability
PAR	photosynthetically active radiation
S	sulphur
s	second
SE	standard error
SR	stocking rate
t	tonnes
μ	micron, 10^{-6}m
VAM	vesicular-arbuscular mycorrhiza
Zn	zinc

CHAPTER 1

Introduction

1.1 The new technology of pasture improvement in the tropics and subtropics

The knowledge about planting and managing forages that increase production and stabilize the environment in the tropics and subtropics has undergone a major revolution in the past two or three decades. There have been advances in selecting and commercializing élite plant germplasm that is reliably adapted to specific but diverse farm situations; scientists understand better the bases of successful adaptation to the particular stresses of different environments, and this leads to a definition of which varieties perform best and how their management may be optimized so that the goals of farmers are realized in terms of better livestock production and survival, better yields of crops which follow pastures, and better retention of soil in the face of erosive rains.

Production benefits

The surest test of the validity of scientific advances in agriculture is their long-term adoption in farm practice. Scientists are often optimistic about the utility of their findings, as demonstrated in the sheltered environment of government-supported research stations. These benefits need to be devalued to take account of the riskiness of technological change in a varying economic and climatic environment, the skills of the farmer in applying technology, and the failure of technology to meet the special needs of particular farm enterprises.

It is therefore heartening to note the advance of pasture improvement in many areas of the tropics and subtropics. In Thailand pasture seed production, principally of *Stylosanthes hamata* cv. Verano and *Brachiaria ruziziensis*, reached 800 t year^{-1} in 1992 (C. Manidool, personal communication); the demand for *B. ruziziensis* seed was such that the free market price at Khon Kaen was $US10 kg^{-1} relative to a government-controlled price of $US3 kg^{-1}. In Queensland the increase in area of planted pastures has averaged *c.* 100 000 ha year^{-1} from 1960/1990 and plantings have recently exceeded 380 000 ha year^{-1} (Walker and Weston 1990). Additionally the area of pasture naturalized by exotic species has grown to 5.0 Mha, including 2.3 Mha of legume-based pastures. Other examples could be cited, especially from Latin America. However there are many regions of the world where the pace of adoption is slow, and some of the reasons for this are discussed in this book.

The integrated grazing of natural pastures with a smaller area of improved legume-based pasture is an attractive option for many farmers, since the benefits of investment in one field are spread over a wider area. At Carimagua, Colombia (lat. 5° N, 2160 mm annual rainfall, 175 m altitude) cattle graze native pastures dominated by the tall *Trachypogon vestitus*, *Paspalum pectinatum* and other species. The performance of a cow/calf operation based on savanna alone, continuous mating and mineral supplementation (treatment 1) was compared with performance where 20% of the savanna was converted to legume-based planted pasture (CIAT 1991d). Animals in this system also had continuous mating (treatment 2), seasonal mating (treatment 3) or had an intensive management system applied where nursing dams had access to planted pastures early in the rainy season when they were mated, and in mid-dry season when pregnant cows grazed the improved pastures; the pastures were used for fattening steers at other times (treatment 4). Access to improved pastures led to substantially improved conception rates (Table 1.1) and to increases in the net output per animal unit (AU) and the stocking rate (SR). The internal rate of return on the investment increased from 3.0% in treatment 1 to 21.1% in treatment 4.

Similar increases are evident for the rate of fattening on planted pastures from many other studies. At Narayen, Queensland (lat. 26° S, 710 mm annual rainfall) cattle grazing native pastures based on *Heteropogon contortus* in open *Eucalyptus melanophloia* woodland gave an annual live weight gain (LWG) of *c.* 95 kg hd^{-1}, where cattle on fertilized pastures of *Cenchrus ciliaris*–*Macroptilium atropurpureum* gained *c.* 160 kg hd^{-1}; when this is coupled with higher SR, output per unit area was increased by a factor of 4.5, or 6.5 if N fertilizer were applied (Mannetje and Jones 1990). In the humid tropics of coastal Queensland (lat. 15–19° S, 1500–5000 mm annual rainfall) the application of new technology based on well-adapted species, identification of soil nutrient deficiencies and their rectification, appropriate SR and weed control measures has led to the successful and sustainable development of large areas of commercial pasture whose performance has now been proven over 20 years (Teitzel 1992). Grass–legume pastures continuously grazed at 2.5

Table 1.1 Performance of alternative cow–calf systems at Carimagua, Colombia (from CIAT 1991d).

Treatment	Relative stocking rate (%)	Conception (%)	Weaning weight (kg)	Sales kg/AU year	IRR* (%)
(1) Native pastures	100	47	133	34	3.0
(2) Access to planted pasture	131	83	148	81	13.3
(3) as for (2) + seasonal mating	134	84	144	85	14.9
(4) as for (3) + intensive management (see text)	191	89	142	69	21.1

*IRR, internal rate of return.

steers ha^{-1} produced an LWG of 450 kg ha^{-1} year^{-1} or more, whilst grasses fertilized with N grazed at 5 steers ha^{-1} gave an LWG of c. 900 kg ha^{-1} year^{-1}. This topic is expanded in Chapter 6.

This book is about growing forages and crops in farm systems. Semi-arid and arid zone pastures are thereby excluded, and attention is focused on planted pastures. The justification for these may lie primarily in the animal production they yield, but there are many situations where crop yields benefit from their rotation or association with forages. These benefits derive from changed organic matter (OM) status of the soil and biological nitrogen (N) fixation (Chapter 2), improved soil structure and reduced soil erosion (Chapter 3), greater efficiency of resource use and control of pollution (Chapter 4), better crop protection (Chapter 5) and more flexible and diversified production (Chapter 7). These relationships can be applied in the context of different farming systems: tree crops and alley cropping with shrub legumes (Chapters 8 and 9) or of pastures with annual crops, where some gains in crop yield are illustrated in section 10.1; the use of forage crops in conjunction with annual crops is explored in Chapter 11.

Species adaptation

Farmers expect to plant productive forages that establish reliably, show persistence of yield, and are easily eradicated when land is again prepared for crops. The plants that are ecologically successful in farm practice are able to replace themselves through continued tillering or natural seedling survival, are resistant to environmental stresses and successfully compete with species which would otherwise invade the pastures. It is convenient to consider environmental stresses as climatic, edaphic or biotic (Humphreys 1981). Signal advances have been made in selecting and breeding élite germplasm that exhibits these qualities. There is now a suite of forage legumes and grasses well adapted to each humid and subhumid situation where crops are planted. Twenty years ago this may have been partly true for forage grasses, but there were many farm situations for which legumes were unavailable: flooded places, dry situations, heavily grazed places and soils of low pH with high levels of Al or Mn. These constraints have largely been overcome (Cameron et al. 1993) and examples are given in section 10.3 of commercialized plants suitable for particular ecological niches.

Biological N fixation is the main rationale for growing legumes in rotation or association with crops. The rate of N fixation in legumes effectively nodulated with *Rhizobium* or *Bradyrhizobium* is primarily dependent on the rate of plant growth, which determines the availability of assimilate to the root nodule. Tropical legumes grow as fast or faster than temperate legumes, so that the amount of N fixation in tropical forage systems is high; this varies from negative fixation in drought years to 560 kg ha^{-1} year^{-1} for *Leucaena leucocephala*, as discussed in section 2.5.

New varieties of plants provide the simplest means of changing levels of farm production, and represent the innovation most readily adopted by farmers. The availability of good quality seed or planting material is a basic requirement, and greater emphasis has been given to seed production potential in selection programmes. Most of the recent releases of pasture legumes are well synchronized in their flowering and capable of high seed yields in excess of 1 t ha^{-1} (Humphreys and Riveros 1986). Many of the improved pasture grasses have lower harvest yields, and seed availability and price is a significant constraint to farmer adoption.

Soil fertility management

The role mineral nutrition plays in successful pasture development has been accorded greater recognition in recent years, and the correct bases for the diagnosis of particular mineral deficiencies have been developed. The once-prevalent blanket NPK fertilizer recommendation has been replaced by a knowledge that the application of N fertilizer is inimical to legume performance and that deficiencies of other nutrients such as S or minor elements may be the predominant cause of pasture failure or of weed invasion.

There are two different approaches to this question. One is the low-input option of seeking plants which are effective in extracting nutrients from low fertility soils or which are efficient in exhibiting a high ratio of the amount of growth per unit of mineral uptake. Legumes such as *Stylosanthes scabra* and *Lotononis bainesii* exhibit these capacities. The alternative is to rectify the mineral deficiency with appropriate fertilizer application. This will lead to forage of higher nutritive value and to faster rates of N fixation in legume-based pastures, as outlined in section 10.4 and Figure 10.11. Teitzel (1992) gives an example of how specific fertilizer recommendations for pasture establishment can be linked to a land classification based on vegetation and soil origin.

Where a key nutrient is determining the level of legume output, guidelines may be laid down for the appropriate SR and expected level of production. There have been sufficient grazing experiments carried out in northern Australia to show how output from legume-based pasture, expressed as safe SR, is modified by annual rainfall, pasture type and available soil P (Miller *et al.* 1990). Low levels of P supply may be augmented by P application, resulting in a higher safer SR (Table 1.2), or production may be increased by clearing open woodland and replacing native grasses with planted exotic species. SR varies from 0.1 to 1.9 b ha^{-1} according to the level of these variables.

Grazing management

The primary goal of management is to effect a synchrony between the pasture available and the forage requirement of the animal. This is achieved mainly through adopting an SR (or a cut-and-remove system) that sustains animal production in the long term through:

- adequate forage allowance (the forage available per head);
- plant growth and persistence;
- maintenance of cover as a defence against erosion; and
- desirable botanical composition, especially in relation to legume content (Humphreys 1991).

Continuous stocking, perhaps with variation in the seasonal SR of particular fields, is the norm for pastures in the tropics and subtropics. Rotational grazing systems have either reduced or have not affected animal production, but have reduced returns to the farmer. Tropical grasses with the C_4 photosynthetic pathway have a higher fibre content and a lower leaf/stem ratio than temperate grasses, and maximum opportunity for selective grazing is necessary if individual animals are to acquire a satisfactory diet and show good levels of survival, LWG, reproduction and milk output.

Continuity of forage allowance is of primary concern to the farmer, who seeks to minimize animal stress and to maintain animal production. This is achieved through:

Table 1.2 Safe stocking rates (300-kg steers ha^{-1}) in relation to available soil P and annual rainfall in northern Australia on (a) uncleared open woodland with native pastures oversown with a legume and (b) cleared land planted to legumes and grasses (from Miller *et al.* 1990).

(a)

Soil P (p.p.m.)	Average annual rainfall (mm)		
	600	750	1000
2	0.1	0.1	0.2
4	0.2	0.2	0.4
6	0.3	0.3	0.6
8	0.4	0.5	0.8
10	0.4	0.6	1.0

(b)

Soil P (p.p.m.)	Average annual rainfall (mm)		
	750	1000	> 1250
4	0.7	0.9	1.0
6	0.8	1.2	1.5
8	1.1	1.5	1.8
10	1.2	1.6	1.9

- varying SR seasonally to match forage availability, through policies of time of mating and of purchase and sale of livestock;
- providing a sequence of feeds of differing seasonal utility – this may require the planting of shrub legumes;
- modifying the environment in which pastures grow through fertilizer or irrigation practice;
- conserving or purchasing feeds.

1.2 Processes of land degradation in cropping systems

Land degradation occurs when the flow of outputs from the farm system exceeds the inputs, when non-replaceable resources are lost, the biotic environment in which crops are grown changes, and the level of continuing crop production falls. This is associated with soil erosion and with adverse changes in the physical or chemical characteristics of the soil that modify its capacity to deliver nutrients and water to the crop; coincidentally the incidence of pests, disease or weeds may increase. Changes in the water table or the exposure of sodic subsoil may lead to problems of salinity.

Trends in crop yields

The first cost to soil fertility is the effects of clearing forest land and converting it to crop land (Colour Plate 3). The earlier focus scientists made on the loss of nutrients in the vegetation biomass (Nye and Greenland 1960) has been modified by the recognition that the bulk of the N in tropical forest ecosystems is in the soil (Sánchez 1982, and as illustrated subsequently in Table 2.1; Cerri *et al.* 1991). Perhaps 20–25% of the N present is lost when tropical rain forest is slashed and burnt, but if the forest is cleared with bulldozers the losses are greater, since topsoil is displaced and accumulated outside the area that is cropped. In the traditional slash-and-burn system of shifting cultivation or swidden a long cycle of forest (or 'bush') fallow leads to a replenishment of nutrients; cycles of 15–30 years of forest (or 10 years in special circumstances), followed by short duration of cropping (one or two crops) are often conservative (Kellman 1979 for Mindanao; Turenne 1977 for Guyana; Toky and Ramakrishnan 1981 for Assam; Dove 1983 for Indonesia). The system degrades when the pressure of population on land resources shortens the fallow cycle or from low input when sedentary cropping replaces the swidden system.

Decreasing crop yields with time are inevitable unless fertilizer inputs, incorporation of legumes in the cropping cycle and a system which maintains OM are adopted. Newton and Jamieson (1968) used an exponential model to illustrate declining crop yields at a humid lowland site at Keravat, Papua New Guinea (lat. 4° S):

$$Y = S\,[1 - a\,(1 - e^{-kt})]$$

where yield (Y), seasonal factors, including pests and diseases (S), and number of crops (t) are related by constants a and k. For sweet potato a and k had values of 0.472 and 0.365 respectively. Yields declined by a factor of eight after 10 successive sweet potato crops over 5 years. This type of result has been duplicated with varying magnitude throughout the tropics and subtropics wherever low input continuous monocropping has been practised. In northeast Thailand cassava yields decrease from initial values of *c.* 25 t ha^{-1} to 6–10 t ha^{-1} after a few years, as discussed in section 10.4 (Gibson 1987); in south Venezuela the yield of a second crop of cassava decreased to 65% of a first crop (Uhl and Murphy 1981), whilst at Ile-Ife, Nigeria cassava yields decreased by a factor of four to eight by the fourth season (Obi 1989). The transition from shifting to permanent cultivation incorporating fertilizer use may require heavy inputs of lime and OM; in the absence of these, crop yields in the Kilombera Valley, Tanzania decreased to zero after 6 years (Vieweg and Wilms 1974). Finally, a study of southern Queensland heavy textured soils used for winter cereal cropping over 20-70 years showed an annual decline in yield of 1.8-8.5%, and an annual decrease in crop N uptake of 1.7–2.4% (Dalal and Mayer 1986a).

These gloomy illustrations should be placed in the perspective that viable long-term cropping systems have also been demonstrated, and sound fertilizer practice combined with rational crop rotation and management of crop residues may stabilize yields. At Kumasi, Ghana the long-term yields of cassava and of maize increased with time when crops were appropriately limed and fertilized with NPK (Ofori 1973). At Yurimaguas, Peru (6° S, 2100 mm annual rainfall, 180 m altitude) in the Amazon headwaters Sánchez et al. (1982) demonstrated on a well-drained ultisol that sedentary cropping was sustainable. Over a period of 8.5 years yields from a fertilized rice–maize–soybean rotation increased, whilst yields were well maintained in a rice–peanut–soybean rotation. There was an average annual grain production of 7.8 t ha^{-1}, with 21 crops harvested over the period. This provides an attractive alternative to shifting cultivation, since less land is needed to be placed under crop in order to meet food demands. Subsequently Sánchez and Benites (1987) moved their advocacy to a lower input system incorporating the vine legume *Pueraria phaseoloides*. The thesis developed in this book is that in many farm situations an alternative which sustains crop yields and protects the environment is the incorporation of forages in the cropping system.

Soil erosion
Consequences of soil erosion

Soil erosion is a primary problem for farmers throughout the tropical world and a secondary but significant problem for the urban communities they support. Soil erosion has become an increasing constraint to farm production as the pressure of rising population has brought more marginal, sloping land with greater erosion hazard into cropping systems, especially if lighter soils of intrinsically lower fertility are involved in this expansion. The reduction in crop production arises from:

- the loss of nutrients that are concentrated in the upper layers of virgin soils and which are necessary for good crop performance;

- the increasing aridity of the farm environment as the capacity of the soil to accept rainfall is reduced and as less soil water is stored at depth for subsequent crop use;
- the effects on farm practice as the loss of OM, the deterioration of soil structure and the physical barriers of gullies (Plate 1.1) make tillage and crop planting more difficult.

Similar effects occur on pastoral lands if overgrazing and excessive clearing of trees lead to erosion. Concurrently the lands below the farm suffer losses of production due to:

- the altered hydrological characteristics of the catchment area, which then delivers greater runoff to lower lying areas, and which results in the incidence of flash flooding increasing and the stability of groundwater supplies decreasing;
- the deposition of unwanted sediments, often of lower fertility.

The costs to the whole community are expressed through:

- the damage to public structures – roads, bridges and riverside installations – and to housing and buildings by increased stream flooding;

Plate 1.1 Gully erosion from farming steep slopes near Asebeteferi, Ethiopia.

- the reduced storage capacity of public reservoirs as siltation occurs;
- the poorer quality of water delivered from the catchment area and its effects on the quality of both human and aquatic life;
- the altered stream bed characteristics;
- the environmental pollution of dust.

This subject is discussed in more depth in sections 3.3 and 3.4, and at this juncture some examples of the magnitude of soil erosion and its effects are given to place the topic in perspective.

Extent and seriousness of soil erosion

The varying occurrence of soil erosion is related to local environmental factors such as the nature of rainfall and wind, the erodibility of the soil and the steepness of the topography, and land use factors such as the intensity of cropping and the husbandry used. In Latin America Posner (1982) quotes various authorities which have stated:

- 42% of Mexico and 77% of El Salvador suffer from accelerated erosion;
- 30% of the savanna of Bogatá, Colombia has severe erosion;
- 60% of the crop land in Jamaica and 50% of the crop land in the Dominican Republic are on steep slopes and suffer from erosion.

Low (1967) estimated rates of erosion in Peru from a consideration of rainfall characteristics and topographical features (the 'orographic' coefficient). In this country with its predominantly Andean hilly terrain there is a special vulnerability to erosion (Table 1.3). If the level of tolerable soil loss is of the order of 10–13 t ha^{-1} yr^{-1} nearly half of the lands of Peru are deteriorating. El-Swaify and Dangler (1982) quote figures for the annual suspended sediment load of the main rivers of the tropics and subtropics. When this load is adjusted for the size of the catchment area to estimate annual soil erosion from the field, the worst

Table 1.3 Estimated rates of erosion and their extent in Peru (from Low 1967).

Rate of erosion (t ha^{-1} year^{-1})	Area (km$^2 \times$1000)	Percent of total area
0–10	216	17
10–15	671	52
15–20	216	17
20–30	95	7
30–70	83	7

regions are those of the Kosi in India and the Yellow River in China, followed by the Damodar and the Ganges in India. It is estimated that 150 Mha in India are subject to erosion, of which 69 Mha are critically affected.

Most of tropical Africa lies in high-risk erosion classes. Annual soil erosion of the Ethiopian highlands is estimated to be well in excess of 1×10^9 t (Plate 1.1)(Brown 1981). Gully erosion is severe in the coastal sediments cropped in southeast Nigeria (Lal 1990) and rainfall intensity in West Africa is unusually high relative to other tropical regions (Wilkinson 1975a; Kowal and Kassam 1976); erosion is especially severe on the structurally unstable and predominantly low-activity alfisols in West Africa (Lal 1990). These examples will be multiplied in subsequent discussion.

The level of erosion varies enormously in different cropping systems. Posner (1982) illustrates this diversity (Table 1.4) from the literature of erosion measurement in sloping lands of tropical America, indicating that erosion in some cropping systems may be completely controlled with appropriate mulching techniques, or may exceed 100 t ha^{-1}.

The severe erosion problems associated with cultivating hilly land, especially when a highly erodible soil type is employed, are indicated by a study at Chiang Rai, Thailand (lat. 20° N, 900 m altitude) of Anecksamphant et al. (1990). This hilly, rolling country has slopes ranging from 20 to 50%. The experimental site had replicated plots laid down with a slope length of 36 m, and additional plots of natural forest and of bare soil were added for comparison. The farming treatments were:

T1 Farmer practice, planting upland rice up and down the slope
T2 Alley cropping using mixed *Leucaena leucocephala* and *Cajanus cajan* as hedgerows, with clippings returned to the inter-hedge cropping area
T3 Contour grass strips and coffee intercropped with rice
T4 Hillside ditches
T5 Coffee intercropped with rice

In the first year 1649 mm rainfall fell, almost all in the five months June–October. The levels of soil erosion (Figure 1.1a) and of runoff (Figure 1.1b) were unacceptable in all but the forested area. Current farm practice (T1) showed total soil loss of 120 t ha^{-1}, but the protective treatments also failed in the year of establishment, the best of them (hillside ditches, T4) leading to 62 t ha^{-1} of soil loss. Upland rice showed an average yield of 1.04 t ha^{-1}, with reduced yields in the T3 and T4 treatments, due to loss of crop area.

There are two especially significant features of this study. First, the erodibility of the soil can be a dominant factor in determining erosion hazard. This soil was a Ustic Kandihumult, a deeply weathered

red-brown silty clay loam derived from granite. This highly erodible ultisol showed much greater soil loss than that from a more resistant alfisol at Chiang Mai derived from shale (Anecksamphant *et al.* 1990). Second, farming steep slopes requires intensive conservation measures which exceed the technology applied in this study, at least in the establishment year. A yield of *c.* 1 t ha^{-1} of rice is unlikely to generate the investment available to farmers that is required to protect their land, and higher value cropping systems and better technology are required if any alternative to closed forest systems is to be contemplated.

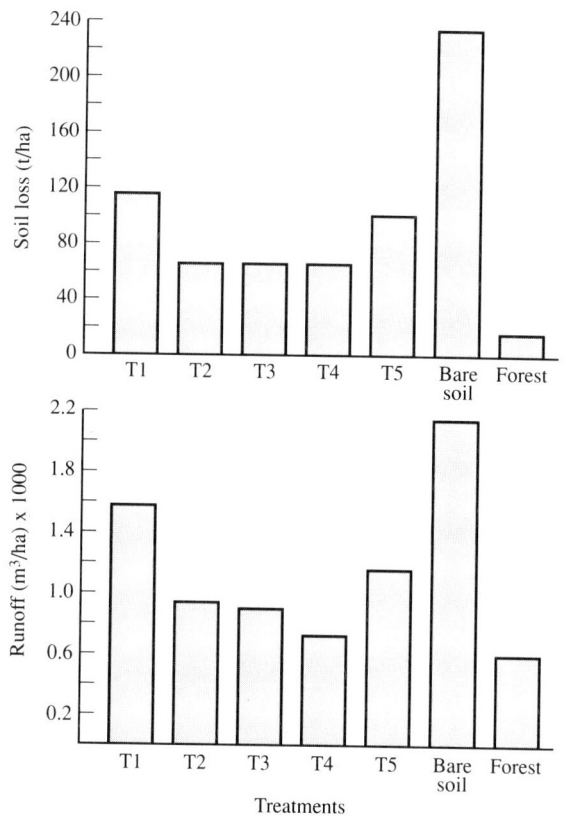

Figure 1.1 Soil loss (a) and runoff (b) from bare soil, forest and different cropping systems (T1 current practice, T2 alley cropping, T3 grass strips, coffee intercropped, T4 contour ditches, T5 coffee intercropped) at Chiang Rai, Thailand (from Anecksamphant *et al.* 1990.)

Table 1.4 Rates of erosion from various cropping systems of tropical America (from Posner 1982).

Location	Slope (%)	Cropping system	Erosion (t ha^{-1} year^{-1})
Humid tropics			
Trinidad	25	Corn relay cowpea	3
Trinidad	35	Pineapple on terraces	5*
Puerto Rico (Mayaguez)	40–45	Annual crops	44
		Sugarcane–burned	19
		Sugarcane–with mulch	2
		Pasture	5
Jamaica (Smithfield)	30	Yams–traditional culture	134
		Yams with contour mounds	27
		Yams on terraces	17
Peru (San Ramon)	30	Potato–fallow–potato	16
		Corn–beans–potato	119
Wet-dry tropics			
El Salvador	30	Corn relay beans	100
		Corn relay beans on terraces	30
High elevation tropics			
Colombia (Chinchina)	22	Pasture	7.1
	45	Young coffee trees	1.8
Panama (Boquete)	35	Carrots and beans	80
Peru (Huancayo)	25	Potato-fallow	5

*Total soil loss per hectare during 18 storms.

Loss of plant production

The loss of potential for plant production due to erosion is illustrated by three studies. The basic factor is the high concentration of nutrients and OM in the surface layers and the poorer fertility of subsoil. In the first of these examples, Pressland and Cowan (1987) compared the growth of grasses and of crops grown on different layers of a light red soil (Gn 2.12 in the Northcote system) at Charleville, Queensland (lat. 26° S) which had been cleared of the woodland species *Acacia aneura* (mulga). Six weeks of seedling growth revealed considerable differences in the production of the introduced grass *Cenchrus ciliaris* and the native grasses *Aristida armata* and *Thyridolepis mitchelliana* growing in the

0–5 cm or 5–10 cm soil layers (Table 1.5). *C. ciliaris* was much more productive than the native grasses when growing in the more fertile surface soil, but its dry matter (DM) yield was reduced by a factor of five when growing in the 5–10 cm soil layer, and it failed to flower. The allocation of assimilate to the root system, which is not accessible to the grazing animal, was markedly increased under low fertility conditions; carbohydrate accumulates wastefully in the roots when nutrients are not available for leaf expansion and growth. The yield of native grasses was also reduced when grown on the 5–10 cm soil layer, but the reduction in growth was less than in the case of *C. ciliaris*, and growth was marked by the greater differentiation of many small leaves. This soil is deficient in P, and the upper 0–5 cm contains three times the P available in the 5–10 cm layer; the latter also has a lower pH, lower exchangeable K, and higher Cl and conductivity.

In Malaysia, maize cob yield on a marginal soil was reduced to 88, 36 and 10% of yield potential as erosion removed respectively 7.5, 15 and 30 cm of topsoil. On a deeper, higher fertility soil the removal of 30 cm of topsoil reduced maize yield to 50%. In West Africa, yield of maize decreased to *c.* 45% of yield potential as 12.5 cm topsoil was progressively removed, whilst the yield of cowpea decreased to *c.* 37% (both studies cited by El-Swaify and Dangler 1982). These examples show the disastrous losses of crop production attributable to erosion.

Vulnerability of the tropics to erosion

Figure 1.1 indicates the high level of soil loss on a particularly erodible soil. Many of the soils developed in subhumid regions are more vulnerable to erosion than the highly weathered soils of the

Table 1.5 Characteristics of grasses growing on differing soil layers at Charleville, Queensland (from Pressland and Cowan 1987).

Species	C. ciliaris		A. armata		T. mitchelliana	
Soil depth (cm)	0–5	5–10	0–5	5–10	0–5	5–10
Plant characteristic						
Total DM (g)	20.2	4.1	7.1	4.5	7.7	4.3
Height (cm)	99	43	56	51	35	35
Leaf no.	95	33	64	94	100	132
Seed no.	368	0	468	195	172	82
Root wt (% total)	36	82	47	56	72	56

humid tropics. El-Swaify and Dangler (1982) suggest a series of decreasing vulnerability from vertisols to inseptisols and oxisols to ultisols. This question is elaborated further in section 3.3. A similar range of vulnerability to erosion occurs amongst soil classes which occur in temperate zones.

In wet–dry climates the opening rains of the season often fall on bare unprotected soil, and this feature is common to Mediterranean climates as well as monsoon climates. The essential feature that distinguishes the tropics and subtropics as zones of special erosion hazard is the nature of rainfall. Soil loss is closely related to the kinetic energy of falling rain. The intensity, the amount of rain, and the median drop size are greater in the tropics than in temperate zones.

Hudson (1981) gives a simple example of this dichotomy. Rainfall of low intensity causes little erosion and a threshold value of 25 mm h^{-1} has been postulated as indicating the commencement of soil loss. In many temperate climates 5% of rain falls at erosive intensities, whereas it would be common in tropical climates for 40% of rain to be erosive. It might also be expected that the kinetic energy of tropical rain falling at an intensity of say 60 mm h^{-1} would be at least 28 $J\ m^{-2}\ mm^{-1}$ relative to 24 $J\ m^{-2}\ mm^{-1}$ for temperate rain falling at say 35 mm h^{-1}. If a comparison were made of a temperate site with annual rainfall of 750 mm and a tropical site with annual rainfall of 1500 mm, the annual erosivity values would be calculated as:

$$\text{Tropical site } E = 1500 \times 0.4 \times 28 = 14\ 400\ J\ m^{-2}$$
$$\text{Temperate site } E = 750 \times 0.05 \times 24 = 900\ J\ m^{-2}$$

The vulnerability of different districts of Zimbabwe to erosive rainfall has been predicted from studies of rainfall intensities and measurements of soil erosion, which have utilized the local relationships between an erosivity index of 30 min peak storm intensity and mean annual rainfall (Stocking and Elwell 1973a). Seven classes of erosivity were mapped (Figure 1.2) and these may be used to modify land use according to the degree of hazard specified, as subsequently modified by soil and slope factors.

This brief summary of the problems of soil erosion, which are expanded in sections 3.3 and 3.4, is intended to place this topic in the context of the central role it occupies in any consideration of the sustainability of agricultural systems in the tropics. The challenge is for scientists to interact with farmers in devising the appropriate technology which contains a set of options that are in tune with farmer goals and have in-built economic incentives leading to land care. These options are elaborated in a consideration of the objectives of crop/pasture/animal interactions (Chapters 2–7) and of their

16 Introduction

application in the management of crop/pasture systems (Chapters 8–11).

1.3 Deterioration of pastures

The success of pasture management and of the technology employed in pasture improvement is gauged by whether the trend in productivity is positive, neutral or negative. Maintenance or increase of productivity occurs in the long term for some pastures, as described in the humid Queensland tropics by Teitzel (1992). Similarly in the subhumid subtropics at Rodds Bay, Queensland (lat. 24° S, 810 mm annual rainfall) suitably fertilized legume-based pastures showed increasing

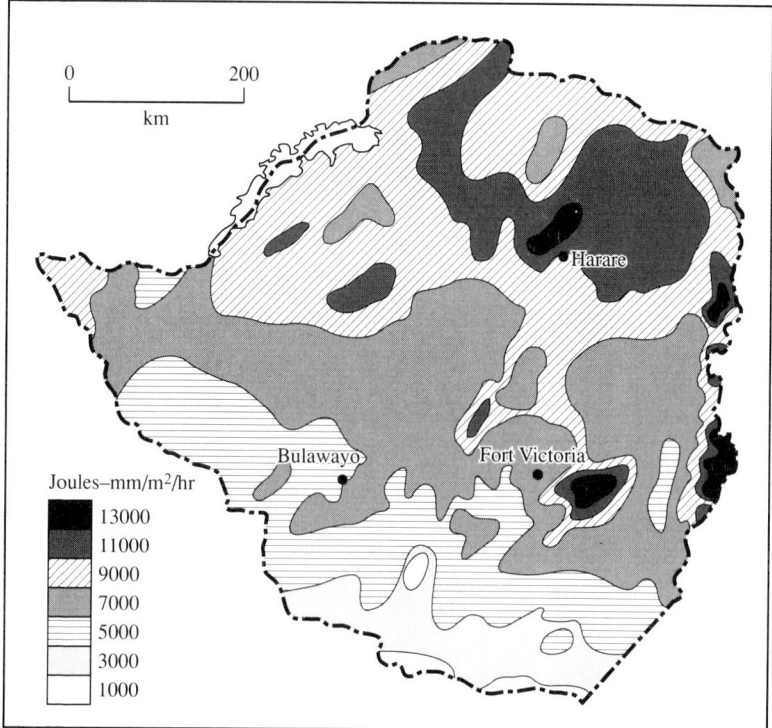

Figure 1.2 Distribution of erosivity over Zimbabwe according to the energy of rainfall (from Stocking and Elwell 1973a).

productivity over 7 years, and safe SR had doubled by year 5 as the effects of the legume on the N status of the system became evident (Shaw 1978).

On the other hand the productivity trend line becomes negative when the legume is lost through overgrazing or the ravages of pests or disease, when weeds invade the pasture, or when no legume is planted and grass pastures gradually run down the soil fertility initially available, as discussed in section 2.4 and Figure 2.8.

Decreasing productivity as pastures age

A common problem of grass-based pastures is the gradual immobilization of N in the system as the years pass since the pastures were first planted. This is well illustrated by a study at Gayndah, Queensland (lat. 20° S, 730 mm annual rainfall) where pastures of *Panicum maximum* var. *trichoglume* were planted in successive years on a black earth soil and reserved for grazing in winter and spring at 2.4 weaner steers ha^{-1}. Data were available for six successive years on pastures of varying age (Robbins *et al.* 1987). The mean LWG (Figure 1.3a) showed a linear decrease of 9.4 kg ha^{-1} for each year since planting; animal production on 5–year-old pasture was only 75% of animal production on 1–year-old pasture. Most of the reduction in LWG for winter occurred up to age 3 years, whilst LWG in spring deteriorated most after 3 years since planting.

This problem was not associated with a change in botanical composition, as occurs for many pastures which become weed-infested with age. The SR was not excessive for this environment, since pasture yield was adequate and little affected by the age of pasture, although pasture growth of aged pastures was restricted in the spring. Decreasing animal production appeared to be related to factors of pasture quality. The percent green leaf on offer (Figure 1.3b) decreased with age of pasture in the late spring (November) and the N concentration in this material (Figure 1.3d) was less on old pastures throughout the season. This affected the N concentration of the diet (Figure 1.3c), which was suboptimal, especially in the period July to October. Availability of S showed similar trends to those for N, and may also be implicated in the syndrome. The increasing amount of N immobilized in decomposing litter (Robbins *et al.* 1989) may account for the decreased pasture quality which affected LWG.

Rudder *et al.* (1982) report similar findings on a commercial property, whilst Myers *et al.* (1986) quote an instance where grass yield on a fertile clay soil decreased from 16 t ha^{-1} in the first year to c. 4 t ha^{-1} within 3 years of sowing; when adjusted for seasonal factors the decrease was at least a factor of two. Net mineralization of N is slow in undisturbed perennial pasture and the microbial population competes with the plant for available N, as discussed in Chapter 2, and the shoot/root ratio decreases. There are various options for countering this decrease in productivity: incorporation of a well-adapted legume, increased level of N fertilizer application,

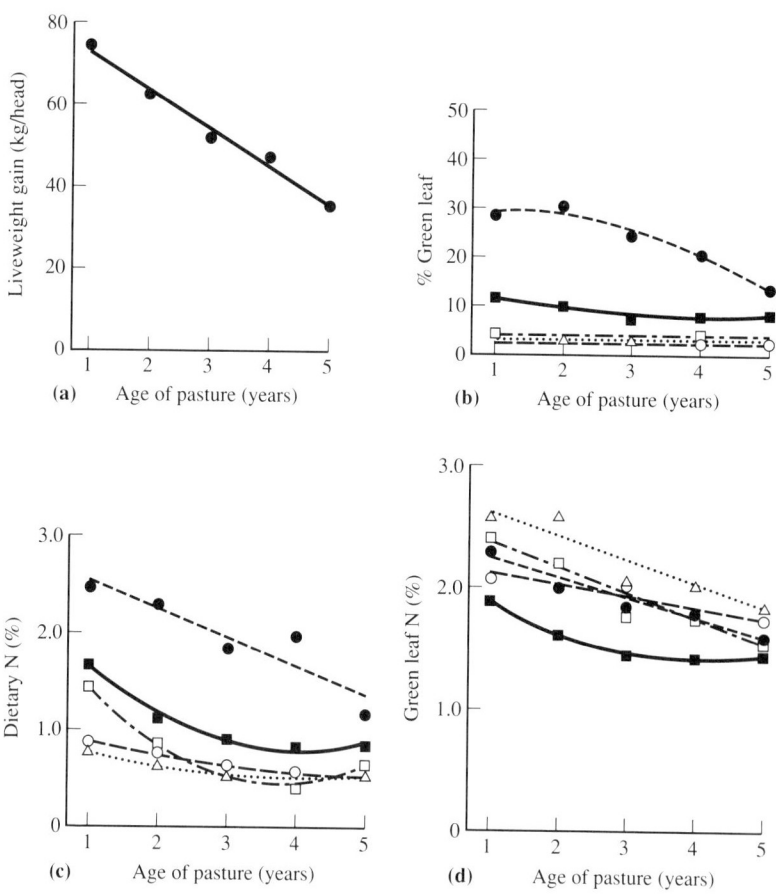

Figure 1.3 Age of *Panicum maximum* var. *trichoglume* pasture since planting and its relation to (a) LWG, (b) % green leaf, (c) % dietary N in oesophageal-fistulated steers and (d) % N in green leaf for samplings in June (■), July (○), September (△), October (□) and November (●) (from Robbins *et al.* 1987).

or the ploughing up of the pasture and the introduction of a cropping phase.

Deterioration of botanical composition

Pasture productivity is related to botanical composition in some circumstances (Humphreys 1991).

- Dominance of the germplasm that grows the fastest leads to higher plant yields.
  This objective needs to be qualified by consideration of the seasonality of growth and whether low nutritive value is associated with high DM yield and high fibre content, as occurs in C_4 grasses.
- Animal production is favoured by the presence of plants of high nutritive value, the absence of unpalatable plants which do not contribute to the feed ingested, and the absence of toxic plants which damage animal health.
- Maintenance of adequate legume content in the pasture sustains the N economy of the pasture and usually enhances nutritive value.
- Plants providing good ground cover, such as sod-forming grasses and creeping legumes, minimize soil erosion and thereby contribute to the long-term productivity of the system.
- Species diversity may provide complementarity of seasonal growth rhythms, which contributes to the continuity of forage supply. Mixtures also provide insurance against natural disasters, the possible occurrence of nutritional imbalances, the evolution or incidence of new pests and diseases, and any adverse effects of changed property management.

Problems associated with weed invasion are discussed in Chapter 5. An illustration of the dependence of rate of LWG on the legume content of a pasture and of the consequences of loss of legume is shown by a study (CIAT 1989) at Quilichao, Colombia (lat. 3° N, 1840 mm annual rainfall). Pastures of *Brachiaria dictyoneura* and *Desmodium ovalifolium* initially contained 60% legume (Figure 1.4b) but the pasture was attacked by a root-feeding scarabeid beetle which killed most of the legume, as indicated by the vertical dashed line at 800 days grazing in Figure 1.4b. The rate of cattle growth (Figure 1.4a) was initially nearly 0.7 kg hd^{-1} day^{-1}, but as the legume content decreased the N concentration of the grass on offer decreased, and after 4 years LWG decreased to *c.* 0.25 kg hd^{-1} day^{-1}. The formulation of a robust technology for the maintenance of legumes in pasture swards is a continuing challenge to scientists; when technology fails due to

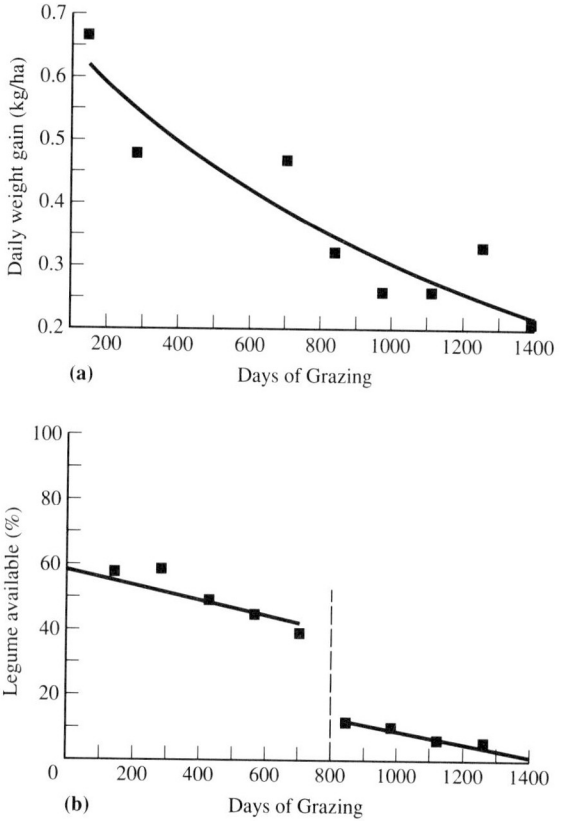

Figure 1.4 Age of *B. dictyoneura* and *D. ovalifolium* pasture and its relation to (a) LWG and (b) % legume (from CIAT 1989).

pest attack the introduction of a cropping phase may be used to control the pest in question.

The problems of pasture degradation in the Amazon basin are described by Toledo and Serrão (1991).

1.4 Concepts of sustainability

'Sustainable' is a term that has become fashionable in the past decade. It is used in so many different senses that it is necessary to clarify the contexts in which it is used in this book. The World Commission on Environment and Development (1987) stated: 'Sustainable development is development that meets the needs of the present without compromising the ability of future generations to meet their own needs'. The Commission emphasized that sustainable development contains a commitment to the essential needs of the world's poor and the idea of limitations imposed by technology and social organization on the capacity of the environment to meet needs. The definition lacks precision since needs are in many senses limitless and future needs cannot be defined by this generation.

It should be noted that sustainability is not a concept bounded by the understanding of physical and biological processes, but incorporates notions of social equity and the involvement of the farm in the ecosystem. Ison (1990) suggests that sustainability may be seen as a process (rather than an outcome) of change, involving the building of sustainable relationships between people and people, and between their environment; development is primarily a learning process which reflects self-development.

This process occurs in the context of the maintenance of a productive agriculture that is essentially in balance with the nutrient and hydrological cycles in which it is cast (Williams and Chartres 1991). A central characteristic of farming systems is their subjection to episodic stresses of differing amplitude (Lynam and Herdt 1989), and a property of the system which measures sustainability is its capacity to recover rapidly and to demonstrate elasticity. This resilience in sustainable systems will maintain performance above a boundary line, defined in terms of productivity or of some property of the environment that is significant for human welfare. Contrasting responses of sustainable and non-sustainable systems are shown in Figure 1.5 (Williams and Chartres 1991).

The Standing Committee on Agriculture (1991) of the Australian Agricultural Council suggests that there are five main principles against which institutional policies might be judged:

- Farm productivity is sustained or enhanced over the long-term.
- Adverse impacts on the natural resource base of agriculture and associated ecosystems are ameliorated, minimized or avoided.
- Residues resulting from the use of chemicals in agriculture are minimized.
- Net social benefit derived from agriculture is maximized.
- Farming systems are sufficiently flexible to manage risks associated with the vagaries of climate and markets.

These principles are encapsulated in the statement: 'Sustainable agriculture is the use of farming practices and systems which maintain or enhance the economic viability of agricultural production, the natural resource base, and other ecosystems which are influenced by agricultural activities'. This definition is accepted with the reservation that economic viability embraces the concept of the well-being of rural communities.

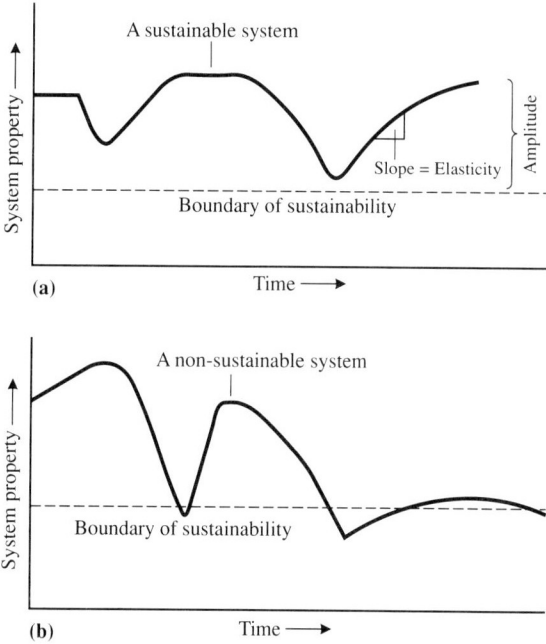

Figure 1.5 The amplitude and elasticity of response of a system property to applied stress for (a) sustainable and (b) non-sustainable systems (from Williams and Chartres 1991).

CHAPTER 2

The maintenance of soil fertility
I. Nitrogen and organic matter

2.1 Introduction

The function of soil nitrogen and organic matter in sustaining crop production

The limits to sustained crop production in the tropics and subtropics are set by many factors such as light, temperature and intrinsic soil characteristics which are not readily changed by management. The availability of soil N is the first constraint which can be readily modified by farmers, and the main rationale for growing forage with or in rotation with crops is the additional atmospheric nitrogen that forage legumes can fix in forms that are available to companion grasses and associated crops (e.g., Sinha and Chatterjee 1966).

This generalization needs to be taken in the context that organic matter (OM) will provide the main pathway for the mineralization of N, that the soil N status is linked to its OM content, and that effective crop responses to improved N availability hinge upon the beneficial influence of OM on (1) moisture status, (2) the availability and retention of other nutrients, (3) biological activity, (4) soil physical attributes, (5) the minimization of soil erosion, and (6) the deactivation of agricultural chemicals.

Since N and OM are inextricably linked in systems of farm production, this chapter enunciates some basic concepts of soil fertility about their value and their management, which are expanded in the later sections of the book that deal with particular crop/pasture systems. The effects of OM on soil structure and erosion are reserved for Chapter 3.

2.2 Organic matter and plant production

Organic matter as a source of nutrients

Components of organic matter

The cycle of OM accretion, storage and decomposition in agricultural ecosystems can be considered simply in the following compartments:

- living microflora and fauna forming the biomass;
- fresh debris and readily decomposable material;
- humus, considered as humic substances showing resistance to decomposition, and non-humic components, mainly polysaccharides.

Scientists disagree as to which components actually comprise OM; it is usual to exclude the macrofauna and the macroflora but to include the microbial biomass in soil OM. The humic substances may be considered as fulvic acid (soluble in acid and alkali), humic acid (soluble in alkali, insoluble in acid)) and humin (insoluble in acid and alkali) (Vaughan and Ord 1985; Brady 1990).

The size of the organic pool and its degradation under some systems of management were referred to in Chapter 1. The components of this pool are illustrated from a study of rainforest and its replacement by pasture at Manaus (lat. 3° S) in the Central Amazon basin of Brazil (Cerri *et al.* 1991). In this district aerial forest phytomass is *c.* 200–225 t C ha^{-1}, and *c.* 240 t C ha^{-1} occurs in the top 5 m of soil. The whole above-ground residues under the rain forest totalled 6.5 t C ha^{-1} (Table 2.1). The top 0–0.2 m of the soil, which had a higher C concentration than the rest of the soil profile, contained 11 t ha^{-1} of alkali-extractable C and 44 t ha^{-1} of resistant humin. When the forest is destroyed the surface C increases, and contains much charcoal, wood fragments and loose soil aggregates; there are only small immediate changes to soil organic C. However, planting the area to *Brachiaria humidicola* pasture did not arrest the initial decrease in soil organic C, since forest OM decomposed more rapidly than the rate of pasture OM accretion over the first 18 months. Soil humus C was restored to its initial level after 8 years of well-managed pasture.

Nutrients in organic matter

The nutrient composition of OM reflects the type of plant material from which it is derived. Most of the N which becomes available to plant roots arises from mineralization of OM, as outlined in section 2.4. OM mineralization also supplies most of the S. About 95% S in soil is organic, depending upon the sulphate content, whilst 2–3% soil S is in the microbial biomass (Scott 1985); proximity to industrial areas and the extent to which they contribute to atmospheric pollution influences the

Table 2.1 Distribution of organic C (t ha^{-1}) in plant residues and soil humus (0–0.2 m layer) at Manaus, Brazil under differing management conditions (from Cerri et al. 1991).

Fractions	Native rain forest	Burned forest	Pasture (18 months)
Leaves	2.3	0	3.1
Wood fragments	2.7	1.5	0.5
Litter roots	0.2	0.2	0.2
Charcoal	0	2.5	0.5
Aggregates	0.1	1.8	0
Unidentified fragments	1.2	3.6	1.4
Whole residues	6.5	9.6	5.7
> 2-mm fragments	5.1	4.9	3.7
< 2-mm fragments	11.0	7.0	9.0
> 2-mm roots	2.0	1.8	1.1
Alkali-extractable humus	11.0	10.0	6.0
Unextractable humus	44.0	40.0	39.0
Whole 0–0.20-m soil	73.1	63.7	58.8
Total residues and soil	79.6	73.3	64.5

level of inorganic S accretion. The extent of organic P supply (20–90%) will vary with soil conditions (Tate 1985); the microbial P pool is small but labile.

The nutrients in plant litter entering the OM material are illustrated from an early study (Jaiyebo and Moore 1964) at Ibadan, Nigeria (8° N, 1300 mm annual rainfall). The native bush fallow system produced litter of higher nutrient content (Table 2.2) than from a stargrass (*Cynodon nlemfuensis*) pasture, but a leguminous cover of *Pueraria phaseoloides* was superior.

It should be recognized that the addition of massive amounts of OM to the soil increases the demand for other nutrients. These may be met from soil reserves, or may require increased fertilizer use to maintain a balanced ecosystem. The ratio of N:S in soil OM is c. 10:1.3 whilst the N:P ratio varies from 10:0.3 to 10:3. (Probert 1982). An increase of 100 kg N ha^{-1} in the OM component implies a content of 13 kg S ha^{-1} and from 3 to 30 kg P ha^{-1}, depending on soil conditions. Similarly, as discussed subsequently, the addition of carbonaceous materials with a high C:N ratio will immobilize much N from the microbial mass in its decomposition and will increase the shortfall between the N requirement for plant growth and soil N availability, which needs to be met from enhanced N fixation by legumes or the use of more N

Table 2.2 Nutrients (%) in litter from different covers at Ibadan, Nigeria (from Jaiyebo and Moore 1964).

Nutrient (%)	Grass	Legume	Bush
N	0.85	2.00	1.80
P	0.05	0.11	0.08
K	0.4	0.8	0.6
Ca	0.5	1.5	1.7
Mg	0.2	0.6	0.3

fertilizer. It is foolish to regard OM as a panacea for all agricultural ills, since there may be significantly negative aspects to OM accretion.

Another controversial question is the direct effect of organic substances in the soil on plant growth. This can be dismissed as of little importance to crop yield, on the grounds that excellent growth is achieved when plants are grown in wholly inorganic nutrient solutions; there is no essentiality for the direct uptake of organic substances from the roots. On the other hand it is known that the availability of organic substances does influence the processes of growth (Lee and Bartlett 1976), protein synthesis, and reproduction. Humic substances have been shown to increase ion uptake and to affect cellular membranes; the growth of microorganisms may also be stimulated. Compounds applied to leaves have increased their fresh weight. Humic substances enhance the production of adventitious roots in nutrient solution; this may be an indirect effect of restricting the precipitation of iron phosphates (Vaughan and Malcolm 1985). The scientist is faced with some religious fervour in the community on this issue which further research will no doubt illuminate.

Ion exchange capacity

Augmenting OM in soil has a beneficial effect on nutrient availability and retention through its increase of the soil colloid bearing negative charges (Waring 1991). The cation exchange capacity (CEC) of humus may exceed 3000 mEq kg^{-1}, (Simpson 1983) or 2–50 times that of an equivalent weight of clay. From 20 to 70% of the CEC of many soils is due to OM (Vaughan and Ord 1985). In sandy soils the retention of nutrients, especially of mobile ions such as K$^+$ which are readily leached, depends upon the OM content of the soil. The buffering of the soil against sudden changes of pH also depends upon OM status. Biological activity continuously produce acids and bases; the acidity of root exudates would lower pH if buffering mechanisms were not available.

The effect of OM on CEC status is illustrated by further data from the study of Jaiyebo and Moore (1964) at Ibadan presented above in Table 2.2. Soil was maintained in a bare condition, or a grass mulch of *Imperata cylindrica* maintained; these treatments were compared after 6 years with plots growing *C. nlemfuensis*, *P. phaseoloides*, or regenerating bush. These covers enhanced the OM content of the latosol (Table 2.3), which was reflected in substantial increases in the values for exchangeable Ca, Mg and K, especially in the legume and the bush fallow treatments, and this improvement in soil fertility resulted in higher yields of maize. The mulch of *I. cylindrica* increased the C:N ratio of the soil from 9.0 to 10.7.

Humic substances in soil form stable complexes with some metals (Beckwith and Butler 1983; Vaughan and Ord 1985) which influence their availability. This is due to functional groups such as carboxylic acids, C=O structures and phenolic, alcoholic and enolic hydroxyls. Complexes with fulvic acid are labile and able to transfer nutrients such as Fe to plants, whilst complexes with humic acid are largely insoluble. The latter is beneficial where toxic levels of minerals are involved, as discussed in the next section.

Soil acidity and the alleviation of mineral toxicity

A significant role for OM in soil fertility is its effect on soil acidity and on the availability of elements such as Al which are often present in deleterious amounts.

In many tropical countries large tracts of ultisols and oxisols, which have low pH values and high content of Al and/or Mn, exhibit low levels of crop production. The N fixation of crop and pasture legumes

Table 2.3 OM, exchangeable cations and CEC of soils (0–0.1 m layer) maintained under different covers at Ibadan, Nigeria (from Jaiyebo and Moore 1964).

Nutrient	Bare	Grass mulch	Grass	Legume	Bush
OM (%)	1.4	2.3	2.8	3.4	4.4
Exchangeable cations (mEq/100 g)					
Ca	3.3	2.2	5.0	5.0	7.0
Mg	0.6	1.2	1.4	1.4	1.8
K	0.1	0.4	0.3	0.3	0.4
CEC	4.6	5.6	7.4	7.2	9.3

is also drastically reduced in the presence of excess Al; the growth of plants dependent upon sources of combined N, as may be supplied by fertilizer N, is less affected by Al toxicity than is the growth of legumes dependent upon symbiotic N fixation (Carvalho et al. 1981). Management of Al toxicity is therefore fundamental to the maintenance of N availability in the ecosystem, quite apart from its direct effects on production.

Aluminium occurs in many forms, some of which are less toxic to plants than others. The phytotoxicity of Al is often measured by its effects on root elongation, since while excess Al decreases the absorption and translocation of nutrients to shoots, it exerts a particular constraint on the processes of cell division and elongation in roots. It is the monomeric forms of Al that reduce root growth, and the polymeric forms that occur in soil solution do not damage root activity (Blamey et al. 1983; Alva et al. 1986). Techniques are now available for the analysis of inorganic monomeric Al (Kerven et al. 1989) and these give an indication of potential phytotoxicity that is superior to the conventional analysis of the degree of Al saturation in soil solution.

The addition of OM to highly acid soils increases the pH (Hue and Amien 1989), helps to precipitate Al in forms unavailable to the plant, and forms Al complexes with OM which do not damage plant growth. This also alters the balance of other cations in the soil solution in ways which may be favourable for plant growth. The type and level of OM influence the extent of these benefits. These effects are illustrated by a study (Bessho and Bell 1992) contrasting the effects of leaves of the shrub forage legume *Calliandra calothyrsus* and of barley (*Hordeum vulgare*) straw on a highly weathered acid soil. A red podzolic soil (Epiaquic Haplustult) containing 1.86% organic C was mixed with the organic materials and basal fertilizer and incubated for 4 or 10 weeks.

The addition of OM slightly increased soil pH (Table 2.4; Bessho and Bell 1992); legume leaves had a greater effect than barley straw, a phenomenon that was evident at the lower levels of application of 5 and 10 t ha^{-1}, which are more within the reach of field practice than the higher levels tested. The difference in content of bases in the two types of OM may partly account for this dichotomy. Legume had a high content of Ca and Mg whilst barley straw was rich in K and Na; Ca and Mg have stronger affinities with exchange sites than the monovalent ions K and Na and would therefore better raise base saturation. Similarly the reduction of monomeric Al concentration (Table 2.4) was related to level of OM addition, with legume having a better effect than barley, which actually increased monomeric Al at up to 20 t ha^{-1} application. Barley treatments had a higher ionic strength. There was a concomitant increase in the level of the inoffensive organic complexed Al in both treatments, and this increase was of lesser magnitude than

Table 2.4 Chemical properties of solution phase of soil incubated with varying amounts of legume or barley (from Bessho and Bell 1992).

Plant material	Amount (t ha^{-1})	Soil pH	Al concentration (μM)	
			Monomeric	Organic
Control ($-$F)*	–	3.86	86.4	3.4
($+$F)	–	3.83	176.5	2.8
Legume	5	3.91	123.4	6.5
	10	4.00	77.7	9.3
	20	4.09	42.0	7.8
	40	4.32	14.2	17.5
	80	4.92	0.4	22.9
Barley	5	3.87	192.7	0.5
	10	3.90	175.6	5.8
	20	3.95	164.8	1.8
	40	4.07	84.6	15.6
	80	4.34	44.6	11.0

*Basal fertilizer (F) added ($+$) or not ($-$).

the reduction in monomeric Al. It is expected that the small increase in pH caused precipitation of Al.

A secondary consideration is the effect of OM addition on the balance of exchangeable cations. The sum of exchangeable bases and Al increased with increasing level of each OM, which would result in additional charge on the variable charge colloids. The percentage of Al saturation (Figure 2.1; Bessho and Bell 1992) decreased substantially. The addition of legume increased the proportion of Ca and Mg exchangeable cations; by contrast the application of barley straw substantially increased the %K and slightly increased %Na.

The alleviation of Al toxicity was measured by planting germinated seeds of mung bean (*Vigna radiata*) in the treated soil and monitoring root length after 48 h. Relative root length (RRL) was plotted as a percentage of the best treatment, and Figure 2.2 (Bessho and Bell 1992) indicates that barley straw up to 10 t ha^{-1} had no significant effect in ameliorating Al toxicity, but was successful at very high rates of application. On the other hand each increment of legume addition from zero to 40 t ha^{-1} improved relative root length.

Rather massive amounts of organic residues are needed to change soil pH, but this study shows how readily decomposable leaves, high in N, Ca and Mg, can be beneficial in improving soil conditions and reducing Al activity. The increased presence of basic cations in soil

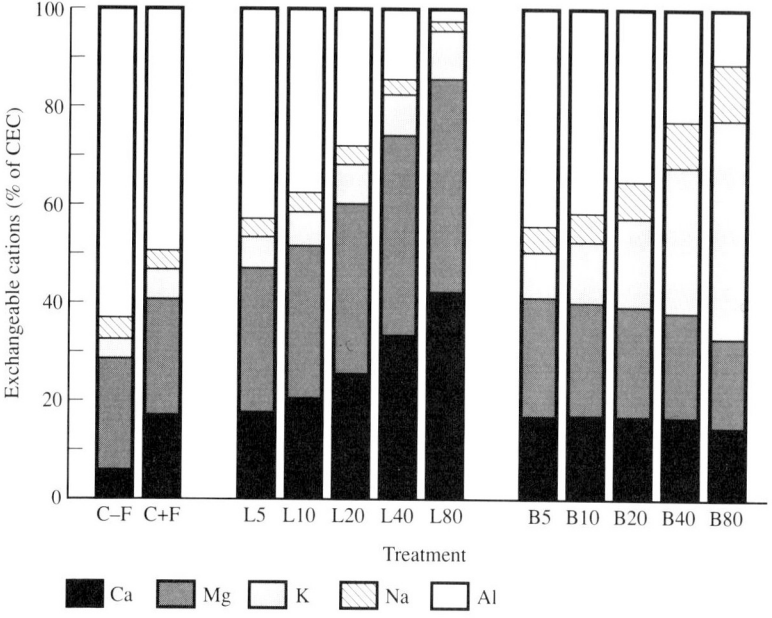

Figure 2.1 Relative amounts of exchangeable cations (Ca, Mg, K, Na, Al) in soils incubated with varying amounts of legume (L 5, 10, 20, 40 or 80 t ha^{-1}), or barley (B) (control C, − or + fertilizer (F)) (from Bessho and Bell 1992).

solution are expected to alleviate Al toxicity to plant roots. In this particular acutely acid soil it would have been necessary to add 14 t ha^{-1} of legume leaves, 42 t ha^{-1} of barley straw or 0.75 t ha^{-1} CaCO$_3$ to lower the monomeric Al activity to 10 mol L^{-1}; this figure is regarded as a critical level for phytotoxicity and one which reduces RRL by 10%. The findings of this laboratory study have been quoted at some length to illustrate the complexity of responses, which are also supported by field experience in which the accretion of organic materials has alleviated Al toxicity (Vaughan and Ord 1985).

One negative aspect of the use of pasture legumes is the soil acidity engendered under long-term legume-based pastures. The continued fixation of atmospheric N and the formation of ammonium and its oxidation to nitrate leads to a drain of hydroxyl ions, and a reduction in soil pH (Haynes 1983). In temperate pastures clover dominance over a period of 30–40 years may lead to a reduction of more than one unit of pH over a 60 cm depth of soil profile (Bromfield *et al.* 1983). We do not have good studies of tropical experience in this respect, but we may expect a similar long-term vulnerability to arise under

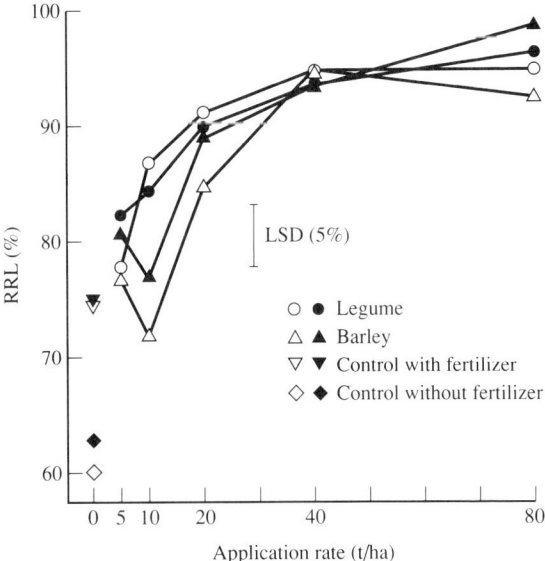

Figure 2.2 Effects of incubation of soil for 4 (open symbols) or 10 weeks (closed symbols) with varying amounts of legume or barley, control (with and without fertilizer on relative root length (RRL) of mung bean (from Bessho and Bell 1992).

legume-based pastures. This constitutes a further rationale for a ley system of short-term leguminous pastures interrupted by sequences of food or cash crops.

Organic matter as an energy source for biological activity

The living soil

Most organisms which inhabit the soil are heterotrophic, and depend for their activity upon energy sources derived from autotrophic green plants and algae. The nitrifying bacteria use CO_2 as a carbon source but it is the oxidation of ammonium and nitrite that provides energy. Biological activity therefore requires sources of labile carbohydrate, which are provided by plant litter, the incorporation of crop residues in soil, and by the decay of roots.

The rhizosphere, or the zone of soil influenced by the root, is a rich source of available carbohydrate (Pera *et al.* 1983). Microbial populations are 10–100 times more numerous in the zone within 1 mm of the root than they are in the overall soil volume (Thompson

1991). Exudates from roots, mucilaginous sheath, and the sloughing of epithelial root cells (Sparling 1985) account for this phenomenon. The free-living N-fixing bacteria are promoted by this source and the N-fixing *Azospirillum* which is associated with grass roots also depends upon root exudate (Umali-Garcia et al. 1980; Swift and Sánchez 1984). The vesicular-arbuscular mycorrhiza (VAM) inhabiting the roots, which play such a significant role in increasing the uptake of nutrients such as P and Zn, send out hypae several centimetres from the root surface, and are abundant in tropical pasture plants (Medina et al. 1988).

The soil biomass has a diverse population of bacteria, actinomcyetes, fungi, protozoa, algae and microfauna such as nematodes. Only a small proportion of these is biologically active at any time, because of a deficiency of energy substrate. These organisms show remarkably rapid shifts in density in response to changing conditions and a resilience in adversity; complete fumigation of the soil which kills most of the organisms is followed within a few days by a resurgence of activity with a new compositional pattern that continues to break down OM and release soil N. The maintenance of microbial biomass following the repeated use of the herbicides paraquat, diquat and glyphosate to control weeds in a no-tillage system is quite satisfactory, as illustrated in Table 10.7 (section 10.3; Thompson 1991).

The microbial biomass typically represents c. 2% of the organic C present, but this figure depends upon the balance of organic substances. These are predominantly polysaccharides (Cheshire 1985), but the small amounts of monosaccharides, the free sugars and the oligosaccharides, which are more labile, contribute disproportionately to microbial activity. The quality of the organic substances present determines not only the level of activity but the type of organisms present. Scientists used to emphasize the role of bacteria in the decomposition of crop residues, but in recent years it has become accepted that fungi play the primary role in the decomposition of complex OM (Swift and Sánchez 1984). Some scientists believe the intervention of many litter-feeding animals may be delayed until microbial action has broken down the more intransigent substances so that fungal hyphae and bacteria with a low C:N ratio of 5–15 may be used by mesofauna such as earthworms. On the other hand we should not undervalue the role of earthworms in comminuting litter to smaller fragments of material, more accessible to microorganisms.

Earthworm activity

Geophagers, or soil-eating organisms, are especially numerous in humid tropical and subtropical regions. At Lamto, Ivory Coast (1250 mm annual rainfall) the earthworm *Millsonia anomala* may consume 20–310 times its own weight of moist soil each day (Lamotte 1982), which may lead to an annual turnover of as much as 1000 t ha^{-1} year^{-1}. The breakdown of soil fractions, the microbial activity in the gut of the earthworm and the mucus it deposits may lead to the mineralization of as much as 800 kg OM ha^{-1} year^{-1}. The decomposer industry in the tropics is further reviewed by Morris *et al.* (1982) and Lamotte and Bourlière (1983).

The significance of leguminous cover crops for the development of microbial biomass and earthworm activity is illustrated by a study from Ibadan, Nigeria (8° N, 1300 mm annual rainfall) of an alfisol (Oxic Paleustalf) in which maize was grown either alone or in strips of a *Psophocarpus palustris* leguminous cover (Mulongoy 1986). Earthworm activity was pronounced under the live leguminous cover, as occurred in other experiments at Ibadan (Jaiyebo and Moore 1964; Lal *et al.* 1975), and the earthworm casts contained high amounts of organic C, total N and available P and showed a higher cation-exchange capacity and a low C:N ratio (Table 2.5). Microbial biomass C was also augmented in casts and in soil below the legume, when compared with soil under maize monoculture.

A further comparison between legume plots intercropped with maize and maintained totally under legume indicated greater earthworm activity in the absence of maize (Table 2.6). The earthworm casts represented a useful source of available N as indicated by the flush of mineral N following incubation. The percent stress-labile N (PSN), which is the percentage of total soil N mineralized after the soil has been fumigated (Ayanaba *et al.* 1976), is regarded as a measure of the quality of soil organic N that is a better index of soil N reserves than

Table 2.5 Characteristics of soils (0–0.1 m depth) and earthworm casts under *Psophocarpus palustris* at Ibadan, Nigeria (from Mulongoy 1986).

Soil characteristic	Under legume		Soil in bare maize plot
	Casts	Soil	
pH	6.1	5.5	5.6
Organic C (%)	3.36	1.80	1.73
Total N (%)	0.51	0.21	0.18
Bray-1 P (μg g^{-1})	30.5	5.3	9.6
CEC (mEq/100g)	17.3	3.54	2.20
C:N	6.59	8.57	9.61
Biomass C (μg g^{-1})	447	194	142

is total N. The values of PSN were high in earthworm casts (Table 2.6) and this further supports their value in augmenting soil fertility.

Other mesofauna are also significant contributors to OM degradation. In drier savanna woodland conditions termites may fulfil a role similar to that of earthworms in more humid regions, but may feed primarily on surface litter. In semi-arid northern Queensland, termites may represent between two and five times the biomass of cattle grazing the pastures (Holt and Easey 1984).

2.3 Organic matter accretion and decomposition

The balance between additions of OM to the soil and the losses of OM from the soil by decomposition is clearly of prime significance for plant production and also for the environment, since global warming is related to increasing levels of CO_2 in the atmosphere. Many studies do not distinguish the processes of accretion and of disappearance, and a simplistic appreciation of net changes in OM may overlook changes in the components of OM that are occurring; for example, additions of C to the refractile humin component may have a half-life of c. 2000 years (Vaughan and Ord 1985) whilst losses of lower molecular weight substances from the substrate may have contributed immediately to

Table 2.6 Characteristics of soils and earthworm casts under *P. palustris*/maize intercrop or *P. palustris* continuous cover at Ibadan, Nigeria (from Mulongoy 1986).

Attribute	Intercropped with maize		Legume cover		LSD (5%)
	Topsoil	Casts	Topsoil	Casts	
Biomass C ($\mu g\ g^{-1}$)	192	447	186	510	116
Mineral N flush ($\mu g\ g^{-1}$)	21	186	20	170	26
PSN* (%)	1.20	4.34	1.43	4.46	1.43
No. casts m^{-2}	NA	485	NA	876	
DM (g)	NA	285	NA	528	

*PSN, Percent stress-labile N.

microbiological activity. The equilibrium OM and N content of soils in natural systems depend upon the balance between the processes of accretion, which are often fast in the tropics, and the processes of decomposition, which may also be faster than in temperate regions. Decomposition is reduced under conditions of excess water, cool conditions, or high levels of acidity, and these factors determine the equilibrium value.

Organic matter accumulation

Usually OM accumulates under pastures and declines under cropping. Russell and Williams (1982) estimate that in Australia gains of C to soil under pasture are c. 7.7 Mt year^{-1} whilst losses of C from soil under crop are c. 4 Mt year^{-1}. Whilst it is expected that soils in undisturbed natural savanna will attain an equilibrium C value, managed grasslands are accumulating C to levels which greatly exceed the original value. At Beerwah, Queensland (lat. 27° S, 1630 mm annual rainfall) soil organic C increased under fertilized leguminous pasture from 0.56 to 1.99% in 8 years (Bryan and Evans 1973). Further examples are given in section 10.2. At Marondera, Zimbabwe (lat. 18° S, 1650 m altitude, 950 mm annual rainfall) soil organic C increased at 0.03% annually year^{-1} under a grass ley, (Table 10.2), whilst at Ibadan, Nigeria (lat. 8° N, 1300 mm annual rainfall) the annual rate of increase in soil C varied from 0.04 to 0.16% according to the type of pasture cover. The most extreme example is the change in the soil under a *B. humidicola* pasture at Manaus, Brazil where in this humid tropical environment the rate of soil C accretion was as high as 5 t ha^{-1} year^{-1} (Table 2.1; Cerri *et al.* 1991).

Rate of plant production

The rate of primary plant production is the first determinant of OM accumulation, although the latter is modified by edaphic and physiographic conditions, as expounded later in this chapter. Earlier studies of production under the auspices of the International Biological Programme grossly underestimated the role of tropical savanna in these processes, and it is now believed (Long *et al.* 1989; Jones *et al.* 1992) that primary productivity is between two- and five-fold greater than the previous mean figure for net photosynthesis of c. 800 g m^{-2} year^{-1} for tropical grasslands (Lieth 1978; Jordan 1981). A study from the United Nations Environment Programme (Figure 2.3; Long *et al.* 1989), reflects a truer picture, since it takes additional account of the continued death and decomposition of pasture shoots, roots and rhizomes, from which total net production may be calculated. Net primary productivity was 1242 g m^{-2} year^{-1} at Nairobi, Kenya (lat. 1° S, 950 mm annual rainfall, 1500 m altitude), 1741 gm^{-2} year^{-1} at a saline site at Montecillos, Mexico (lat. 19° N, 700 mm annual rainfall, 2220 m altitude) and 2220 gm^{-2} year^{-1} at Klong Hoi Kong, Hat Yai, Thailand (lat. 6° N, 2100 mm annual rainfall).

Previous estimates of productivity were based either on the difference between the maximum and minimum biomass present, or the summation of positive changes in the biomass present. Errors then arose because of the following:

- Senescence is continuous in tropical savanna. The rate of leaf senescence is often highest when tropical grasses are growing actively (Humphreys 1966; Wilson and Mannetje 1978; McIvor 1984), and in annual pastures plants disappear and decompose during the vigorous growth of a closed canopy (Rickert and Humphreys 1970). Measurements therefore need to be made of the accumulation of litter and of the decomposition of plant material. At Klong Hoi Kong shoot and root material turned over on average once every 2.7 and 7.7 months respectively.
- Many studies have concentrated on monitoring herbage shoots and neglected roots and rhizomes. The above-ground biomass represents a variable fraction of the total biomass present; in swards of *Paspalum notatum* the C present in roots and stolons may greatly exceed the C in shoots (Hirata *et al.* 1986). It is difficult to measure changes below ground, and Figure 2.3 indicates that this component has a varying significance not simply related to changes in the above ground biomass; net production of shoots was highest at Klong Hoi Kong, but the production of roots and rhizomes was proportionately greater in the saline environment of Montecillos. Similarly, high levels of *Echinochloa polystachya* shoot production at Manaus, Brazil were associated with low levels of root production (5% total biomass) at a seasonally inundated site (Long *et al.* 1989).
- The reflectance ratio of red/near-infrared light is a powerful tool for estimating changes in the green biomass of growing crops by the use of remote sensing. This technique may be misleading when applied to tropical savanna, since the red/near-infrared ratio fluctuates widely according to the amount of dead material present in the herbage (Jones *et al.* 1992).

Management of plant production

The net gain or loss of C to the soil system is next dependent upon the management of the pasture or the ways in which crop growth and crop residues are manipulated.

Organic matter accretion and decomposition

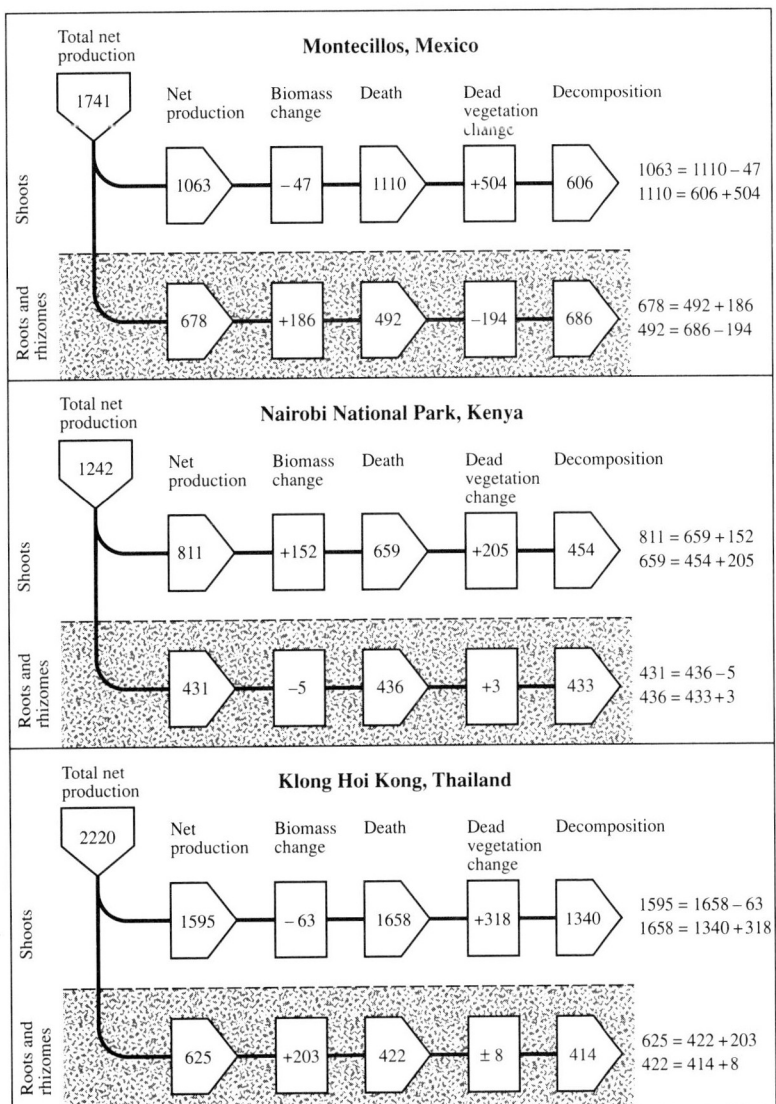

Figure 2.3 Net primary production (g m^{-2} year^{-1}) of grassland sites in Mexico, Kenya and Thailand, showing values for shoots and for roots and rhizomes of fluxes (arrowed boxes) and of net changes in conserved quantities (rectangles) in net production, death and decomposition, together with the method of calculation (from Long *et al.* 1989).

- Litter accumulation is negatively related to the SR of the pasture (Mears and Humphreys 1974) and the allocation of C from leaves to roots and stems is also negatively related to grazing pressure (Humphreys 1991). Gains of soil C are not expected from overgrazed pastures, and the high figures for net primary production in Figure 2.3 were obtained for pastures from which large grazing animals were excluded, although other fauna and other decomposers were present. This topic is further discussed in section 10.5.
- Burning of grasslands and of crop residues limit the opportunities for accretion of C to the soil mass. Most tropical grasslands have resilient growth responses to fire, but frequent burning not only converts above-ground C to CO_2 but reduces the subsequent growth of roots and of shoots. At the Klong Hoi Kong site production of *Eulalia trispicata* grassland, in the year after a fire which followed a long period of protection, was 91% of the unburnt site, but a second fire combined with the occurrence of drought led to a net loss of roots and rhizomes of 229 gm^{-2} year^{-1} and to greatly reduced shoot production of 406 gm^{-2} year^{-1} (Kamnalrut and Evenson 1992).
- Choice of species for planted pastures will influence the level of C addition and also its ease of incorporation in the soil C pool. In Sri Lanka some farmers recondition their soils with grasses before replanting tea. A series of field trials compared different grasses in terms of their shoot DM production and their root residues at the end of the reconditioning period. The best grasses were *Tripsacum laxum*, *Cymbopogon confertiflorus* and *Panicum maximum* (providing cut material was retained), whilst *Eragrostis curvula*, sugar-cane and *P. maximum* (cut-and-remove practice) were less successful in this system (Sandanam *et al.* 1982). The cost in nutrients needed to convert organic materials to humus should enter this assessment, as mentioned in section 2.2.
- Inclusion of legumes in the pasture provides both readily decomposable material and a source of N, which may otherwise constrain production. Fertilizer inputs (section 10.4) may be needed to develop a viable system, whilst intensive systems involving irrigation provide more rapid OM accretion than occurs under dryland conditions.

Organic matter depletion

Many models of the relationship between OM content and time have been developed: single exponential, double exponential, asymptotic, linear, quadratic and power (Wieder and Lang 1982). Two-pool models, in which one OM pool decomposes rapidly and the other pool, containing the more refractile components, mineralizes more slowly, have merit (Skjemstad *et al.* 1988; Melillo *et al.* 1989). An exponential equation embracing the C values in the virgin soil, at equilibrium and at the point of measurement and a rate constant for decomposition is illustrated in section 10.2 and Tables 10.4 and 10.5 for data of Dalal and Mayer (1986b) on the effects of cropping on OM loss on the Darling Downs of Queensland. The main factors involved in OM decomposition appear to be as follows.

Soil temperature and moisture

Plants grow faster in warm conditions than in cold weather and provide greater additions of OM; the rate of decomposition of OM has a similar positive relationship with temperature. This is illustrated simplistically by comparing the decomposition of identical standard ryegrass material labelled with ^{14}C and buried in soils at Ibadan, Nigeria and at Rothamsted, UK. The retention of C with time fitted a double exponential function (Figure 2.4), a two-compartment model in which 70% of the material decomposes at a more rapid rate than the remaining material. The decomposition coefficient for each compartment was four times greater at Ibadan than at Rothamsted, and the adjustment of the time scale of Figure 2.4 by a factor of four put the values on the same curve. The mean temperature at Ibadan is 17.2 °C greater than at Rothamsted.

The rate of decomposition is also positively related to soil moisture. At Gayndah, Queensland (lat. 26° S, 730 mm annual rainfall) Robbins *et al.* (1989) measured the decomposition of *P. maximum* var. *trichoglume* litter in nylon bags buried in soil. Although N release was proportional to the amount of N present, the rate of DM loss was independent of the initial N content of litter. Net release of N occurred when the N concentration of residues reached *c.* 0.6–0.7% N, i.e. a C:N ratio of 70–80:1 and not a ratio of 20–30:1 (Mulder *et al.* 1969) was critical for decomposition of this tropical grass. The best relationship with DM decomposition was the moisture content of soil in the 0–10 cm layer, and this relationship was improved in combination with average daily temperature. These two components gave a linear relationship with rate of disappearance, which accounted for 60% of the variation about a mean of 4.9 mg g^{-1} d^{-1}. The cycle of wetting and drying increases N mineralization from soil humus (Birch 1958), and the amount of nutrients mineralized upon wetting is positively related to the duration and severity of the previous drying period.

Aeration

Aerobic environments provide the highest levels of decomposer activity. Accumulation of OM occurs under anaerobic conditions (Swift *et al.* 1979), and swamp grasslands have higher contents of OM than grassland developed on well-drained soils. This mainly arises from the better moisture availability favouring pasture growth and OM accumulation and the inhibition of decomposers in anaerobic conditions, although nutrient and pH factors are also involved. The composition of the decomposer industry adjusts to anaerobic conditions but these are less efficient overall than anaerobic decomposers. The development of acidity slows down the rate of organic transformations (Beckwith and Butler 1983).

Soil particle size and density

Changes in the total OM of soils are a cruder indicator of the effects of management than are changes in the more labile forms of OM, which both reflect and determine biological activity and nutrient availability. Earlier segments of this chapter discussed the difficulty of defining what is meant by OM and the need to distinguish biomass, fresh debris and readily

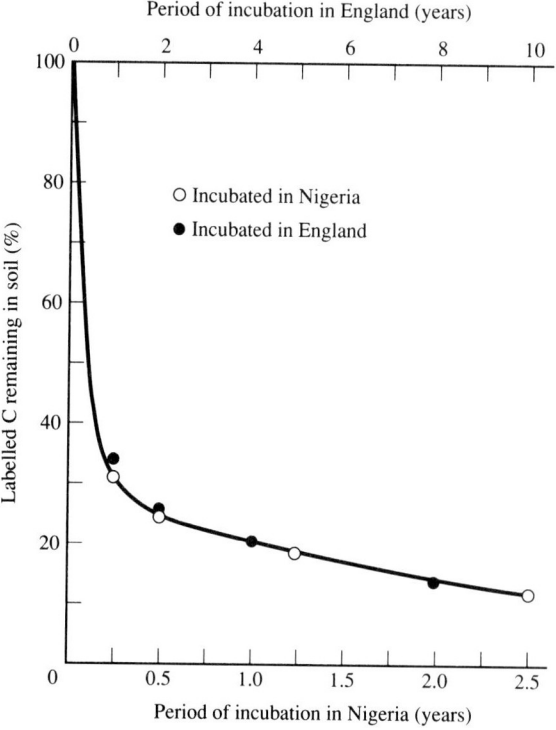

Figure 2.4 Decomposition of uniformly labelled ryegrass in Rothamsted, England, and in Ibadan, Nigeria (from Jenkinson and Ayanabe 1977).

decomposable material, and the humic substances that are fairly resistant to breakdown, and which were designated conventionally as fulvic acid, humic acid and insoluble humin (Vaughan and Ord 1985). An important consideration is the degree to which they are associated with the clay fraction of the soils, where they are relatively protected from microbial and enzymic attack.

It is therefore of interest to consider the size distribution of particular soils, since this modifies the mineralization of OM and indicates also the location of OM fractions within the soil, which are all undergoing mineralization simultaneously but at very different rates. Skjemstad et al. (1988) refers to the three most significant moderators of rates of OM turnover as:

- molecular recalcitrance, as brought about by a highly cross-linked aromatic structure high in free radicals;
- direct association of substrate with inorganic ions such as Ca, Fe, Al, polycations of Fe and Al, and with hydrated clay surfaces which form complexes;
- physical separation of substrate from the soil microflora and fauna, associated with the small pore size of the clay particles (Ladd et al. 1977).

Soil particles may be considered as sand-size (0.02–2 mm), silt-size (2-20 μm) and clay fractions (<2 μm). When these fractions are separated and the changes in the OM content of each studied, the sand-size fraction is usually the one which exhibits most rapid turnover of OM. This is illustrated for a Langlands-Logie clay soil (Typic Chromusterts) from southern Queensland (Dalal and Mayer 1986c). The soil in virgin sites contained 2.2% organic C and 49% clay. Sites with varying periods of cultivation, which extended to 40 years, were examined, and the decrease in organic C in the three size fractions with age of cultivation investigated. The rate of decrease in the sand-size fraction (Figure 2.5; Dalal and Mayer 1986c) was clearly greater than for the other fractions; the rate constant averaged 0.109 t C ha^{-1} relative to values of 0.078 and 0.039 for the silt and clay fractions respectively. The rate of organic C loss was substantially greater from the sand-size fraction than from the total soil. The sand-size fraction consists mostly of plant debris and fungal hyphae; these are readily lost from the soil by oxidation, but also disintegrate upon cultivation to enter the silt-size and clay-size fractions. In the latter it is protected against enzymic and microbial attack (Dalal and Mayer 1986b). However, in some soils high rates of loss occur from the silt-size fraction, indicating that whilst particle size is a significant factor determining rates of turnover, the nature and extent of organic C association with particular inorganic soil constituents modify this generalization. There may be a 'priming action' of newly added residues containing labile C

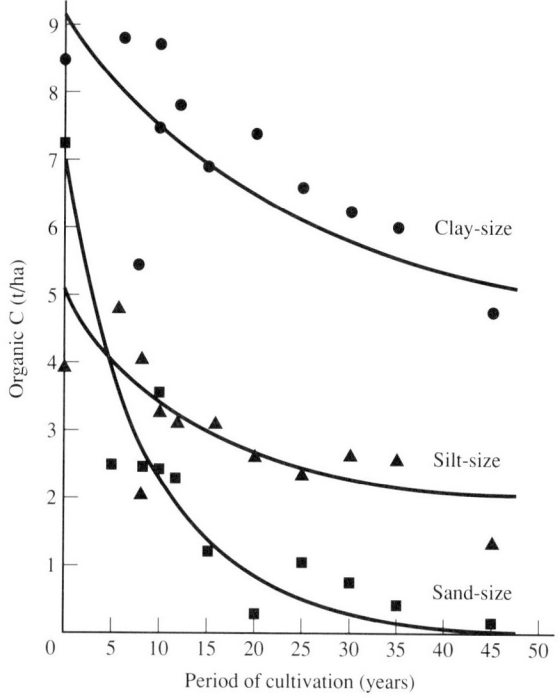

Figure 2.5 Changes in amounts of organic C in sand-size, silt-size and clay-size fractions with period of cultivation of a Langlands-Logie soil in southern Queensland (from Dalal and Mayer 1986c).

which increases the decomposition of native and otherwise refractile C (Vaughan and Ord 1985).

An alternative approach in identifying the labile OM is to consider the density rather than the size of soil particles. Greenland and Ford (1964) drew attention to the significance of the OM in the 'light' fraction (usually of density <2 Mg m^{-3}) as an early and more sensitive indicator of the effects of soil management on OM turnover than that derived from studying total soil OM. The 'light' fraction accumulates in soils under pasture and decreases rapidly upon cultivation. It contains much structural litter which has not undergone much humification. Various densimetric techniques have been developed to separate soil particles without changing their basic structure. Dalal and Mayer (1986d) used a bromoform–ethanol mixture with an added surfactant and applied a density fractionation procedure to soils with a differing history of cultivation. Their study has been simplified (Table 2.7) to consider

Table 2.7 Density fractionation of various soils in southern Queensland and organic C distribution (from Dalal and Mayer 1986d).

Soil	Organic C (%)		Light fraction % soil weight	% soil organic C in light fraction
	Total soil	Light fraction		
Waco	1.63	12.8	2.6	20
Langlands-Logie	2.23	20.5	2.8	26
Cecilvale	1.73	16.3	2.2	21
Billa Billa	1.48	15.5	3.2	34
Thallon	0.77	7.3	1.8	18
Riverview	1.28	15.6	2.4	29

only the light fraction (< 2 Mg m^{-3}) and the aggregate of other density fractions. In these soils the light fraction accounted for only 1.8–3.2% of soil weight, but contained 18–34% of total soil organic C. The loss of organic C from the light fraction was 2–11 times faster than that from the heavy fraction in fine-textured soils, but this did not apply in the coarse-textured Riverview soil. Dalal and Mayer (1986d) found that the rate of loss of light fraction OM was significantly related to the reciprocal of clay content.

Other soil factors involved are the ratio of organic C/urease activity, which is a measure of OM accessibility to enzyme degradation, the degree of soil aggregation, and the exchangeable Na percentage, which reflects clay mineralogy, montmorillonitic clays (low in K) protecting organic C better than illitic clays (high in K) (Dalal and Mayer 1986b). Allophane, which is especially prevalent in soils of the wet tropics, is an even more effective protector of OM (Sánchez 1976).

Cropping practice

Cropping practice is a major determinant of the rate of OM breakdown, and the augmentation of OM through the return of crop residues or the maintenance of 'live' mulch is discussed further in section 10.3.

- Tillage causes breakdown of OM, since reduction in particle size and the increased aeration of the soil exposes molecules to greater attack by microbial enzymes and other decomposing agents (Swift and Sánchez 1984). Tillage is consequently a means of increasing the availability of nutrients to the crops subsequently planted or to perennial pastures which benefit from renovation. The alternative view, which has been illustrated in Chapter 1, is that continuous

cultivation leads to a rundown in OM (Dalal and Mayer 1986a), associated with decreasing crop yield and increasing soil loss. Long bare fallows are especially damaging.
* Systems of minimum tillage have therefore been developed which are soil conserving and in which a slower rate of OM mineralization is a concomitant of lower temperatures under mulch and reduced intensity of cultivation (Uren 1991). The retention of crop residues on the soil surface provides better protection against erosion and a slower rate of OM mineralization than the incorporation of material in soil. This is illustrated by a study on the Darling Downs, Queensland, in which ^{14}C-labelled wheat straw was either spread uniformly on the soil surface or was intimately mixed with the top 10 cm of soil. In the first 1.5 months recovery of added C was greater in surface straw (Figure 2.6; Cogle et al. 1987), and thereafter decomposition occurred at rather similar rates for the next 12 months. This difference is attributed to the greater positional availability of carbon to soil organisms where straw was incorporated, and this also led to greater denitrification.

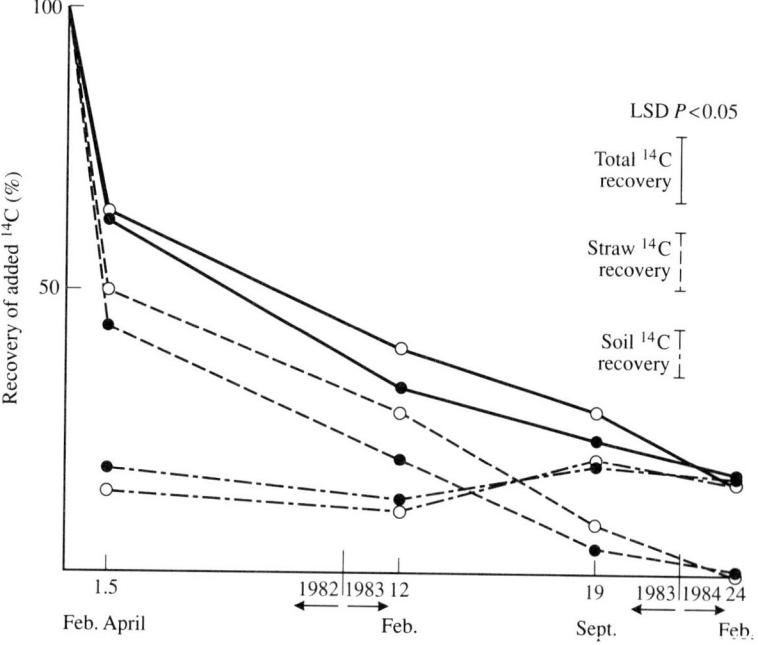

Figure 2.6 Percent recovery ^{14}C-labelled wheat straw following field incubation on the soil surface (○) or incorporated in the soil (●) (from Cogle et al. 1987).

- Incorporation of cereal residues of low N content can lead to a greater N demand on the agricultural system, as mentioned in section 2.2, and as elaborated subsequently in section 2.4. An alternative is the increased incorporation of legumes in crop rotations. These provide residues of lower C:N ratio which may augment N supply as well as OM. One problem is the more rapid decay of surface residues, which reduces their effectiveness in ameliorating the soil microclimate (Figure 10.7) and reducing soil erosion. In the absence of comparative crop data an illustration of the persistence of pasture residues in a ley system at Katherine, Australia (lat. 14° S) is given. Residues of perennial grass *Urochloa mosambicensis* and of the legume *Stylosanthes hamata* cv. Verano were applied to the soil surface at the rate of 3 t ha^{-1} and watered every second day. After 60 days the loss of legume DM was 50% (Figure 2.7; J. P. Dimes personal communication) and that of the grass 35%.
- Forage legumes and grasses used in ley systems differ in their rate of litter breakdown. This is illustrated by a study at Carimagua, Colombia (lat. 5° N, 2160 mm annual rainfall, 175 m altitude) in which litter was collected on three occasions from pure stands and buried in nylon bags in May, July and October. The half-life of litter was calculated from an exponential function, and was greatest during the dry season. The legumes *Stylosanthes capitata* and *Arachis pintoi* produced litter with faster decomposition rates

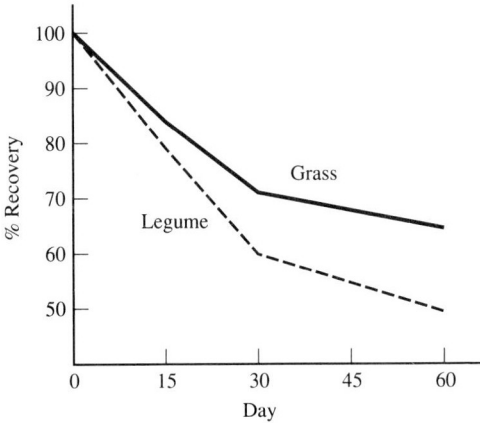

Figure 2.7 Percentage DM recovery of surface applied grass (*Urochloa mosambicensis*) and legume (*Stylosanthes hamata*) at Katherine, Australia (J.P. Dimes, personal communication).

Table 2.8 Half-life (days) of litter from forage species buried on different occasions at Carimagua, Colombia (from CIAT 1991a).

Species	Time of litter distribution		
	May	July	October
Legumes			
Arachis pintoi	50	45	150
Centrosema acutifolium	122	120	250
Desmodium ovalifolium	136	169	251
Pueraria phaseoloides	111	119	218
Stylosanthes capitata	26	70	85
S. guianensis	79	105	250
Grasses			
Andropogon gayanus	89	102	222
Brachiaria decumbens	99	118	244
B. dictyoneura	116	123	328
B. humidicola	91	173	248

than those of *Centrosema acutifolium*, *Pueraria phaseoloides* and *Stylosanthes guianensis* (Table 2.8; CIAT 1991a). The grasses had rather similar rates of decomposition during the wet season, but *Brachiaria dictyoneura* was slowest to decay during the dry season. These differences were best explained by the ratio of lignin: N concentration which was positively correlated with litter half-life.

- The practice of burning crop stubbles leads to a gradual decrease in soil OM. Loch and Coughlan (1984) monitored the properties of a cracking clay soil (Udic Pellustert) at Warwick, Queensland (lat. 28° S) on which the management of wheat stubbles had been varied for 5 years. Plots were either tilled or managed with zero tillage and herbicides, and stubble was either retained or burnt. Organic C in the top 0–10 cm layer was not significantly affected by tillage treatment, but was reduced by burning of stubble (Table 2.9). Zero tillage had other effects: soil dispersion ratio was reduced, and this was found to be solely due to reduced Na exchange capacity. The concentration of chloride was less under zero tillage, and increased leaching of chloride was consistent with the effect of this treatment in increasing moisture accumulation in the soil profile. Deep infiltration under zero tillage might be due to better stability of the surface soil, as indicated by its reduced dispersion ratio, or to the maintenance of soil cracks not removed by tillage. The problem of nitrate leaching in zero-till systems is mentioned in section 10.3.

2.4 Soil nitrogen

The opening affirmation of this chapter declared availability of soil N to be the first constraint on production which is susceptible to management by farmers. Most of the N (c. 90%; Anderson and Vaughan 1985) in soil is bound in OM, and the rate of mineralization of OM primarily determines the supply of N for plant growth. More direct sources of N from fertilizer application, return of animal excreta and biological N fixation assume differing significance according to the type of agricultural system. The overriding significance of N supply for pasture growth is illustrated by a study (Graham *et al.* 1981) of pasture productivity in relation to the age of *Cenchrus ciliaris* pastures established after clearing woodland dominated by *Acacia harpophylla* and *Eucalyptus cambageana* in central Queensland (lat. 23-25° S, *c.* 600–750 mm annual rainfall). Sites varied from 2 to 34 years since clearing, and total soil N in 31 virgin sites adjacent to the pastures ranged from 0.094 to 0.197%. Soil from eight paired sites, which varied in N status, was collected and *C. ciliaris* was grown in pots

Table 2.9 Effects of management practices on soil properties at Warwick, Queensland (from Loch and Coughlan 1984).

Soil property	Tillage		No tillage		F value	
	Burn	Retain stubble	Burn	Retain stubble	Zero tillage	Stubble retention
Organic carbon (%) 0–10 cm	1.60	1.67	1.59	1.72	n.s.	11.39**
Chloride (μg g^{-1}) 60–90 cm	360	444	185	115	6.02*	n.s.
Ratio CEC:% clay 0–4 cm	0.87	0.87	0.82	0.81	197.1**	n.s.
Exchangeable Na (%) 0–4 cm	4.3	4.4	2.6	2.2	53.6**	n.s.
Dispersion ratio 0–4 cm	0.33	0.35	0.27	0.30	19.71**	n.s.

n.s., not significant ($P > 0.05$), *$P < 0.05$, **$P < 0.01$.

to compare production from the virgin woodland areas and the areas long established to grass pastures. The uptake of N in grass shoots was positively associated with soil total N%, and DM yield from the pasture areas (Y_d) was then plotted as a percentage reduction relative to DM yield from the virgin woodland areas (Y_v) against the reduction in total soil N% associated with pasture development. This showed a linear relationship ($r=0.76$; Figure 2.8) in which 50% reduction in total soil N% reduced plant growth by 52%.

The availability of nitrogen for plant growth

Nitrogen transformations

Soil N is mostly unavailable for uptake by plant roots; for example, Queensland soils usually contain 5–12 t N ha^{-1}, with kraznozems containing in excess of 20 t N ha^{-1}, but mineral N content, excluding fixed ammonium (NH_4^+), is normally less than 1% of these values (Vallis 1979; Myers et al. 1986). Plant roots take up the great part of

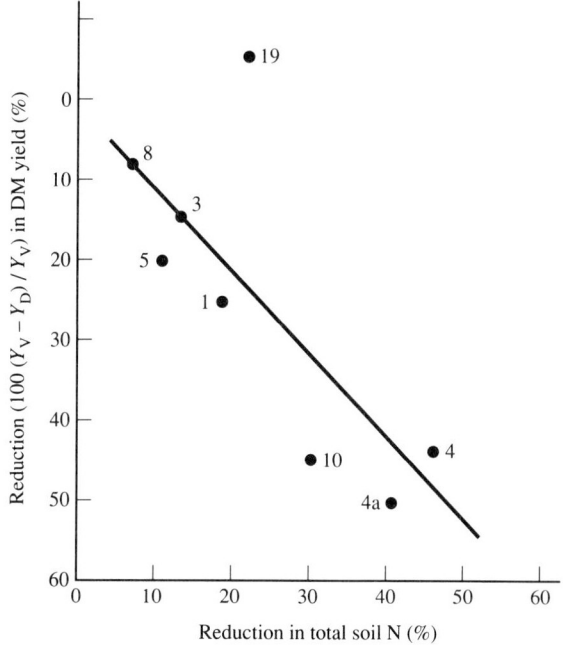

Figure 2.8 The relation between the percentage reduction in DM yield of *C. ciliaris* and reduction in percentage total soil N (from Graham et el. 1981).

their nitrogen as nitrate (NO_3^-); (NH_4^+) is also a direct source and is especially significant in very acid soils where ammonium is not readily nitrified and there is a shortage of microbial nitrifiers (Figure 2.9). The normal pathway for N uptake is the transformation of NH_4^+ to nitrite (NO_2^-) by microbial transformation with *Nitrosomonas*; nitrite rarely accumulates in soils, but is transformed to nitrate (NO_3^-) by *Nitrobacter*.

An understanding of plant N uptake therefore depends upon monitoring a number of dynamic processes (Lathwell and Bouldin

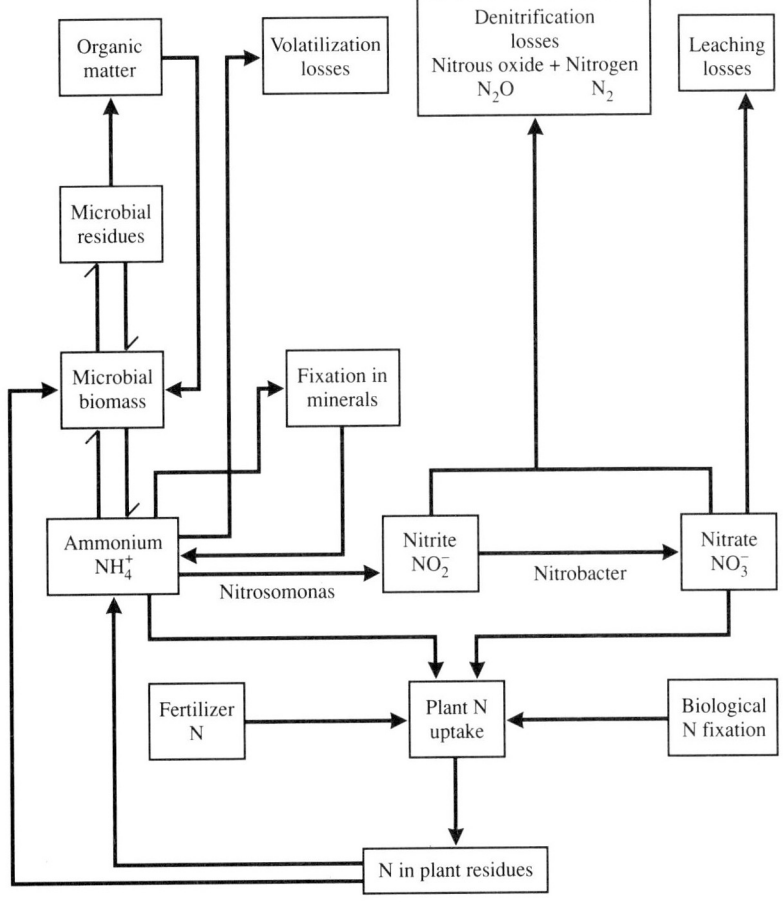

Figure 2.9 Some pathways of N transformation in the soil–plant system.

1981). The pools of N which are involved in the soil–plant system are set out simplistically in Figure 2.9; what should be recognized is that the size of the pools is secondary to the rate of movement between pools, although small changes in a key pool such as the availability of labile carbohydrate can cause rapid shifts in the microbial biomass which results in rapid release of NH_4^+. Grazing of forage, which is mentioned in the next section, introduces a further cycle in which N is immobilized in animal tissues or returned as excreta to the soil as OM and ammonia or lost to the atmosphere. Biological N fixation is shown in Figure 2.9 as contributing directly to plant uptake, as occurs within the *Bradyrhizobium* nodules on plant roots, but it may also be directed to OM, as discussed subsequently.

The central stabilizing factor is the pool of OM, which contains material of differing enzymic and microbial availability. Whilst OM is being decomposed, the processes of mineralization (through microbial transformation of OM to ammonium) and immobilization of N (through the conversion of inorganic forms of N to organic N) are occurring simultaneously; and scientists often monitor the net balance of these two sets of processes which determine the N available for plant uptake. During the decomposition of plant residues plant N is degraded and transformed to organic N in cells and extracellular metabolites or appears as ammonia. Factors affecting the gains and losses of soil OM were discussed in the previous section 2.3, and much of that material is pertinent to the N cycle. The flow of N to and from OM is essentially mediated through living organisms, whose actual biomass contains only a small fraction of the organic N but may have a low C:N ratio of *c.* 6; a representative range of values for grassland soils is 2–15 g N m^{-2} in microbial biomass relative to 90–1600 g N m^{-2} in soil OM (Woodmansee *et al.* 1981). The dead microbial residues are an especially labile fraction. The different components of OM and their varying resistance to mineralization were described earlier in the Chapter; reviews are available from Van Veen *et al.* (1984), Anderson and Vaughan (1985), Myers *et al.* (1986), Skjemstad *et al.* (1988), and Steele and Vallis (1988). It is also possible that endomycorrhiza may transform N direct from OM to the plant root system.

The binding of ammonium by clay minerals and the formation of ammonia–organic matter complexes creates a further N pool. Native fixed ammonium accounted for 1–13% of total N in the 0–15 cm layers of soils in southern Queensland, and 1–20% of total N in subsoil (Black and Waring 1972). It is especially prevalent in soils with the micaceous clay minerals illite and vermiculite. Availability to plants of native fixed ammonium is very low. Recently fixed ammonium, which usually occurs following fertilizer application, has moderate availability (Black and Waring 1972).

The losses of N from the soil-plant system are:

- Volatilization losses. Direct losses of ammonia from plant residues occur, in addition to the expected losses from fertilizers, urine and dung. Hot, dry conditions favour losses of ammonia, and this problem is exacerbated on soils of high pH and low cation exchange capacity. Deficits of up to 10% of ^{15}N occurred from surface-applied litter of *Chloris gayana*, which may have been associated with ammonia loss (Moore 1974).
- Denitrification from nitrate (Figure 2.9) following the transformation processes from ammonia results mainly from biological activity. It is most acute under anaerobic conditions, as occur in poorly drained or flooded soils, and is positively associated with soil temperature (Steele and Vallis 1988). A small level of nitrous oxide is now known to occur during nitrification. Additionally, if nitrite accumulates after fertilizer application where high pH inhibits its conversion to nitrate there may be gaseous loss from chemical denitrification.
- Leaching losses depend upon the degree of soil drainage, and are most acute when rainfall is greater than evapotranspiration, in irrigation systems, and when high nitrate availability, as may occur through heavy fertilizer N application, exceeds plant N demand. Nitrate level of groundwater and of water entering streams used in human consumption is clearly an issue of environmental quality.

The 'priming' effect of adding organic residues

Plants given fertilizer N may take up more N from soil than is taken up from the soil by unfertilized plants (Jenkinson *et al.* 1985). This effect was first demonstrated with ^{15}N-labelled fertilizer, and may arise in different ways. An augmented N supply to the plant may cause increased root exploration of the soil mass, resulting in greater uptake of native soil N. Alternatively there may be substitution between the different pools of N in the soil, and microbial immobilization of N through the processes of OM decomposition may change the level of plant N uptake, perhaps depending upon the quality of the plant residues present. In ley systems the farmer is interested in the effects of legume residues on the behaviour of the plants which are subsequently grown.

The differing effects of forage legumes and of pulse crops on the performance of the succeeding crop is well illustrated by a study in which a varying number of crops of soybean (*Glycine max*) and of Siratro (*Macroptilium atropurpureum* cv. Siratro) were grown and the tops returned to the soil (Yaacob and Blair 1980). The culture of legumes influenced the total N of the sandy loam (ex granite) used, and this varied from 0.041 to 0.086% with one to six crops of soybean, and 0.046 to 0.110% with one to six crops of Siratro. A standard amount of ^{15}N-labelled legume residue, which contained 1.48% and 2.61% N in soybean and Siratro respectively, was then mixed with the soil or not added; the latter gave an indication of N uptake from the native soil N unamended with further legume residues. Seedlings of *C. gayana* were next grown following two cycles of wetting and drying to commence the processes of residue breakdown. The recent addition of legume residues increased both available soil N and grass growth, which was also positively associated with the number of legume crops previously grown in the soil. The total uptake of N in *C. gayana* shoots and roots (Table 2.10) was also increased according to the number and type of legume crops grown, Siratro giving a superior result compared to soybean. The N recovery in the plants grown in pots recently amended with legume residues was substantially greater than expected from estimates of the ^{15}N recovered from the added residues. This indicated a positive 'priming' effect of added legume residues on the release which led to additional recovery of N in *C. gayana* and of native soil N, and which varied from 11.8 to 96.1 mg N per pot; this represented 14–39% of the N uptake by plants. The effect was greater from Siratro residues, which were mostly leaf with a C:N ratio of 16.1, than from soybean residues, which contained more stem and pod material and which had a C:N ratio of 28.4. Yaacob and Blair (1980) suggest that additions of residues may have a positive, neutral or negative effect on plant N uptake, depending upon the balance of immobilization and mineralization, as

Table 2.10 Effects of added legume residues on the additional release of N (mg/pot) from native organic matter (from Yaacob and Blair 1980).

Legume cropping history of soil (no. crops)	N recovery from unamended soils (1)	N from residues (2)	Expected recovery (1) + (2)*	Recovery in amended pot	Additional recovery from native organic matter
Soybean					
1	25.6	21.8	47.4	59.2	11.8
3	29.8	21.7	51.5	63.5	12.0
6	42.6	24.8	67.5	98.3	30.8
Siratro					
1	32.0	35.8	67.8	91.3	23.5
3	37.8	110.7	148.5	244.6	96.1
6	111.8	144.9	256.7	298.0	41.3

*Expected recovery assuming added residues had no stimulating effect on native soil N.

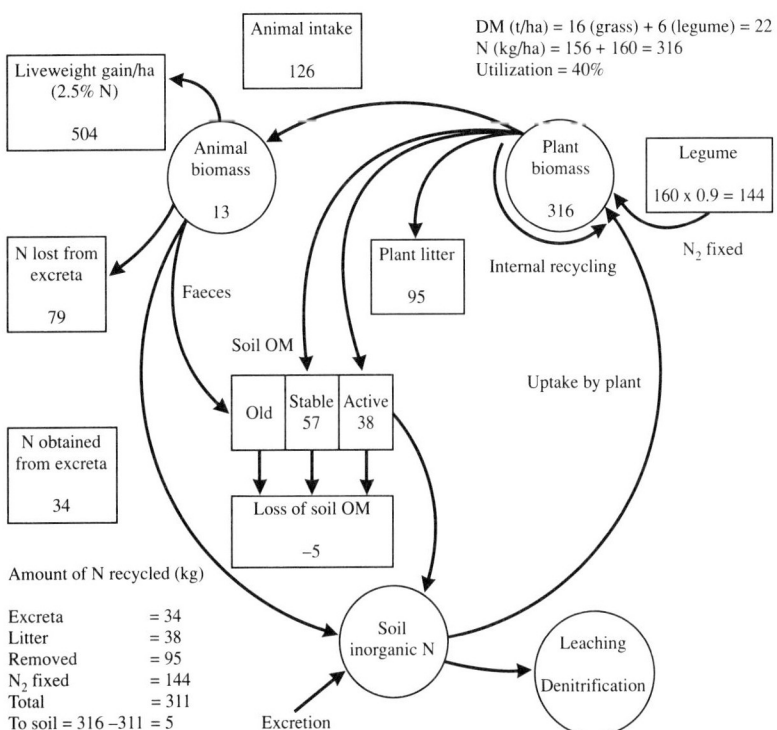

Figure 2.10 Annual N cycle in a grazed *Brachiaria decumbens–Arachis pintoi* pasture at Carimagua, Colombia (from CIAT 1991b).

influenced by the nature of the residues and the condition of soil N present.

Gains and losses of nitrogen in the agricultural system

The nitrogen cycle under grazing

An example (CIAT 1991b; Thomas 1992) of the annual gains and losses of N in a grazed grass/legume pasture is available from the llanos at Carimagua, Colombia (lat. 5° N, 2160 mm annual rainfall, 175 m altitude). This pasture was estimated to produce 16 t DM ha^{-1} of *Brachiaria decumbens* and 6 t ha^{-1} *Arachis pintoi* (Figure 2.10), totalling c. 316 kg N ha^{-1}. The legume was estimated to contribute 90% of its N uptake from *Bradyrhizobium* atmospheric N fixation, or 144 kg ha^{-1}, and the balance of pasture N uptake arose from the soil and from internal plant N recycling. Utilization of the pasture by grazing cattle was controlled to 40%, or 126 kg N ha^{-1} consumption. Liveweight gain of 504 kg ha^{-1} accounted for 13 kg N ha^{-1}, giving c. 90% return of N ingested to the pasture as faeces and urine. Most of this (79 kg N ha^{-1}, a high estimate) was lost from the system, presumably as ammonia or leached nitrate, leaving 34 kg N ha^{-1} returned to the soil. Plant litter production of 95 kg N ha^{-1} contributed 38 kg N to the active OM pool, available for plant uptake, and 57 kg N to replenish the stable OM pool. Losses from denitrification also occurred, but the net loss from the soil OM was only 5 kg N ha^{-1}, indicating the great benefits from the N fixation of a well-adapted legume.

It is difficult to find comparable (and courageous) studies of such detail from the literature of tropical pasture science; Steele and Vallis (1988) provide two examples from Queensland, whilst Robbins (1984, quoted in Humphreys, 1991, pp. 33–34), reports a study from a lower rainfall site. A basic assumption is that grazing accelerates the N cycle. Under lenient grazing or in the absence of grazing the increased accumulation of litter and of root growth immobilizes much N in their decomposition, leading to pasture run-down, as discussed in Chapter 1. Grazing cycles N in forms more available to the plant, but the losses through volatilization of ammonia and leaching of nitrate from areas where excreta are concentrated need to be compensated from soil OM reserves, from biological N fixation, or from N fertilizer. A cut-and-remove system clearly deprives the soil–plant system of nutrients, unless the excreta of the fed animals are carefully returned to the area. In Zimbabwe Swift et al. (1989) calculated that the collected excreta from 14 ha of savanna are needed to support production of 2 t ha^{-1} maize grain on 1 ha of arable land, and this is further discussed in Chapter 6.

N losses of 70% from excreta were estimated in the Carimagua example above. At Katherine, Australia (lat. 14° S, 950 mm annual rainfall) urine labelled with ^{15}N was applied during the dry season, and losses accumulated to 46% of the urea N applied (Vallis et al. 1985); greater losses are expected under hot, dry conditions than from cool, moist soil. In the dry tropics at Townsville, Queensland, losses were 16–32% (Vallis and Gardener 1984). There is therefore considerable variation expected in the efficiency of N cycling under grazing.

Nitrogen losses under cropping

The run-down of soil fertility and the decrease in production associated with continuous cropping received attention in Chapter 1. The losses of organic C are illustrated subsequently (Tables 10.4 and 10.5 in section 10.2) for a study of cereal cropping on the Darling Downs of Queensland, and references to their relationship to the soil 'sand-size' fraction and to the labile OM in the 'light' fraction were made in Table 2.7 and Figure 2.5 in section 2.3. These changes are mirrored in the losses of soil N, and the average annual loss of N from the profiles of six soils varied from 31 to 51 kg N ha^{-1}, which could be mainly accounted for by crop removal (Dalal and Mayer 1986e). The half-life of loss of N varied from 2.5 to 11.8 years of cultivation, whilst the overall rate of loss (the rate constant k_c) ranged from 0.06 to 0.27 year^{-1}. The proportional loss of N was less in soils of higher clay content; the protection of OM by the clay fraction was discussed earlier. Similarly, the loss of N was greatest in the sand-size fraction, accounting for 89% N loss in one soil, and in the light fraction ($<$ 2 Mg m^{-3}), where labile OM was concentrated (Dalal and Mayer 1987a). Cultural practices which increase the amount of OM in the light fraction may be expected to enhance the rate of N cycling through the microbial biomass and to augment the availability of the nutrients present.

These findings suggest that changes in total soil N are less sensitive indicators of the effects of cultural practices on soil fertility than other measures. Microbial biomass and the mineral N flush were referred to in Tables 2.5 and 2.6 in section 2.2, and are also discussed in section 10.3 and Table 10.7. In Dalal and Mayer's (1987b) study the 'stabilized' microbial biomass accounted for 2–5% of total N in the soils of the Darling Downs they monitored. Microbial biomass decreased with age of cultivation as did the N mineralization potential, N_o. This index is used by many scientists (Campbell *et al.* 1981) to predict the capacity of the soil to supply N to plants and is measured by incubating soil under standard conditions for up to 30 weeks and estimating the rate of N mineralization. This value accounted for 11–23% total N in different soils, and represents soil organic N fractions which differ in their susceptibility to mineralization.

2.5 Biological nitrogen fixation

It is clear from the above discussions that soil N status in many tropical farming systems will only be maintained if changed practices involving increased use of fertilizer N or the stimulation of biological N fixation are adopted. Increased application of fertilizer N requires the use of fossil fuel, which exacerbates the foreign exchange difficulties of developing countries that are not rich in resources of oil or natural gas, and which adds to environmental problems such as global warming. Biological N fixation is an environmentally 'cleaner' solution, and may sometimes be accomplished in lower input agricultural systems appropriate to the economic and social conditions which apply.

Sources of nitrogen fixation

Legume symbiosis

The symbiosis of host legumes with rhizobial bacteria which convert atmospheric N_2 to forms which the plant can use is the greatest source of biological N fixation.

Rhizobial bacteria are free-living in soil, and plants form nodules containing the bacteria as sessile lateral outgrowths of the root; plants such as *Sesbania spp.* (Alazard et al. 1988) or *Macroptilium lathyroides* which grow in flooded situations also form nodules on their stems. There is some difference of view about the naming of these bacteria; recent practice designates the type of bacteria that are slow-growing when cultured and which mostly inhabit tropical plants as *Bradyrhizobium*, whilst the bacteria that are fast-growing when cultured retain the original genus name of *Rhizobium*. This is not a simple division according to the provenance of the host legume; for example, the rhizobial bacteria for the tropical shrub legume *Leucaena leucocephala* are fast-growing. A common method of infection follows the curling of root hairs in the presence of these bacteria and the subsequent development of an infection thread; in *Stylosanthes* there is direct infection of the 'crack' where a lateral root meets a main root (Chandler et al. 1982). Nodules show varying shape, size and differentiation depending upon the host plant and its tribal affinity (Corby 1988).

The host plant supplies carbohydrate to the root system and there is an energy cost to the subsequent N transformations in the nodule; Warembourg and Roumet (1989) estimate this is c. 4 mg C per mg N fixed, and Henzell (1988) regards the cost of N fixation as having only a minor effect on the yield of plants in the field. Ammonia is the primary product of N_2 fixation; this is released from the bacteroids and assimilated into glutamic acid and glutamate in the surrounding cells (Peoples and Herridge 1990). In most of the tropical forage and food legumes the ureides allantoin and allantoic acid are the principal N compounds exported from the nodules to the xylem sap, and the degree to which legumes are relying on soil N and on nodule N can be estimated from ureide analysis.

Nodulation is widespread throughout the Leguminosae; the non-nodulating legumes are usually confined to some woody shrubs, especially within the group occurring in wetter tropical regions (Corby 1990). The flora of Zimbabwe is taken as an illustration. The number of legume species forming nodules relative to the total number of species examined for the three sub families were 410/412 for Papilionoideae, 62/69 for Mimosoideae, and 41/58 for Caesalpinioidiae (Corby 1974). These legumes supply N for the growth of plant organs which then enter the cycle of the agricultural system, as illustrated in Figure 2.10.

The growth of the legume and its capacity to supply carbohydrate to the nodule is the first determinant of the amount of N_2 fixed, provided the host and bacterium can form an effective symbiotic relationship. The factors determining legume growth include the genetic capabilities of the legume and the bacterium, the levels of radiation, temperature and rainfall, soil factors which determine nutrient availability, and biotic factors such as grazing, pest and disease attack and the interference of non-legume plants. This is further modified by the degree of dependence on soil N, which is susceptible to management. Management of the legume to maximize N fixation is discussed subsequently.

The amount of N contributed by the legume to the soil–plant system is sufficiently great to modify agricultural output, but is insufficient in the humid tropics and subtropics for associated grasses to express their full growth potential. High input systems will therefore continue to incorporate N fertilizer application, as discussed in Chapter 6. A brave generalization is made that 15–40 kg N fixation occurs for each 1000 kg DM of legume shoots produced. The output of N fixation of tropical and temperate legumes is similar.

There is now a vast literature relating to N fixation of tropical legumes. The estimates are sometimes crudely based on the difference in productivity of pastures based on grass–legume mixtures and on grass alone, on the equivalent benefit of N fertilizer relative to the inclusion of legumes in a sward, on changes in total soil N, or on estimates of N balance. Some estimates are based on the level of acetylene reduction as a measure of nitrogenase activity, on studies of the natural abundance of or enrichment with ^{15}N, or on the ureide content of legume cell sap (Peoples and Herridge 1990).

The following field illustrations, based on methods of estimation of varying reliability, are grouped into sites with good rainfall or subhumid sites receiving less than 1000 mm annual rainfall. Some of the early claims should have converted scientists who were sceptical about the benefits of tropical legumes. Jones (1942) recorded 180 kg N ha^{-1} year^{-1} for the first 5 years, and 110 kg N ha^{-1} year^{-1} for the last 4 years of a *Neonotonia wightii* ley in Kenya; Hutton and Bonner (1960) estimated N uptake of 560 kg ha^{-1} year^{-1} from *L. leucocephala* in Queensland; and Moore (1962) noted 280 kg N ha^{-1} year^{-1} from *Centrosema pubescens* in Nigeria. There are other reports in the 100–300 kg ha^{-1} year^{-1} range (Bruce 1965; Jones 1967; Jones *et al.* 1967; Whitney and Green 1969; Haggar 1971; Keya 1974; Johansen and Kerridge 1979; Mello 1980; Reynolds 1982). In drier regions much lower levels of N fixation are expected. In a drought year there may even be a net loss of N, as reported by Vallis (1972; -29 kg N ha^{-1} year^{-1} at Rodds Bay, Queensland), which may be more than counterbalanced in moister years ($+10$ to $+30$ kg N ha^{-1} year^{-1}). In the monsoon tropics of Northern Territory, Australia Myers (1976) found more available

N but no change in total soil N under a *Stylosanthes humilis* pasture, whilst at the same site Wetselaar (1967) recorded 73 kg N ha^{-1} year^{-1}. In northeast Australia *S. hamata* contributed 33–38 kg N ha^{-1} year^{-1}, despite some leaching of nitrate (Probert and Williams 1986), whilst Crack (1972) found that N fixation of *S. humilis* increased from 77 to 120 kg N ha^{-1} year^{-1} with increasing rate of P application. These levels of N fixation are sufficient to influence grass growth, the quality of the diet of grazing animals, and any subsequent cropping activity.

Frankia

The significance of rhizobial N fixation by leguminous trees and shrubs is expanded in Chapters 8 and 9. It should also be noted that some non-leguminous trees have N-fixing organisms in their roots. The soil actinomycetes of the genus *Frankia* form an association with *Casuarina* spp. (Dreyfus et al. 1987) which is capable of fixing 40–60 kg N ha^{-1} year^{-1}. This has been demonstrated with *C. equisetifolia* growing on coastal sand dunes in Senegal (Dommergues 1963), a finding subsequently confirmed using ^{15}N (Gauthier et el. 1985).

The actual benefits of field inoculation with *Frankia* are illustrated by a study (Reddell et al. 1988) at Gympie, Queensland (lat. 26° S, 1340 mm annual rainfall), which also raises an important philosophical question. Biological N fixation may be regarded as a process appropriate for low input agricultural ecosystems. Much research effort has been directed successfully to commercializing legumes which are tolerant of low fertility soil conditions and efficient in their exploration for and utilization of scarce nutrients. The alternative approach is to maximize N fixation by removing the constraints imposed to the process by low mineral supply.

Seedlings of *C. cunninghamiana* were planted on a sandy lateritic podzolic soil (ultisol) at a spacing of 2×2 m. They were inoculated with *Frankia* (or not), and split dressings of double superphosphate totalling 150 kg P ha^{-1} were applied (or not) over a period of 18 months. There was also an uninoculated N fertilizer treatment, in which split dressings of ammonium nitrate totalling 160 kg N ha^{-1} were applied. P was a primary limitation to tree growth (Table 2.11); 22 months after planting there was no response to *Frankia* in the absence of applied P, and little response to N fertilizer. *Frankia* inoculation increased volume of wood by 34% when the level of soil P was increased. Actinorrhyzal trees are likely to play a greater role in agroforestry systems.

Azolla

The aquatic fern *Azolla* has a symbiotic relationship with the cyanobacterium *Anabaena azollae* which fixes atmospheric N^2. This fern is widely grown in flooded tropical rice fields and has played a significant part in

Table 2.11 Growth of *C. cunninghamiana* at Gympie, Queensland, as influenced by *Frankia* inoculation and application of P and N (from Reddell *et al.* 1988).

P treatment	N treatment	Height (m)	Basal stem diameter (cm)	Estimated wood volume (m^3 ha^{-1})
No P applied	Nil	2.31	5.3	6.0
	Applied N	2.51	5.8	7.7
	Frankia inoculation	2.34	5.1	5.5
P applied	Nil	3.19	6.8	14.3
	Applied N	3.89	9.0	27.9
	Frankia inoculation	3.47	7.8	19.2
	SE_d	0.12	0.2	1.7

maintaining the productivity of these cropping lands in Vietnam for some centuries (Lumpkin and Plucknett 1980; Martinez *et al.* 1986). It is used as a green manure, as a weed suppressant, as feed for animals (including fish) and as a component of digester systems producing methane.

The fern floats on water, forming a deep mat. It has a heterosporous life cycle. The *Anabaena* inhabit cavities in the leaves enclosed by epidermal cells which form a pore capable of atmospheric gaseous exchange. The cyanobacterium fixes N_2, producing ammonia which is assimilated by the fern or which may leak from the plant. There are various reports of the level of N fixation, most of which indicate fixation in the range 1–2 kg N ha^{-1} day^{-1} (Lumpkin and Plucknett 1980). Since the cyanobacterium and its host normally remain associated throughout vegetative and reproductive development there is limited opportunity for interchanges, but Plazinski *et al.* (1988) have demonstrated that diversification in *A. azollae* exists and might be used in developing new and potentially more productive symbiotic relationships. There are several species of *Azolla*; in Thailand the growth of *A. caroliniana* and *A. microphylla* was superior to that of some lines of *A. pinnata* (Choonluchanon *et al.* 1988).

There are many reports of *Azolla* benefiting subsequent rice production. It may be cultivated in special ponds for year-round production of *Azolla*, 5–10% of the crop area being required for each rice crop, i.e. 10–20% of the area if double cropping is practised. Alternatively *Azolla* may be cultivated in the rice fields and incorporated into the soil between rice crops. This requires that a separate starter culture area be retained for use after the dry season (Lumpkin and Plucknett 1980). *Azolla* clearly has a role in sustaining the yields of irrigated rice and may have marginal significance as a forage for cattle but the subject is peripheral to the main thrust of this book.

Associative nitrogen fixation

The discovery of nitrogenase activity in the rhizosphere and roots of tropical grasses, which was initially identified with the bacterium *Azospirillum lipoferum* (Day et al. 1975), caused considerable excitement amongst scientists in the 1970s. *Azospirillum* was found to be widespread amongst C_4 tropical grasses, appeared to be associated with malate in the root substrate, and the levels of acetylene reduction which were measured suggested the presence of a significant N input. This helped to explain why grass production was sustained in the absence of legumes, and the possibilities for inoculation with *Azospirillum* were canvassed. Additional bacteria with these properties were discovered (Boddey and Döbereiner 1988): *A. brasiliense*, *A. amazonense*, *A. halopraeferans*, the acid-tolerant *Acetobacter nitrocaptans* in sugar-cane, *Bacillus azotofixans*, *Campylobacter nitrofigilis*, *Herbaspirillum seropedicae*, and *Pseudomonas* sp.

Some of the high expectations (Schank et al. 1977) about the level of N fixation were not realized, and in some studies on poor soils an *Azospirillum* response only occurred if some fertilizer N was applied (Smith et al. 1976). The levels of estimated N fixation proved usually to be in the low range of 0–30 kg ha^{-1} year^{-1} (Koch 1977; Bouton et al. 1979; Weaver et al. 1980; Weier et al. 1981; Sanoria et al. 1982; Okon et al. 1983). However, subsequently Boddey and Victoria (1986) using ^{15}N, showed that pastures of *Brachiaria humidicola* and of *B. decumbens* were fixing 30 and 45 kg N ha^{-1} year^{-1} respectively, which represented 30 and 40% respectively of their N uptake, and which must be regarded as influential amounts. *A. brasiliense* shows host specificity (Smith et al. 1984) and inoculation appears to have been sufficiently beneficial to justify commercial adoption in Israel (Boddey and Döbereiner 1988).

Controversy exists concerning the mechanisms of plant response to *Azospirillum*. It is known that *Azospirillum* stimulates the production of growth hormones (Gaskins et al. 1983) and growth responses to inoculation have also been associated with increased plant uptake of soil and fertilizer N, perhaps linked to increased bacterial nitrate reductase activity in plant roots (Boddey and Döbereiner 1988) and to enhanced root exploration of the soil mass. Current scientific fashion emphasizes rhizobial N fixation as a process more likely to modify the sustainability of agriculture than associative N fixation.

Management of biological nitrogen fixation

The farmer can manage the agricultural system to maximize N fixation through manipulation of the following:

- choice of elite legumes well adapted to the local environment;
- inoculation of legumes showing specificity;
- management of soil conditions to remove mineral constraints to N fixation and to divert legume N uptake from soil sources of N;
- establishment of a legume density leading to good growth;
- appropriate management of the interference of non-legumes;
- defoliation practice which maintains N fixation and N transfer as designated.

These objectives will be elaborated for the different types of farming systems described in Chapters 8–11, and in the following section some general principles are enunciated.

Rhizobial specificity and ecology

Most tropical legumes are promiscuous in their rhizobial affinities and will nodulate well with indigenous rhizobial bacteria wherever they are planted. Some legumes are quite specific in their requirements for a rhizobial strain, some strains are more effective in N fixation than others, and some soils have a low rhizobial population. This has led to the manufacture of superior inoculants and to the field inoculation of seeds, which are demanding processes. The practice has only been adopted in a limited number of developing countries.

The failure of nodulation following legume establishment is common in a number of forage species, and field responses occur in demanding tropical forage species such as *Centrosema pubescens* (Vargas and Suhet 1982 in Brazil; Tang and Menéndez 1988 in Cuba; Nurhayati *et al.* 1989 in Java), *Leucaena leucocephala* (Homchan *et al.* 1989a), *Lotononis bainesii, Stylosanthes guianensis* var. *intermedia*, and in the pulse legume *Glycine max* (Nautiyal *et al.* 1988). Frequently scientists have isolated rhizobial lines exhibiting élite performance for N fixation under laboratory or glasshouse conditions. These strains often contribute to nodulation and N fixation in the first 2 or 3 years after establishment, before being displaced by native rhizobia, which are more incursive and perhaps faster in infecting root cells (Henzell 1988). In Australia applied rhizobia initially formed 100% of nodules on *L. leucocephala*, but 2 years later these accounted for only 12–16% of nodules (Bushby 1982). In northeast Thailand *L. leucocephala* inoculant formed > 97% of nodules at all field sites, but after 64–72 weeks it was responsible for 50–90% of nodules and the growth response to nodulation had disappeared (Homchan *et al.* 1989a). For the more promiscuous *S. hamata* and *S. humilis* native rhizobial strains were even more competitive (Homchan *et al.* 1989b), and local strains were especially successful on *S. hamata* cv.

Verano (Ruaysoongnern and Aitken 1980). Even where the inoculant successfully invades the host, responses in N fixation and growth may not occur, as in Hawaii (Singleton and Tavares 1986).

Responses to inoculation are most likely to occur under the following conditions:

- appropriate inoculation technology – the designated strains need to survive environmental stresses such as sunlight and pesticides and be delivered to the planted seed with more than 300 bacteria per seed, covered with an adhesive;
- low soil nitrate availability, as discussed subsequently;
- highly strain-specific legumes;
- a long history of cropping with non-legumes which has depleted the indigenous rhizobial population;
- use of legumes new to the region.

The introduction of legumes from distant sites and their inoculation with bacteria from yet another area may lead to a poor symbiosis. Thus accessions of *Stylosanthes* spp. from alkaline soils only nodulated effectively with bacteria isolated from equivalent conditions (Date *et al.* 1979), and similarly *Desmanthus* spp. with a provenance of alkaline soils required specific inoculation for nodulation on an acidic soil at Gympie, Queensland (Date 1991). This problem of transfer across geographical boundaries is illustrated by a study of *S. guianensis*, which exhibits considerable genetic variation in the effectiveness of symbiotic N fixation (Date and Norris 1979). *S. guianensis* was collected from various locations in the Americas, which have been grouped into phytogeographic regions (Figure 2.11) originally proposed by Good (1964). Some 184 lines of known provenance were then grown and inoculated with the widely used rhizobial strain CB756, which was collected from *Macrotyloma africanum* in Zimbabwe (Edye *et al.* 1974). These lines showed diverse responses (Table 2.12), which have been grouped according to their provenance. Most of the *S. guianensis* accessions from the humid tropical low-altitude environments of the Mexican lowlands and coast, Guatemala, Panama, Venezuela and Guyana and the higher altitude Andean region (numbers 24A, 24C, 25 and 28) were effectively nodulated by CB756. Only eight accessions were collected from regions 24B and 24D and the results are less reliable. However, many of the accessions from South Brazil and the adjacent pampas were ineffectively nodulated with CB756. The authors relate these findings to the variation in *S. guianensis* evident in other work they have done which led to morphological–agronomic grouping of these accessions. This study well illustrates the need to recognize the special character of the host–bacterium relationship and the need for appropriate choice of inoculants.

Management of soil conditions

Two guiding principles for farmers are the need (1) to plant élite legumes well adapted to local soil conditions and (2) to ensure that legumes are more dependent for their N supply on nodule activity than on soil N. Much research effort is directed to extending the range of legumes which are resistant to environmental stresses such as waterlogging, soil acidity, salinity and other mineral toxicities (see Humphreys 1981, chapter 6). The simplest method of agricultural improvement is for farmers to establish new cultivars which perform better in N fixation than the plants previously grown. The level of N fixation will reflect the growth and persistence of the legume in the local soil conditions.

Table 2.12 Response of *S. guianensis* accessions to inoculation with rhizobial strain CB756 according to provenance (from Edye *et al.* 1974).

		Response to inoculation (% accessions)		
No.	Region	Ineffective	Effective	Highly effective
24	Caribbean A Mexican lowlands and coast	5	52	43
	B South Florida, West Indies Bahamas, Bermudas D North Colombia North Venezuela	87	7	0
	C Guatemala Panama	4	70	26
25	Venezuela Guyana	50	50	0
27	South Brazil A Eastern coast B Uplands central Brazil, Bolivia	37	34	29
	C Eastern Highlands D Grand Chaco–Paraguay Argentina	53	34	13
29	Pampas			
28	Andean	0	82	18

Economic conditions may favour the use of fertilizers to increase legume growth and N fixation. This is discussed further in section 10.4 and a case study is illustrated in Table 10.10 and Figure 10.11; Table 2.11 also refers to this problem. The first concern is the deficiency of nutrients that limit legume growth. There is usually a positive association between P and N concentrations in legume shoots, which is additional to but associated with the effect of P supply on legume growth. The response to P fertilizer should therefore be evaluated not only in terms of the additional forage grown, but also in terms of the increased N supply, which benefits both soil fertility and the quality of animal diet. This is illustrated by a study (Blunt and Humphreys 1970) at Mt. Cotton, Queensland (lat. 28° S, 1400

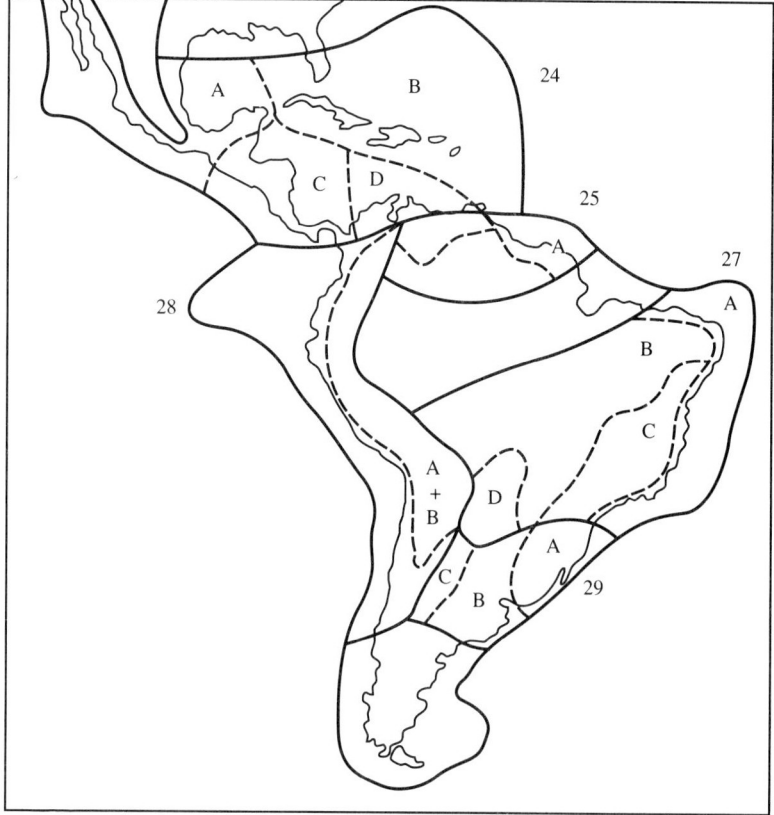

Figure 2.11 Phytogeographic regions of the Americas used in Table 2.12 (from Edye et el. 1974).

mm annual rainfall) where various legumes were grown with *Setaria sphacelata* var. *sericea* on a red-yellow podzolic soil (ultisol). The legumes exhibited differing efficiency of response according to P level. At a low level of P application (11 kg P ha^{-1}) *L. bainesii* and *M. atropurpureum* swards yielded an additional 1.7–1.9 kg N kg^{-1} P applied. *Desmodium intortum* cv. Greenleaf showed its best response (1.0 kg N kg^{-1} P applied) at higher levels of P application (45–89 kg ha^{-1}). In *S. humilis* N assimilation was 17 mg g^{-1} nodule DM day^{-1} with low P supply and 53 mg g^{-1} nodule DM day^{-1} with adequate P supply (Gates 1974).

Information is often available about the nutrients which limit overall plant or crop performance in a district: the need for P, K, S, the response to liming, and the micronutrient deficiencies. Additionally the status of the elements which are specifically involved in N fixation have to be considered, since these may add to the suite of minerals determining plant growth. Reference was made earlier in section 2.2 to the greater effect of Al toxicity on N fixation than on the growth of plants dependent upon soil N (Carvalho *et al.* 1981). In most cases limitations of mineral nutrient supply do not affect the growth and multiplication of rhizobial bacteria in the soil (O'Hara *et al.* 1988), although low Ca availability in the vicinity of the legume rhizosphere limits multiplication. Following entry to the host the bacteria are dependent upon the supply of nutrients in the host tissues. Once they are enclosed by a membrane envelope, the properties of this peribacteroid membrane influences whether the nutrients present in the plant cytoplasm are adequate for the proper functioning of the bacteroids.

The minerals with roles in N fixation that may be regarded as distinctive are Ca, Mo, B, Co, Fe and Cu (O'Hara *et al.* 1988). Ca plays a specific part in nodule formation of species which depend on crack rather than root hair infection, such as *Stylosanthes* and *Arachis*. Mo is essential for N fixation and is a constituent of Mo-nitrogenase. Legumes differ in their sensitivity to Mo deficiency; *Neonotonia wightii* and *D. intortum* exhibited a greater requirement for application of maintenance Mo fertilizer than *L. bainesii* or *S. guianensis* cv. Cook in one study (Johansen *et al.* 1977). B deficiency is known to inhibit nodulation or nodule growth, and this is probably related to the especial sensitivity of meristem activity to B deficiency. Although Co is a widespread dietary deficiency of animals, it is not an essential mineral for plant growth but is required for nodule activity, especially in relation to the synthesis of vitamin B$_{12}$. Fe deficiency is usually evident in plant growth, but Fe may also control nodule function through its constituent role in nitrogenase, leghaemoglobin and in other proteins. A specific role of Cu in nodulation is also suspected. Fertilizer policy for legume forages needs to take these elements into account.

Soil nitrate

Reference is made subsequently in section 10.2 to the higher rates of N fixation that are feasible on low fertility compared to high fertility soils. There are two separate effects operating. On the one hand the nitrate ion inhibits nodule infection; this is a process for which genetic variation in tolerance appears to be available (Murphy and Sherwood 1989). On the other hand the dependence of the legume on N fixation for its own N uptake is reduced under conditions of high soil N availability. This also influences the competitive relationships of grasses and legumes; fertilizer N application favours the C_4 grasses with their higher growth potential. The use of compound fertilizers containing N, sometimes the only type of fertilizer available in developing countries, is inimical to the survival and performance of legumes.

Situations can arise where the earliness of nodulation of legume seedlings is limited by the growth of the seedling and the limited supply of surplus carbohydrate to the young root system, especially on low fertility soils. In these circumstances application of starter N fertilizer dressings at the rate of 10–30 kg N ha^{-1} may favour (Diatloff 1974) or not impede (Whiteman 1972) nodulation.

The presence of a companion grass increases the dependence of the legume on N fixation. This is illustrated later in this chapter. There are seasonal changes in the level of dependence, according to the rate of growth of the companion grass and the effect of temperature on soil nitrate levels (Vallis *et al.* 1977; Cadish *et al.* 1989).

Vesicular-arbuscular mycorrhiza

The occurrence of fungi in the roots and rhizosphere of tropical plants known as VAM is widespread. These organisms depend on plant carbohydrate from the plant root system for their energy, and their presence enhances mineral uptake and rhizobial nodulation. For example, *Glomus fasciculatum* and *Gigaspora margarita* were effective on *Acacia mangium* and *Albizia falcataria* in the Philippines (De la Cruz *et al.* 1988), *G. fasciculatum* on *S. scabra* in Brazil (Purcino and Lynd 1985) and on *L. leucocephala* in India (Punj and Gupta 1988), and *G. intraradices* on several forage legumes in Florida (Medina-Gonzales *et al.* 1987).

The shrub legume *L. leucocephala* is regarded as a seedling which is slow to establish, and careful management is needed to control competing weeds until the shrub overtops them. The rate of seedling growth is improved by early nodulation and by high levels of P supply. It is now clear that *Leucaena*'s reputation in these respects partly arises from the paucity of mycorrhiza in some soils, especially acid soils. This is illustrated by a pot study at the University of Queensland in which a sterilized red-yellow podzolic soil (ultisol) from Mt. Cotton (pH 5.3, limed to pH 5.5, 15 mg kg^{-1} P (bicarbonate extraction)) was planted to *Leucaena* inoculated with *Rhizobium* strains CB81 and NGR8, and grown with different levels of P application, with or without inoculation with magnolioid root pieces of *Leucaena* infected with VAM (Ruaysoongnern 1987). At the lowest level of P application (50 kg ha^{-1}) *Leucaena* growth after 10 weeks (Table 2.13) was poor in the absence of mycorrhizal inoculation, but greatly improved by infection with VAM. This was associated with improved P nutrition, as indicated by the increase in the P concentration of the youngest fully expanded leaves from 0.07 to 0.32%, and with improved N supply, as shown by the substantial improvement in rhizobial nodule mass. The growth response to mycorrhizal presence was still real for these seedlings at the intermediate P level and disappeared if a massive application of P fertilizer was made; the latter still had a positive effect on leaf P and N concentrations and on nodule mass.

Agronomic management of N fixation

Management practices during legume establishment may ameliorate the factors adverse for N fixation, such as high soil temperature (Ranga Rao 1977; Lee and Döbereiner 1982). Legume growth and N fixation are positively related to legume density, at least up to a level which exploits environmental growth factors effectively. The response in these attributes to added fertilizer is greater at high legume density (Rickert and Humphreys 1970; Crack 1972), so that high planting rates facilitate N fixation in the establishment year.

The influence of a competitive grass on legume N fixation is

Table 2.13 Yield, nodulation and P % concentration of *Leucaena leucocephala* as influenced by inoculation with mycorrhiza and by P supply (from Ruaysoongnerm 1987).

Mycorrhiza inoculation	Level of P application (kg ha^{-1})					
	50		150		450	
	−	+	−	+	−	+
DM yield (g/pot)	2	28	21	36	87	81
Nodule DM (mg/pot)	0	300	0	350	1330	2030
Leaf P (%)	0.07	0.32	0.09	0.28	0.17	0.23

illustrated from an intercropping study in north Thailand. The practice of intercropping with annual legumes is elaborated further in Chapter 11. In this experiment (Rerkasem and Rerkasem 1988) ricebean (*Vigna umbellata*) was grown in varying proportions with maize (*Zea mays*) in a replacement series design in which monocultures had densities of 80 000 ha^{-1} for maize and 160 000 ha^{-1} for ricebean. The crops were grown with 20 or 200 kg N ha^{-1} as urea. At the lower level of N fertilizer application the intercrop delivered substantially greater N yield than the monocrops (Table 2.14). The most striking feature of these results is that level of N fixation in ricebean grown in a ratio of 25:75 maize:legume was equal to the amount of N fixation in ricebean monoculture, despite the competition and shading by the maize plants. The percentage dependence of ricebean on N fixation (rather than on soil or fertilizer N) increased with increasing maize competition. This effect also held even at the higher fertilizer N level. This is illustrated in a more sophisticated way from data from a similar experiment in an earlier year (Figure 2.12). The relative concentration of N as ureides in the xylem sap, which is a measure of the origin of N from atmospheric N_2 fixation, is seen to be greater in the two maize–ricebean intercrops with the highest maize densities. The level of dependence on N fixation increased until the commencement of podset, and decreased thereafter. The depletion of the pools of soil N by companion grass growth was shown under this agronomic management actually to enhance biological N fixation.

Table 2.14 N uptake from different sources of maize and ricebean grown in varying proportions at two levels of N fertilizer (from Rerkasem and Rerkasem 1988).

Fertilizer N (kg ha^{-1})	20				200			
Ratio maize:/ricebean	100:0	75:25	25:75	0:100	100:0	75:25	25:75	0:100
N uptake (kg ha^{-1}): Maize								
from fertilizer	6	5	3	–	106	97	46	–
from soil	68	59	40	–	124	86	46	–
Ricebean								
from fertilizer	–	1	2	3	–	15	47	56
from soil	–	6	27	35	–	13	47	66
from rhizobia	–	62	98	97	–	32	60	55
System total:	74	133	169	134	229	242	246	178
% N fixation (ricebean)	–	90	76	72	–	51	39	37

Defoliation of legumes

The potential gains in N fixation through the adoption of the principles described above may be lost in farm practice if inappropriate utilization of the legume occurs. The system of defoliation of the legume needs to take into account its direct effects on N fixation and on the spatial transfer of nutrients. The key principles are as follows.

- Regular or severe defoliation reduces the supply of assimilate to the nodules and reduces rate of N fixation.
- Grazing or cutting increases the proportion of N fixed which is located in the shoots.
- Lenient defoliation which removes flowers or pods increases the diversion of assimilate to nodules.
- Most of the N transfer from the legume to the soil system and to companion grasses occurs through the return of animal excreta

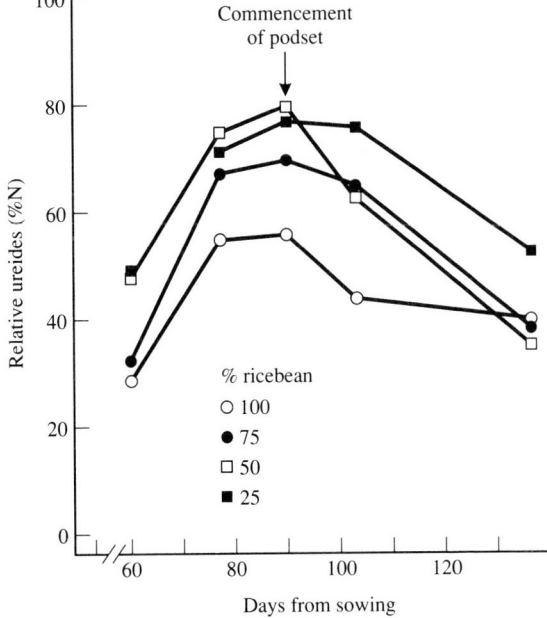

Figure 2.12 Relative concentration of N in ureides of xylem sap of ricebean grown in varying proportion with maize (from Rerkasem *et el.* 1988).

or through the senescence of legume shoots and their addition to surface litter.

The first factor controlling the rate of N fixation is the supply of energy to the bacteroids in the nodules. When the leaf surface is removed by grazing or cutting the reduction in the rate of photosynthesis is reflected first in a shortage of surplus carbohydrate flowing to the root system. The nodules are most reduced by defoliation; this is followed in descending order by roots, stem and inflorescence, with leaf growth following defoliation the organ least affected. Severe defoliation may cause the premature senescence and shedding of nodules, delayed nodule initiation, and a reduction in the size to which nodules grow (Whiteman 1970a). In a study of *Macroptilium lathyroides* in which different defoliation procedures varied the subsequent growth of the plant, Othman and Asher (1987) found that N fixation was positively and linearly related to the dry weight of new shoots. Othman et al. (1988; also illustrated in figure 4.15 in Humphreys 1991) showed elegantly the close relationship between the supply of ^{14}C to the nodules and the nitrogenase activity, as measured by acetylene reduction. Retention of stubble leaves when upper shoots were removed led to more rapid recovery of both N fixation and growth than occurred if stubble leaves were also removed.

There are several implications of this finding. Legumes which are able to maintain a green leaf surface during continual grazing or cutting, as occurs in legumes such as *Arachis pintoi* which maintain some leaves closely appressed to the ground, will maintain high rates of N fixation, despite defoliation. Tropical legumes which are rejected during the growing season when grazing animals positively select green grass, as commonly occurs for many species (for example Böhnert et al. 1985 for *Pueraria phaseoloides* in Colombia), are expected to fix more N than legumes such as *L. leucocephala* and *L. bainesii* selectively eaten during the growing season. The former legumes, if eaten during the dry season when the nutritive value of grass is low, may then maintain the animal in positive N balance and contribute more N to the ecosystem than legumes well accepted by grazing animals when their foliage is green. Shrub legumes which elevate their canopy above the reach of grazing animals are expected to maintain higher rates of N fixation than shrubs managed short and grazed heavily. The formation of a protein bank which is made accessible to the grazing animal in the dry season may meet the dual objectives of maximizing legume N fixation and maintaining continuity of forage supply.

The proportion of N uptake in the legume which moves to the shoots increases with increasing severity of defoliation. The root system of legumes that are spared from grazing or cutting, or which are leniently grazed, is greater than that of legumes frequently or severely grazed, and the proportional location of N follows a similar pattern. For example, undefoliated *M. atropurpureum* cv. Siratro accumulated 53% of whole plant N uptake in its shoots, whereas Siratro defoliated in varying systems accumulated 76–86% of whole plant N uptake in its shoots, according to the severity of defoliation (N. D. Young, unpublished; Table 7.3 in Humphreys 1981). The proportion of N that is available for cycling through the ruminant and the proportion of N which goes directly to soil OM through root and nodule senescence (or excretion) depends upon the farmer's management of the legume.

N fixation decreases as ontogeny becomes strongly advanced. This has been noted in *D. intortum* and *D. uncinatum* (Whiteman 1970b; Gibson and Humphreys 1973) and in *Glycine max* (Lawn and Brun 1974). Figure 2.12 shows the real decline in N fixation in *Vigna umbellata* after the commencement of podset. N fixation is maintained if ontogeny is delayed by the removal of pods and flowers, as occurs through grazing or cutting. The removal of flowers and developing pods from *M. lathyroides* substantially increased the supply of ^{14}C to the nodules and the linked nitrogenase activity (Othman *et al.* 1988). Flowers and pods appear to act as competing 'sinks' with nodules for carbohydrate, and with axillary buds for N supply.

Cut-and-remove systems of forage utilization are often more convenient in tropical farm practice than grazing systems; they also pose a significant threat to the benefits of N fixation unless the return of animal excreta is carefully managed, as discussed in Chapter 6 and sections 10.1 and 10.5. Grass–legume swards with a high legume content commonly contain yellow, N-deficient grass plants if a cut-and-remove system is practised, since little N is available for transfer to the grass from a perennial legume root system and nodule N usually only accounts for 1–3% of the plant N present. A cut-and-remove system of shrub legumes growing with *Panicum maximum* var. *trichoglume* in Indonesia provided little N transfer to the latter (Catchpole and Blair 1990a). Most of the N produced occurs in legume shoots; the proportion contributed as litter is negatively related to stocking rate and 80–95% of the N ingested by grazing animals (Humphreys 1991) is returned as excreta. This will be concentrated where animals congregate under shade or near watering points and will in any event be delivered to the points of excretion on the pasture in luxury amounts. These inefficiencies are still greater than occur under cut-and-remove systems.

2.6 Conclusions

The maintenance of an appropriate level of soil OM is the central concept of this book; the ancillary questions of soil N supply, soil physical fertility and control of soil erosion are inextricably related. A satisfactory content of OM provides the opportunity for farmers to manage the forage–crop system so that productivity and income are maintained. The introductory statement to this chapter is reiterated that OM provides the pathway for the mineralization of N and indicates the N status of the soil, which constitutes the first limitation to forage and crop production. It also represents a source of other nutrients and of energy for biological activity.

Two facets of soil OM require emphasis. As discussed further in Chapter 4, the global environment is made more comfortable for the organisms which inhabit it through the ameliorating effects of soil OM. This is reflected in the deactivation of harmful pesticides and chemicals. OM also contributes to the moderation of global warming. There is a popular perception of ruminants as contributors to the 'greenhouse' effect because of their emission of methane. This negative aspect is minor relative to the amount of C being safely locked away as long-life OM under pastures. Certainly frequent burning and overgrazing change pastures from a C sink to a C source, and cropping may also induce a C loss, but the system is subject to manipulations in which the predominant C_4 tropical grasslands are as significant as forests in reducing net CO_2 emission to the atmosphere.

The second factor meriting mention is the widespread problem of acute soil acidity in the tropics, a central constraint to crop production. The ameliorating effects of OM on the mineral toxicities of Al and Mn, discussed in section 2.2, and the use of organic materials provided by forages to manage these effects, require greater attention than many crop agronomists have been prepared to accede.

Legume N fixation is an unexploited source of gain in many tropical farming systems. In a minor sense this arises from inappropriate management of the outputs of N fixation, as discussed in the last section, but a more cogent reason is the failure of scientists to develop with, and communicate to, farmers legume technologies which are sufficiently robust to attract ready adoption. Some notable exceptions to this generalization are described in Chapters 8–11, where the general principles enunciated in this chapter are developed in more detail for the diverse farming systems of the tropics and subtropics.

CHAPTER 3

The maintenance of soil fertility
II. Soil structure and erosion

3.1 Introduction

Soil structure influences the viability of agricultural systems through its effects on plant growth and on the retention of soil and of soil nutrients. It modifies the physical environment for plant growth by determining the availability of water and of oxygen and the capacity of plant roots to access water and nutrients; soil temperature is also affected by changes in soil structure and is pertinent in some situations. The porosity of the soil influences its acceptance of moisture, the storage of moisture at varying depths and its drainage through the profile.

Many of the characteristics by which we describe soil structure, such as the range and distribution of the size of soil aggregates and the type of clay constituents (for example, swelling or non-swelling) are intrinsic to the soil situation, the parent materials from which it is derived, and the climatic influences which bear on it. Nevertheless many attributes of soil structure are able to be manipulated by the farmer. The OM status of the soil, which has immediate implications for structure, can be rapidly changed in agricultural practice, as discussed in the last chapter. The management of the soil surface, especially in relation to the cover maintained, the organic residues retained, the type of tillage and the crops or pastures grown, has profound effects on the ways in which the properties intrinsic to the soil are expressed. These then influence the level of plant production achieved.

The long-term continuity of agricultural production requires that soil erosion be controlled to an acceptable level, as defined by its effects on plant growth, on the ease of farming, and on the natural resources and public utilities downstream from the farm. The loss of nutrients concentrated in the surface layers of the soil is often the aspect of soil erosion most devastating in its effect on crop production.

Soil structure that is favourable for plant growth depends upon the presence of adequate pores > 75 μm diameter, which ensure sufficient infiltration and aeration, and sufficient pores 30–0.2 μm diameter, which store water for subsequent plant growth (Tisdall and Oades 1982). Rapid infiltration and drainage occur if the pores between the aggregates are large enough. Wetting of aggregates causes slaking and dispersion of the soil aggregates; dispersed clay may block the pores which transmit or store water. The farmer is therefore concerned to maintain soil in a condition where aggregates are stable, which is assisted by biological activity and the binding action of OM and the phytomass. This is a simplistic rationalization for the need to focus on the maintenance of soil structure.

3.2 The nature of soil structure

Soil architecture

The architecture or spatial arrangement of soil particles and aggregates, and their stability in response to wetting and drying, rainfall action, tillage, and other forms of agricultural management is well illustrated by a scheme proposed by Tisdall and Oades (1982) (Figure 3.1). This model was developed from experience with red-brown earths (oxisols) but is applicable to many other soils, and indicates structural units of differing sizes and the main binding agents for each unit.

- Aggregates > 2000 μm diameter. The larger aggregates are porous, and are generally bound together by plant roots and fungal hyphae; consequently their stability depends greatly on agricultural practice. They accumulate under grassland and under cropping systems involving continuous live cover, but are destroyed by tillage.
- Aggregates 20–250 μm diameter. These aggregates are relatively stable under rapid wetting. They consist mainly of particles 2–20 μm diameter which are held together by cements which include inorganic crystalline oxides, aluminosilicates and persistent organic materials. These often reflect intrinsic properties of the soil in question. This is illustrated by a study of Freebairn and Loch (1991) (Table 3.1) in which the aggregation of soils in different parts of Queensland was managed by zero or conventional tillage, with or without stubble retention, and subjected to high-energy rain. The indicator of stable aggregation was the proportion of

The nature of soil structure

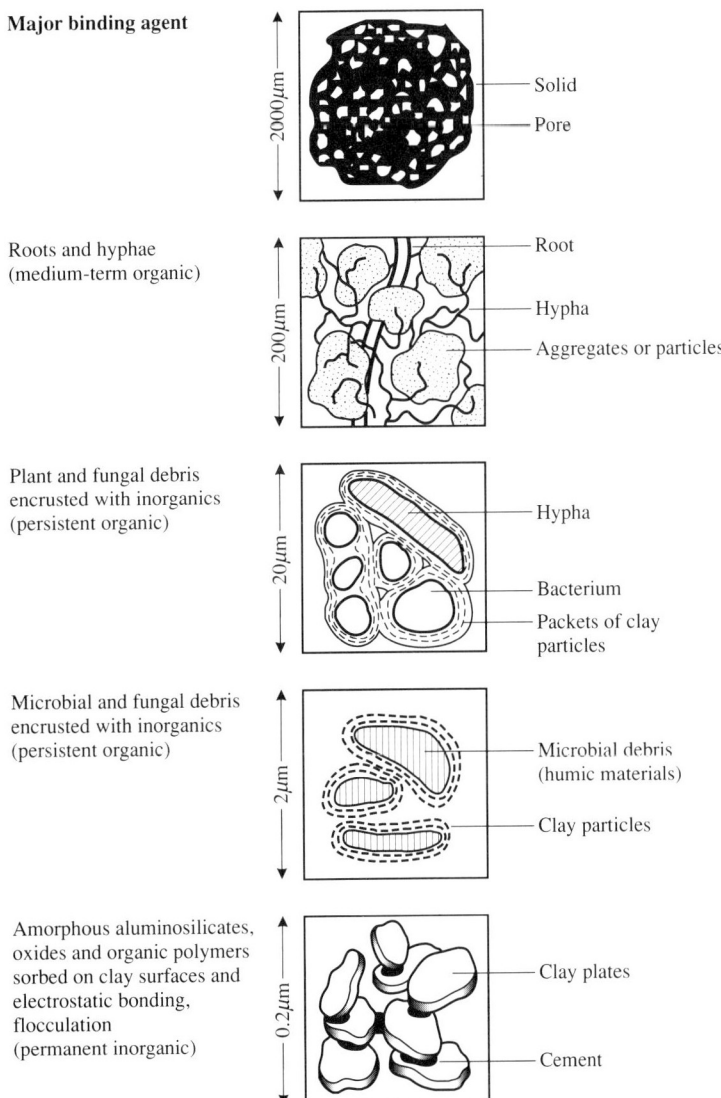

Figure 3.1 Model of organization of soil aggregates, indicating major binding agents (from Tisdall and Oades 1982).

Table 3.1 Percentage of fine particles (< 0.125 mm) at the soil surface after rain of high energy at various sites in Queensland (from Freebairn and Loch 1991).

Tillage	Zero		Conventional	
Stubble retention	+	−	+	−
Site				
Kairi	13.3	16.4	17.6	19.2
Hermitage	31.2	37.1	33.9	36.1
Warra	20.5	NA	23.5	NA
Billa Billa	50.9	56.2	54.6	58.1

particles < 0.125 mm diameter (125 μm). There was a slight trend towards finer particles under conventional tillage or stubble removal, but the largest differences were due to the soil type occurring at the different sites.
- Aggregates 2–20 μm diameter. Particles < 2 μm are the main constituents of these water-stable aggregates; these have persistent organic bonds which are not disrupted by agricultural practice. In section 2.3 reference was made to the effect of clay particles in providing resistance to breakdown of OM. Electron micrographs of young aggregates show bacteria or colonies of bacteria surrounded by a carbohydrate capsule to which fine clay particles adhere. When the bacteria die the products of their decomposition provide an older aggregate with a matrix OM binding clay particles together.
- Aggregates < 2 μm diameter. These water-stable particles are usually based on clay plates held together by physical forces associated with amorphous aluminosilicates, oxides and inorganic polymers; some organic materials may be sorbed onto clay surfaces.

Organic binding agents

It is convenient to classify the organic binding agents mentioned above according to their persistence in soil (see section 2.3).
- Transient binding agents (Tisdall and Oades 1982). These provide the substrate for the activity of microorganisms. They include rapidly available substances such as glucose which provide glues which assist water-stable aggregation but break down within a matter of days or weeks, and polysaccharides which persist for several months.
- Temporary binding agents. Roots and fungal hyphae, especially VAM (see section 2.5), increase within a few weeks or months

according to the growth and activity of the plant root system, and are of course modified by agricultural practice. The superior role of fine grass roots relative to that of legumes in promoting aggregation is referred to in section 10.1, and 'rhizodeposition', or the deposition of organic debris from or associated with plant roots during periods of active growth, has become a focus of scientific interest (Shamoot et al. 1968). These also provide food for earthworms and mesofauna which contribute not only to structural aggregation but have profound effects on infiltration and drainage, as discussed subsequently.

- Persistent binding agents. These are due to refractile, aromatic humic material complexed with iron, aluminium and aluminosilicates which account for the major proportion of OM in the soil, and which are only slowly modified by long-term changes in agricultural practice. The effects of changes in soil aggregation on other soil physical properties and their consequences for plant growth are expanded later.

Slaking and the disruption of aggregates

Pasture and crop practice may be ideally directed to minimizing the breakdown of macro-aggregates. A potent cause of the disruption is the slaking or breakup of macro- into micro-aggregates which occurs when dry soil aggregates are wetted by rainfall or irrigation. This has two primary causes (Emerson 1977). The first is the swelling of dry clay aggregates due to the hydration of the exchangeable cations. The stresses induced depend upon the rate of swelling, which in turn hinges on the clay's permeability and capacity to bend. Different clays respond differently. Kaolinite crystals which are about 50 nm thick provide more porous aggregates than montmorillonite sheets about 1 nm thick, and the latter will take up water more slowly, with illite intermediate in performance.

The second cause is the pressure exerted by entrapped air. The air has an explosive effect on the soil aggregate if it cannot escape as water displaces it. It follows that antecedent soil moisture, and a slowing of the rate of wetting, as may be induced, for example, by the use of surface organic mulch, reduces the amount of slaking. Montmorillonite may bend to accommodate some entrapped air whereas the larger pore size of kaolinite aggregates leads to more disruption. These responses are modified by the presence of the binding agents referred to above.

Formation of surface crusts

The formation of crusts on the surface soil is such a significant aspect of the architecture of soils which affects agricultural performance that it merits a special section. This phenomenon is distinct from the production of claypans, in which the erosion of the topsoil leaves a hard pavement of sodic or clayey subsoil. It is of widespread occurrence (Casenave and Valentin 1989), especially on intensively cultivated soils, and is most commonly caused by the impact of raindrops on a bare soil surface, which produces a thin compact skin seal, perhaps 0.1 mm thick. The dispersion of the aggregates and the splash of fine soil particles washing into and filling in the macropores of the soil surface may also lead to a layer perhaps 1.5–2.5 mm thick with considerably reduced porosity and infiltration (McIntyre 1958a). The formation of these crusts has been described by scanning electron micrographs (Chen et al. 1980). The type of thin surface crusts increases the modulus of rupture, reduces infiltration and increases erosion (Epstein and Grant 1967), whilst the thicker crusts to 5 mm depth also impede seedling emergence.

Bare patches within grassland, which may arise initially from overgrazing or drought and the death of plants, provide runoff to adjacent grassy areas. In the Katherine region of northern Australia (lat. 14° S, 930 mm annual rainfall) the tall grasses *Themeda australis*, *Sehima nervosum* and *Sorghum plumosum* provide basal cover of 10–15% (Mott et al. 1979) on Oxic Paleustalfs, but smooth, bare, crusted areas 5–20 m in diameter are interspersed throughout. Their surface texture contrasts with the crumb structure under the grass tussocks. Micrographs showed the occurrence of a seal about 1 mm deep containing no large pores, and a layer below containing discrete round pores or vesicles. Beneath this the soil was massive and dense without large pores. The depth of the wetting front (Figure 3.2) following the first storm rainfall after the dry season shows the profound difference in acceptance of water according to the amount of foliage cover, and the dry soil environment developed beneath the soil seal. The beneficial effects of OM on aggregate stability and infiltration have been mentioned; later studies of Bridge et al. (1983a) at the same site showed that raindrop impact alone could lead to the formation of these degraded areas even before losses in OM became evident. After burning the vegetation and remaining surface cover, the soil seal developed following structural collapse, and sorptivity and hydraulic conductivity decreased, although soil C content had not been impaired. These changes only became persistent if repeated defoliation that destroyed the vegetation continued.

Tillage commonly reduces aggregation and increases the susceptibility of soils to surface crusting following raindrop impact. This is illustrated by a study in western Nigeria (Wilkinson and Aina 1976) where forest was cleared and the early phases of cropping induced considerable change in soil properties. A sandy loam of the Iwo series (alfisol; Oxic Tropudalf) after 2 years of cropping exhibited

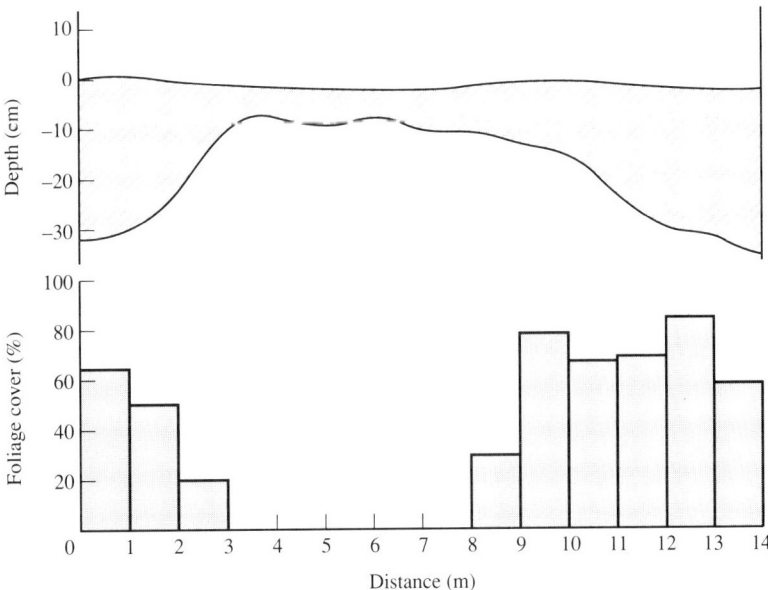

Figure 3.2 Depth of wetting front across a sealed area following storm rainfall at Katherine, Australia (from Mott et al. 1979).

infiltration capacities of 8.4 and 49.5 cm h^{-1} respectively in the presence of a soil crust (c. 1 mm thick) or in its absence. A loamy sand of loose friable and weak granular structure of the Oba series showed infiltration capacities of 7.1 and 21.6 cm h^{-1} in a similar comparison.

This generalization about the dangers of tillage requires qualification; there are hard setting soils, especially in West Africa, for which tillage is beneficial in increasing porosity and infiltration. On untilled soils the presence of a crust is a permanent feature (Hoogmoed and Stroosnijder 1984) of sandy soils of the Sahel zone, the soils become compacted and hard during the dry season, and farmers cannot commence ploughing until the rains have arrived to wet the soil to plough depth. Immediately after wetting, the moisture content of a soil below the crust was 15–20% (w/w), whilst in a non-crusted soil 30–35% (w/w) moisture was observed. Lal (1985) observes that soils with high silt and fine sands are especially vulnerable to compaction and to crust formation and may require periodic tillage unless some special system of surface management which results in high levels of biological activity is used.

The failure of the no-till system on these hard setting soils is illustrated by a study (Lal 1989b,c) of an alfisol (Oxic Paleustalf) at IITA, Ibadan, Nigeria (lat 8° N, 1300 mm annual bimodal rainfall). In this experiment (Lal 1989a) *L. leucocephala* or *Gliricidia sepium* was grown as hedgerows 2 or 4 m apart and the production of maize and cowpeas between the hedgerows contrasted with conventional crop production in plough-tillage or no-tillage systems using paraquat as a herbicide. After 6 years of cropping there was a considerable increase in penetration resistance and bulk density (Lal 1989b), which were greatest in the no-till treatment, whilst soil moisture retention was greatest in the hedgerow systems, and was associated with structural changes in porosity. These changes influenced the rate of infiltration of rainfall (Lal 1989c); the capacity of the soil to absorb water (Figure 3.3) and its hydraulic conductivity were greatly reduced in the no-till system. However it should be noted that the final rate of infiltration in this treatment on this coarse textured soil was still of the order of 10 cm h^{-1}, a high value relative to that exhibited by many other soils. Surface crusting, even in these hard setting soils, becomes

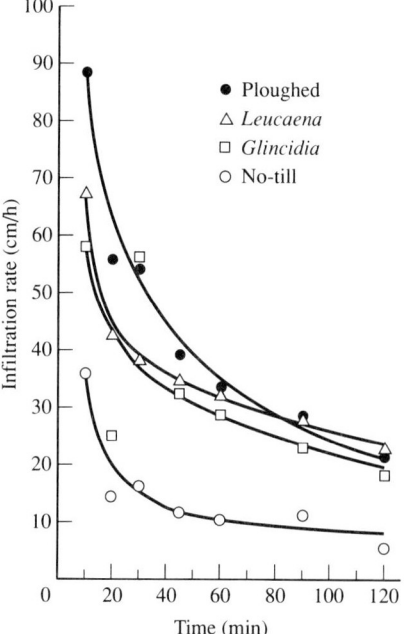

Figure 3.3 Effects of methods of tillage and systems of agroforestry on rate of infiltration of an alfisol in Nigeria (from Lal 1989c).

less of a problem if the level of biological activity is high as occurs with hedgerows and bush fallows; the activity of mesofauna such as earthworms breaks up the mulched surface and provides channels for moisture penetration. The level of crop production in this study did not provide sufficient or stable surface residues which would engender the necessary level of protection.

Physical properties of soils

The physical attributes of soils that are of agricultural moment include aggregation, dispersion and swelling characteristics, which have already been discussed, porosity and bulk density, infiltration rates and hydraulic conductivity, soil strength and penetrability, aeration capacity and rate of gaseous diffusion, which determine susceptibility to waterlogging. Water retention characteristics are a most important physical property and plant yield is linearly related to stored water and to water usage during the growing season; these characteristics are especially related to soil structure, OM content and the related macroporosity and rate of infiltration. Many of these properties are interlinked.

Porosity and bulk density

Porosity and bulk density (the mass of soil *in situ* per unit volume) are in part intrinsic characteristics of the soil, arising from the nature of the parent material and the size fractions which result, but also reflect agricultural practice, especially in relation to OM content. Animal treading, wheel traffic and tillage are the most common management factors influencing the degree of soil compaction. The formation of an impenetrable plough layer is described in Figure 10.5, section 10.3. The static load on the soil is greater for cattle than for sheep (Spedding 1971), and compaction increases with SR. For example, Walker (1980) noted a linear relationship on grass–legume pastures where bulk density (Mg m^{-3} 0–10 cm) = 1.21 + 0.09 SR (steers ha^{-1} for 4 years over the range 1.2–3.3); the more continuous sod cover of N-fertilized *S. sphacelata* var. *sericea* provided better defence against compaction. Similarly Pott and Humphreys (1983) noted that the bulk density (0–7 cm) of a red-yellow podzolic soil (ultisol) varied from 1.2 to 1.4 Mg m^{-3} as sheep SR increased from 7 to 28 ha^{-1}.

The effects of compacting agricultural soils is illustrated by contrasting studies of pigeon pea (*Cajanus cajan*) cropping on a vertisol at a dry inland site at Dalby, Queensland (Kirkegaard *et al.* 1992a) and on an oxisol at a humid coastal site at Redland Bay, Queensland

(Kirkegaard *et al.* 1992b). At both sites combinations of deep ripping and rolling were used to establish three levels of compaction. These treatments led to quite different effects on soil behaviour and crop growth, and indicate the complexities involved in modelling the expected results of compaction. The vertisol (Kirkegaard *et al.* 1992a) was a cultivated black earth (Entic Pellustert) and increasing levels of compaction (labelled C1, C2 and C3, Figure 3.4) produced substantial differences in the bulk density of the surface layers to 20 cm depth (Figure 3.4a, left side). There were similar increases in penetrometer resistance (Figure 3.4c) with increasing compaction, whilst air-filled porosity was negatively associated with compaction in the surface

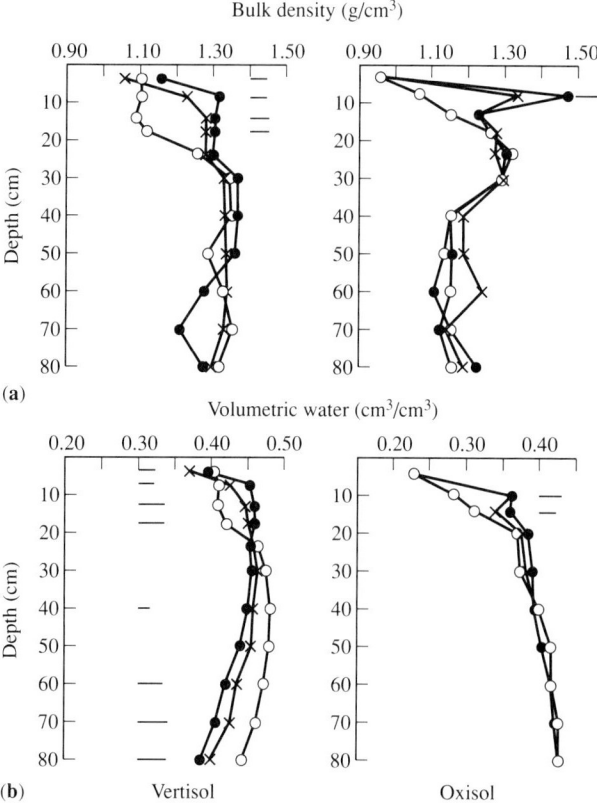

Figure 3.4 Effects of varying levels of increasing compaction (C1 ○, C2 ×, C3 ●) on the physical properties of a vertisol (left side) and an oxisol (right side). Horizontal bars indicate LSD at $P < 0.05$ (from Kirkegaard *et al.* 1992a,b).

layers of the soil (Figure 3.4d); the higher moisture content at depth of the low compaction treatment may have reduced air-filled porosity. Although compaction increased the volumetric water content of the surface layers, the low compaction treatment contained much more water at depth 6 weeks after treatment (Figure 3.4b), indicating the higher infiltration and storage of rain (140 mm) which fell during this period. The higher penetrometer resistance measurements at depths below 30 cm in the compacted treatments are also indicative of reduced infiltration and water storage.

Compaction increased the incidence of branched, contorted and thinner taproots in the top 10 cm of soil. It is likely that air-filled porosity below 10%, as was evident in the 5–20 cm level (Figure 3.4d), restricts root growth and the greater penetration resistance reduced the growth of the root system into the subsoil, which in turn decreased

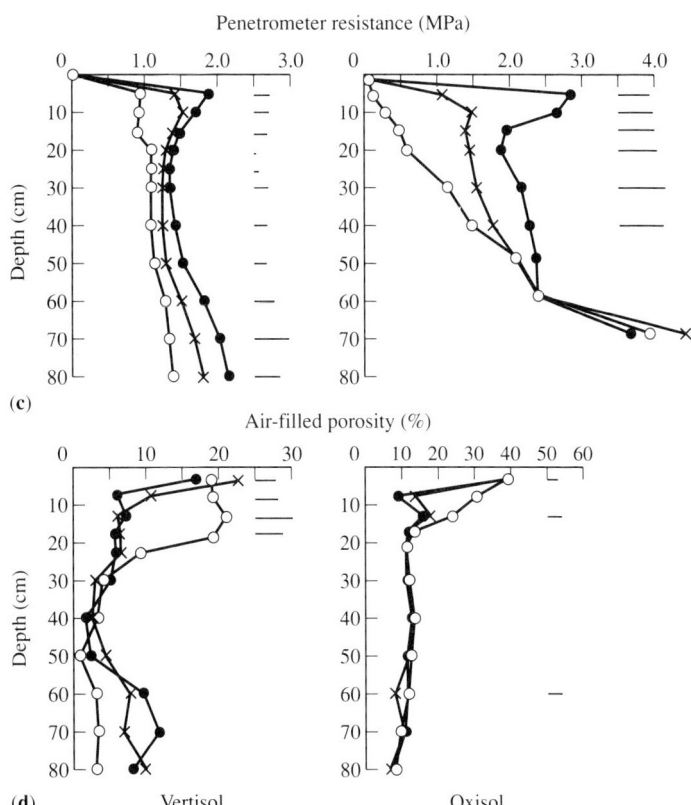

water uptake by the crop. This effect was even more pronounced in a companion experiment with mung bean (*Vigna radiata*). This restriction of root growth was first reflected in reduced growth of the seedling, which at 40 days after sowing had least shoot weight and leaf area in compacted treatments. The second major effect of compaction was reflected later in the growth of the crop, when compacted treatments exhibited more moisture stress and grain yield of *Cajanus cajan* in the rain-grown experiment decreased from 0.64 t ha^{-1} in the C1 treatment to 0.21 t ha^{-1} in the most compacted C3 treatment. This indicates the paramount importance of soil physical characteristics in determining water storage and therefore crop yield.

In a subsequent and wetter year grain yield was independent of compaction; soil strength (penetration resistance) was less in the wetter soil and moisture shortage did not constrain grain yield, which averaged 2.4 t ha^{-1}. This latter result contrasts with experience in cropping systems in temperate regions, where wet conditions increase the negative effect of compaction on crop yield. In these conditions seasonal rainfall is high, evaporation is low, and temperatures are cold; compaction under cold, wet conditions is often linked to increased waterlogging during early growth and reduced surface temperatures during germination.

In the second study (Kirkegaard *et al.* 1992b) the oxisol was a Krasnozem (Rhodic Paleustalf) with a similar high clay content to the vertisol, but the high sesquioxide content and dominance of kaolinitic clays produce a strongly structured, friable, permeable soil whereas the montmorillonitic clays of the vertisol give a narrow friable range. The three compaction treatments produced pronounced differences in bulk density to 20 cm depth (Figure 3.4a, right side) and in penetrometer resistance (Figure 3.4c). Volumetric water content in the surface soil (Figure 3.4b) increased with compaction, but there was no effect of compaction on water storage at depth, which was consistent with measurements of infiltration. Air-filled porosity in the surface soil (Figure 3.4d) was negatively associated with compaction, but levels were above or close to the critical 10% level at all depths.

In this study shoot growth 26 days after sowing was positively associated with compaction, since the increased volumetric water and presumably nutrient contents of the compacted soil favoured growth in the absence of any serious adverse effect of compaction on root growth. However, as the roots grew into the deeper soil layers these differences disappeared and grain yield was independent of compaction and averaged 2.2 t ha^{-1}. It therefore appears that compaction is unlikely to be a serious problem for *Cajanus cajan* yields on these oxisols in a more humid environment than occurs at the vertisol site. It is apparent that seasonal conditions, the hydraulic properties of the soils involved and their susceptibility to structural change by compaction vary the agricultural significance of applying loads or tillage to soils.

Infiltration rate

The acceptance of rainfall by the soil is a key influence on plant growth in most parts of the tropics and subtropics. This is true for humid areas as well as subhumid areas. In well-watered lands SRs of pasture are higher and forage reserves tend to be lower, so that a short drought period of perhaps 3 weeks can cause a discontinuity of forage supply; crops and pastures tend to be more shallow rooted and vulnerable to drought. In all regions the spectre of soil erosion grows if infiltration is low and runoff thereby increased. Soil surface management and the degree to which rainfall is trapped by differences in microrelief are influential in determining the degree of rainfall acceptance, as expanded subsequently, but soil properties determine the longer term situation.

There are various models of infiltration. A widely used model is that of Philip (1957) which distinguishes the sorptivity (S), which describes rainfall acceptance in the early stages of raining, and transmissivity (A) or hydraulic conductivity, which determines the later phases of infiltration and is usually measured until a near-constant value is attained. The cumulative infiltration (i) is then determined from the equation $i = St^{1/2} + At$, where t is the duration of infiltration.

Infiltration is well related to macropore space. This is illustrated by a study in the Katherine district, Australia (Bridge et al. 1983b) which compared the porosity and infiltration characteristics of soils of differing cultural history and pasture treatment. Macropore space, defined as the percentage of soil sectional area occupied by macropores > 0.1 mm diameter, was very low under degraded pasture (Table 3.2), and under a poor pasture of *Stylosanthes humilis* and *Cenchrus ciliaris* (Table 3.2, Figure 3.5b) which contained only 10% legume, had little vegetative cover between grass tussocks, and practically no surface litter. The horizontal cracks in Figure 3.5 are an artefact of the soil removal technique before photography. Native pasture protected from grazing had better pore space, but this was even greater with good interconnections between pores under well-managed sown pastures of *S. hamata* (Figure 3.5a) or of *Alysicarpus vaginalis* with volunteer *Digitaria gibbosa* (Figure 3.5c), which contained c. 80% legume. The acceptance of initial rainfall, or sorptivity, was greatly reduced in the degraded pastures, but well-managed sown legumes led to levels similar to that of protected native pasture, indicating how these pastures may be used to restore good soil structure. On the other hand, levels of hydraulic conductivity were lower than under protected native pasture, perhaps due to the compaction associated with the high

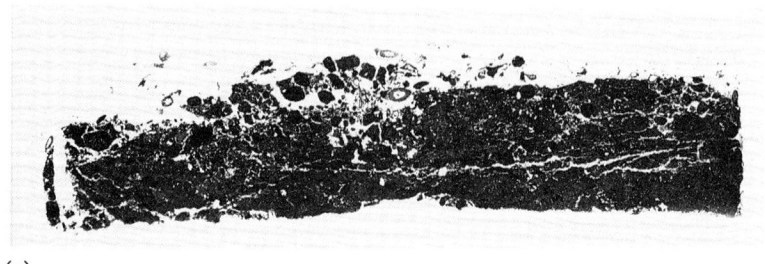

Figure 3.5 Pore micromorphology of surface soil (0–25 mm) at Katherine, Australia, below different pastures. (a) *S. hamata*, (b) *S. humilis* + *C. ciliaris*, (c) *D. gibbosa* + *A. vaginalis* (from Bridge *et al.* 1983b).

wet season SRs employed on these fertilized, legume-based pastures. It is believed these improved pastures give stable soil structure provided a litter layer is maintained.

Macropore space is related to the dispersibility of the fine soil fractions, and this in turn influences the hydraulic conductivity. A study of vertisols in Queensland which had been cropped for varying periods up to 64 years (Cook et al. 1992) found that the size and strength of dry aggregates decreased with cultivation, as did the stability of wet aggregates. This led to increased dispersion and slaking, which reduced hydraulic conductivity. The latter (Figure 3.6) was linearly and negatively related to the content of dispersed clay ($r^2=0.56$). This factor in these vertisols was more influential than OM content, and there was a closer relationship between dispersibility and exchangeable Na content ($r^2=0.88$) than with OM (So et al. 1988). There are many studies which show the importance of sodicity in determining the degree of dispersion and in thereby reducing hydraulic conductivity, which may decrease infiltration and storage of soil water, constraining crop yields. An increase in exchangeable Na may reflect erosion of the surface soil and the exposure of a more sodic subsoil (Cook et al. 1992). One of the benefits of zero tillage may be expressed via increased infiltration leaching Na ions to deeper soil depths, and reduced dispersion ratios in the surface soil (Loch and Coughlan 1984). Soils that develop deep cracks upon drying have a high initial rainfall acceptance.

The final comment in this section emphasizes the role of soil OM in infiltration. On sandy soils there are instances of OM increasing hydrophobicity (Emerson 1977) perhaps associated with algal or fungal seals, but the general case is that increasing OM leads to stable aggregation and increasing infiltration (Stephens 1967; Wilkinson 1975b; Kelly and Walker 1976; Lal 1976; Braunack and Walker 1985). This is illustrated further with respect to the duration of the ley in Table 10.3, section 10.2, and to the effects of different cover crops in Table 10.8, section 10.3. The effect of OM is especially pronounced where a substrate is provided to stimulate the feeding of mesofauna which leave channels through the soil profile. A further example comes from an extension of the study reported in Table 2.10 (section 2.4) in which the effects of number of legume crops on the fertility of a sandy loam derived from granite were measured (Yaacob and Blair 1981). The OM accretion in this experiment is indicated by the increase in soil total N content (Table 3.3) which resulted from legume N fixation and the incorporation of legume residues in the topsoil. This had markedly positive effects on the formation of water-stable aggregates and on the initial infiltration rate (0–5 min), indicating that rainfall acceptance was much improved. There was a linear relationship between initial infiltration (sorptivity) and soil N%. On the other hand

Table 3.2 Macropore space and infiltration characteristics of soils at Katherine, Australia under different pasture types (from Bridge et al. 1983b).

Pasture type	Macropore space (%)	Sorptivity (cm/min$^{1/2}$)	Hydraulic conductivity (cm min^{-1})
Degraded native pasture	2.1	0.35	0.044
S. humilis + C. ciliaris	2.4	0.54	0.029
Protected native pasture	7.4	2.14	0.182
S. hamata	11.0	1.45	0.045
D. gibbosa + A. vaginalis	14.3	2.00	0.074

the hydraulic conductivity of the undisturbed subsoil was determined by intrinsic soil characteristics, and infiltration rate in the 25–30 min period after commencing irrigation was independent of the number of previous legume crops. Some of the beneficial effects of legume leys on subsequent crop yield may be attributed to an improvement in rainfall acceptance and the amelioration of both crop water stress and soil erosion, especially in relation to an improvement in the water retention characteristics of the profile.

Aeration capacity and waterlogging

There is a linear and positive relationship between air-filled porosity and infiltration. Connections between the macropores, as are illustrated in Figure 3.5, greatly enhance the rate of oxygen diffusion. On the

Table 3.3 Effects of number of crops of *Macroptilium atropurpureum* cv. Siratro and *Glycine max* (soybean) on soil properties (from Yaacob and Blair 1981).

Legume	No. crops	Soil N (%)	Aggregation (% > 2 mm)	Infiltration rate (mm min^{-1})
Siratro	1	0.049	47	4.1
	3	0.071	58	10.6
	6	0.108	62	15.4
Soybean	1	0.054	39	8.1
	3	0.062	59	12.3
	6	0.087	77	13.4

other hand the so-called 'dead end pores', which are discrete entities without interconnections, do little to improve aeration or infiltration. Air-filled porosity was discussed in the section on compaction, and treatment differences are illustrated in Figure 3.4d.

In addition to the effects on infiltration and moisture storage the development of favourable conditions for root growth and activity need to be addressed. Poor root growth, root malformation and restriction of the capability of the next system to occupy the soil mass and to extract moisture and nutrients was evident in *Cajanus cajan* growing in a compacted vertisol (Figure 3.4d, left side), and a figure of 10% air-filled porosity was suggested as a critical level below which root growth of many species is restricted. Anoxia not only restricts root respiration but has potent effects on root activity that are evident before the needs for maintenance respiration are exhausted; these are first reflected in a failure of nutrient and water uptake, so that plants wilt. This is associated with the production of toxins such as ethylene, acetaldehyde and alcohol (McManmon and Crawford 1971); under flooded conditions plants are intoxicated and thirsty despite the abundance of water.

Plants vary considerably in their adaptation to soils with impeded drainage or to topographical situations susceptible to waterlogging

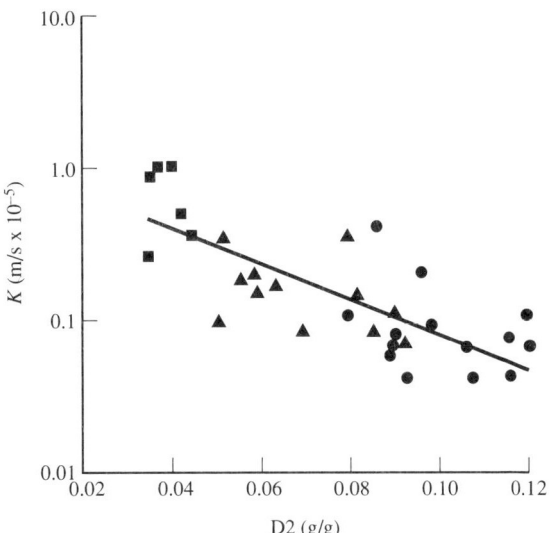

Figure 3.6 Relationship between hydraulic conductivity (K) and dispersibility (D2) of three vertisols cultivated for varying periods in Queensland (from Cook *et al.* 1992).

or flooding. For example *Cynodon dactylon* continued root growth until the oxygen diffusion rate was reduced to 0.15 µg cm^{-2} min^{-1} whereas other grass species ceased root growth at a value of 0.40 µg cm^{-2} min^{-1} (Stolzy 1972). Tropical and subtropical grasses known for their resistance to waterlogging and flooding include *Acroceras macrum, Brachiaria mutica, Echinochloa polystachya, E. pyramidalis, E. stagnina, Entolasia imbricata, Hymenachne amplexicaulis* and *Leersia hexandra* (Humphreys 1981). It is often more difficult to locate legumes which flourish under flooding; some of the more successful include *Aeschynomene americana, Desmodium intortum, Lotus pedunculatus, Macroptilium lathyroides, Sesbania sesban, Vigna vexillata* and *V. pilosa* (Miller and Williams 1981; Whiteman *et al.* 1984).

Soil strength and resistance to penetration

The soil strength and resistance to penetration is an index of the response of the soil framework to applied stress. This has direct effects on plant performance as reflected in processes such as seedling emergence or the growth of roots in the soil layers surrounding them, in addition to indirect effects on moisture and nutrient availability and on aeration. It also bears on management practices such as the ease of tillage and the effect on the soil of wheel traffic or movement of animals. Compressive and tensile strength are determined by the cohesion and the internal angle of friction of the soil. These attributes are modified by the moisture status of the soil at the time and its relationship to the plastic limit. Soil strength is commonly measured with a core penetrometer, and differing degrees of compaction which produced varying resistance to penetration and varying crop responses were illustrated for two soils in Figure 3.4c.

Soil temperature and heat conductivity

The final soil physical attribute to be mentioned is soil temperature and the associated soil thermal conductivity and thermal capacity. These vary less than hydraulic conductivity but are modified by moisture content and by tillage. Soil colour is a significant factor affecting the rate at which soils warm up, dark soils having a higher temperature in the surface layers than light-coloured soils which exhibit a higher reflectance ratio. In the subtropics and the high-altitude tropics cool soil temperatures restrict root growth (Schroder 1970) and the rate of mineralization.

Soils have low thermal conductivity and are excellent insulators. Grassland fires are spectacular and remove all or most of the surface litter. They have little immediate effect on soil conditions below the soil surface, although the removal of vegetation and of litter affects subsequent infiltration, as discussed previously. When Stür and Humphreys (1987) burnt pastures of *Brachiaria decumbens* and of *Paspalum plicatulum* maximum soil surface temperatures varied from 270 to 365 °C, but the highest increase in soil temperature at 1 cm depth was only 3 °C at the time of the fire. In the month after burning the burnt treatments had daily maximum temperatures at 2.5 cm depth *c*. 4 °C higher than in treatments in which the pasture had been cut, especially on clear sunny days, whilst daily minimum temperatures at 2.5 cm depth were *c*. 1 °C lower in the burnt treatments. Following burning the soil is exposed, the surface is black, and the onset of tillering may be delayed.

The management of surface residues varies soil temperature and water loss, and this is illustrated subsequently in Figures 10.6 and 10.7 (section 10.3); surface mulch reduced soil temperatures and improved the seedling emergence of maize, which is restricted by high temperatures.

Effects of soil structure on conditions for plant growth

From this section the significance of variation in soil physical conditions for plant growth may be briefly summarized as follows.

- Creation of water-stable aggregates improves rainfall acceptance, aeration and ease of tillage.
- The absence of soil surface crusts increases rainfall acceptance, seedling root growth and seedling emergence.
- The maintenance of air-filled pore space and of low soil bulk density, which are often associated with low ratios of dispersion and of exchangeable Na, provide good conditions for infiltration and deep moisture storage, aeration which favours root growth, and good access of roots to moisture and nutrients.
- Low soil strength and low resistance to penetration increase root growth, seedling emergence and ease of tillage.
- Soil temperature may be manipulated through the management of surface residues to favour plant growth.

Well-managed pastures favour the development of the desirable physical attributes discussed above but these benefits are transient under tillage (Pereira *et al.* 1954); the major physical benefits are associated with the retention of soil, soil water and soil nutrients, whilst the principal rationale for the pasture ley hinges upon the levels of OM and N accretion which result.

3.3 Soil erosion

Processes of soil erosion

Soil erosion arises from the detachment of soil particles and their transport by water or wind to other locations. Water is the predominant agent of erosion in cropping areas and is therefore the focus of this section. Soil is moved by:

- the splashing of raindrops striking the soil surface;
- the energy of particles displaced by splashing which strike other particles;
- the movement of this displaced soil by flowing water;
- the further entrainment of soil particles from the soil surface by runoff water; and
- the dislodgement and transport by flowing water at high velocity to cause gully erosion.

Splash erosion

Soil erosion is directly linked to the kinetic energy of falling raindrops. The energy expended when rain hits a soil surface is much greater than the energy involved in overland flow (Hudson 1981). Studies in which surface flow has been maintained but the energy of raindrop impact reduced have shown a diminution of soil loss directly commensurate with the reduction (Young and Wiersma 1973). The energy is the product of the mass of rain and the square of its terminal velocity. The energy impact is mostly used in detaching soil particles and breaking down soil aggregates; the explosive effect of entrapped air which causes slaking when dry aggregates are moistened, and the formation of surface crusts by the washing in of fine particles to soil cracks were described in section 3.2. There is a secondary effect as the soil that is displaced into the air causes disruption and entrainment of the particles with which it collides upon landing. Splashed soil may move sideways 1.5 m or more on level surfaces and reach a maximum height of > 70 cm (Lal 1990).

There have been many attempts to derive a useful index which would estimate the kinetic energy of rainfall and its erosive action. In the USA Wischmeier and Smith (1958) found best correlation with the maximum intensity of rain during a 30 min period multiplied by the energy per millimetre rain whilst in Zimbabwe a stronger relationship was found if rainfall of low intensity (< 25 mm h^{-1}) were excluded (Hudson 1981). The intensity of rainfall is a more potent factor than the total amount.

The major component of intensity which causes this effect is the variation in the size of raindrops. Large raindrops have a greater

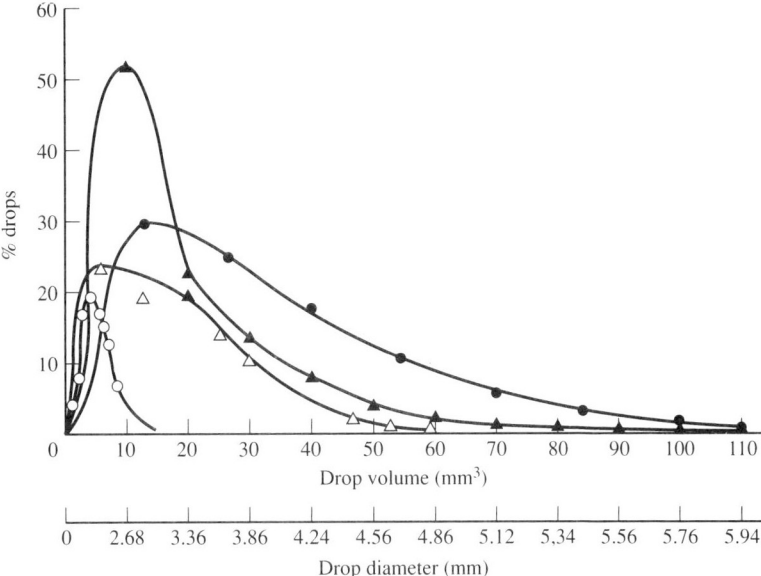

Figure 3.7 Drop size distribution of rainstorms at Samaru, Nigeria (●, heavy rainfall, 41 mm, 1230 drops cm^{-2}; ▲, storm and drizzle, 6 mm, 530 drops cm^{-2}; △, typical rain 20 mm, 95 drops cm^{-2}; ○, drizzle, 7 mm, 135 drops cm^{-2}) (from Kowal and Kassam 1976).

terminal velocity than small raindrops; drops with diameter of 2 mm arrive at the soil surface at 6.5 m s^{-1} whilst drops of 4.5 mm diameter arrive at 9 m s^{-1} (Lal 1990). In Chapter 1 the higher intensity of tropical rain relative to that of temperate rain was mentioned; in south central USA median drop diameter of rains varied from 1.4 to 2.7 mm, whilst at Ibadan, Nigeria a common range was 2–4 mm (Lal 1990). A concrete illustration of tropical rain characteristics (Figure 3.7) of different rains at Samaru, Nigeria (lat. 11° N, 1100 mm annual rainfall; Kowal and Kassam 1976) shows the continuous series of drop sizes, and the greater proportion of large diameter drops in rain of high intensity. The kinetic energy load varied from 31.6 to 38.4 J m^{-2} mm^{-1} rainfall, which are higher values than occur at many sites. In this region peak intensities of 120–160 mm h^{-1} are not uncommon and the turbulence of these rains makes large drops unstable.

Movement of soil downslope by splash erosion has been reported on slopes up to 30°, but this topic is controversial. In some studies overland flow predominates as the main agent of soil transfer and slope makes little difference to the fate of soil splashed (Lal 1990). The presence of surface water after rain to a depth equal to raindrop diameter reduces the direct impact of raindrops on the soil, but the increased turbulence of the water due to rain splash causes entrainment of particles. There may be fluctuations in the rate of soil splash associated with the disruption and reformation of the soil surface skin (McIntyre 1958b). Splash erosion lessens as deep water develops.

Soil splash decreases exponentially with increase in soil cover. The projected foliage cover is the significant measurement in determining the extent of raindrop impact on bare soil; ground cover is the significant measurement for estimating effects on surface runoff. The height of cover is influential, and water shed from shrubs and trees may reach damaging terminal velocities.

Runoff and antecedent soil moisture

The obverse aspect of infiltration, which was discussed in section 3.2, is runoff. Rainfall acceptance depends upon intrinsic soil properties, soil conditions and surface cover at the time of rain, and the amount and intensity of rainfall. The first factor to be considered is the dryness of the soil when the rain commences.

Naturally the significance of antecedent soil moisture in determining runoff depends upon the sorptivity characteristics of the soil in question. In western Nigeria Wilkinson (1975a) found that on an Iwo soil with high equilibrium infiltration rates of about 20 cm h^{-1} under bush fallow the moisture content at the time of the storm had little if any influence on runoff. Storms following two or more rainless days were as likely to produce runoff as storms following one or less rainless days. Similarly Stocking and Elwell (1973b) in Zimbabwe did not find antecedent moisture was a significant variable in their multiple regression analysis of factors associated with erosion.

The situation is quite different when dealing with soils of low rainfall acceptance. Agricultural usage which has the effect of keeping the soil relatively dry will increase sorptivity and reduce runoff. This is illustrated by a study (Freebairn and Wockner 1986a) at Greenmount on the eastern Darling Downs, Queensland of a vertisol (Udic Pellustert) with a self-mulching, cracking character that contained 62% clay. Small experimental contour bay catchments managed with different cropping systems and stubble retention led to variations in the cover present. Runoff from four storms which occurred when different levels of soil moisture were initially present was related to the degree of cover (Figure 3.8). The percentage runoff was linearly and negatively related to cover, but the overriding factor in determining runoff was antecedent soil moisture. On a soil such as this, which has an infiltration rate of only c. 4 mm h^{-1} when wet, runoff was 75–80% if a storm occurred when the profile was very wet, or 0–22% if the profile was dry. Ive et al. (1976) also developed functions that relate runoff to antecedent soil moisture for a Tippera clay loam at Katherine, Australia.

Runoff, rainfall intensity and cover

Rainfall intensity is a major factor determining runoff, which in turn will determine the degree of soil loss as modified by cover. In the same study at Greenmount on the eastern Darling Downs, Queensland (Figure 3.8) the effect of rainfall events of differing intensity was monitored. The treatments which produced differing levels of cover involved zero tillage, stubble mulching, stubble incorporation, bare fallowing and summer cropping with sorghum, sunflowers or maize. The exit water from 1 ha catchments defined by contour banks was sampled for suspended sediment; soil movement down the slope and deposition in the channel above the bank also occurred and probably represented about 80–90% of soil loss. The time of concentration of the catchments is the time taken for water from the farthest point to reach

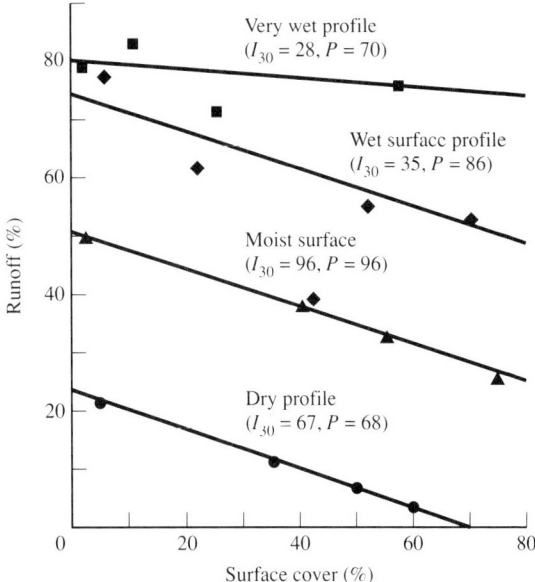

Figure 3.8 Surface cover and runoff from a Greenmount black earth, Queensland for selected storms under varying antecedent soil moisture conditions. Values in parentheses indicate maximum 30 min rainfall intensity (I_{30} mm h^{-1}) and rainfall P (mm) (from Freebairn and Wockner 1986a).

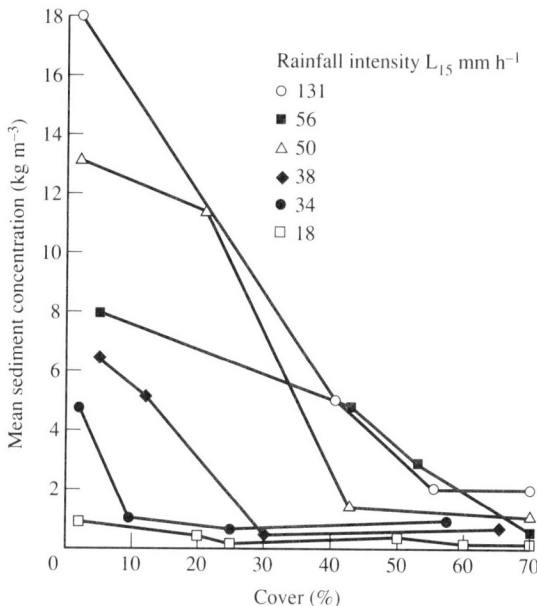

Figure 3.9 Flow-weighted mean sediment concentration and cover for a Greenmount black earth, Queensland for storms of differing maximum 15 min rainfall intensity (from Freebairn and Wockner 1986b).

the outlet and is a key indicator of the influential duration of a storm, since after this period runoff from all parts of the catchment area is present at the outlet. This was 12–15 min from these catchments, so the highest intensity rains for a duration of 15 min were considered.

The selected storms (Freebairn and Wockner 1986b) gave a range of intensities of 18–131 mm h^{-1} over 15 min, and sediment concentration in the runoff (Figure 3.9) was positively associated with rainfall intensity, especially at low levels of cover. The effect of rainfall intensity on soil loss was much less influential if cover was greater than c. 50%, which probably demonstrated the combined effects of surface protection from raindrop impact, greater surface storage of water, and more tortuous flow through the uneven surface.

This study was carried out on land of 5–7% slope, with c. 60 m between contour banks constructed with a gradient of 0.3%. It illustrates the failure of mechanical measures to control soil erosion; the banks acted as soil collecting agents from the inter-terrace areas in which both inter-rill and rill erosion were evident. It was only under agricultural practice which maintained cover of 20–30% during the summer erosive period as a defence against episodic rainfall events of high intensity that tolerable levels of soil loss were maintained.

Cover is a crude index of vulnerability to erosion, since plants which differ in structure but which offer the same projected foliage cover have differing effects on runoff and soil removal, and the presence of particular plants may indicate differences in pasture use which have changed soil characteristics. This is well illustrated by a study (A.J. Pressland, personal communication) at Charters Towers, Queensland (lat. 20° S, 640 mm annual rainfall). Differing amounts of cover occurred at each of two sites on an eroded red duplex soil with neutral reaction, and runoff and soil loss were measured in these different conditions. The native tall bunch grass *Heteropon contortus* at the Silver Valley site, which was located about 150 m from the second site, was the dominant grass in a pasture grazed by cattle at conservative SRs which nevertheless exhibited a gradient of cover from zero to 100%. The creeping sod-forming grass *Bothrichloa pertusa* has been introduced to the region, and colonizes areas which are heavily grazed. Its creeping habit confers grazing resistance and its prolific seeding capacity assists its incursiveness; one measurement of soil seed reserves recorded 63 000 seeds m^{-2} for *B. pertusa* and 4600 seeds m^{-2} for *H. contortus*. It had invaded a site at Kirk River nearby. The two grasses show a different relationship between yield of shoots and pasture cover, measured as the projected foliage cover plus litter, i.e. the degree of absence of bare ground. A yield of 3150 kg ha^{-1} of *B. pertusa* led to 95% ground cover, whilst 7700 kg ha^{-1} of *H. contortus* was needed to produce 90% cover (Figure 3.10a). Additionally at the same level of cover *B. pertusa* gave less runoff than *H. contortus* (Figure 3.10b); at 20% cover surface runoff was 14% in the former but 22% in the latter. The situation for soil loss (Figure 3.10c) was different. At high levels of cover the concentration of sediment in the runoff water and the total soil loss was much less from the sod-forming *B. pertusa* than from the bunch grass *H. contortus*, but at low levels of cover greater soil loss occurred from *B. pertusa* than from *H. contortus*. This no doubt reflected the greater decline in soil structure and the increase in detachability associated with the heavily overgrazed conditions indicated by the presence of *B. pertusa* with low ground cover.

Runoff, entrainment and deposition

The concentration of sediment in runoff depends upon the balance between three processes (Rose 1988).

- Detachment of soil particles from the soil surface by splash erosion into the water flowing overland.
- Entrainment of sediment by overland flow scouring out more soil particles from the soil surface. This factor tended to be neglected in early studies of soil erosion.

98 *The maintenance of soil fertility II. Soil structure and erosion*

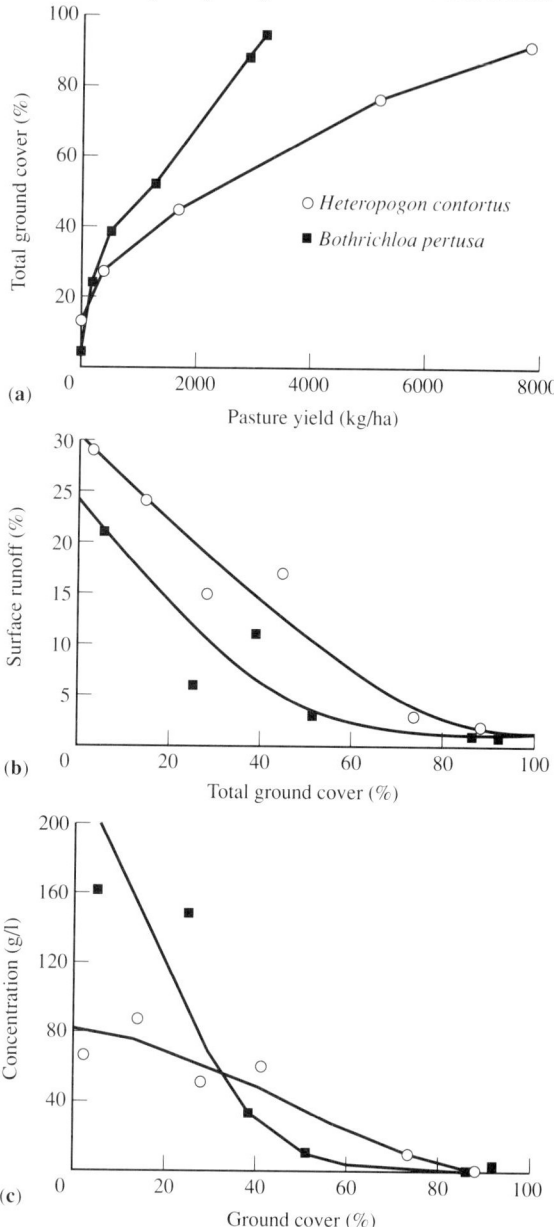

Figure 3.10 Relationships between (a) cover and pasture yield, (b) cover and runoff and (c) cover and sediment concentration for the bunch grass *Heteropogon contortus* and the sod-forming grass *Bothrichloa pertusa* near Charters Towers, Queensland (A.J. Pressland, personal communication).

- Deposition of sediment on the soil surface as gravity settles out material of varying fineness and as the instantaneous rate of flow exceeds its capacity to transport the sediments being moved. A soil particle of 0.01 mm may require a flow of 60 cm s^{-1} to detach it, but is not deposited until the flow velocity drops below 0.1 cm s^{-1} (Morgan 1979).

The transporting capacity of runoff varies with the fifth power of its velocity (Morgan 1979). The standard expression of velocity of flow is that of Manning, who related velocity (v) to the depth of flow or hydraulic radius (R), the slope of the surface below (S), and a coefficient of roughness (n), which depends upon the nature of the surface and its cover and assumes some turbulence of flow:

$$v = \frac{1.009 \; R^{2/3} \; S^{1/2}}{n}$$

The development of conservation measures is therefore often related to reducing the velocity of flow to levels which minimize sediment concentration; this depends upon a high coefficient of roughness, as occurs with good plant cover, shallow depth of flow associated with high rainfall acceptance by the soil, and directing runoff into gently sloping situations.

The scouring action of flowing water is increased by the presence of sediments, since these are more abrasive than water. This is illustrated by a further study of the two clay soil sites used by Freebairn and Wockner (1986a,b) on the eastern Darling Downs, Queensland. Soil loss could be in the order of 200 t ha^{-1} from a single high intensity runoff event with bare soil, indicating the episodic character of erosion against which farmers need to take continual action to protect their land. The soil exposure fraction was varied by the use of stubble from previous crops, and the value of efficiency of entrainment (κ) was calculated for all major runoff events over a 6-year period. Efficiency of entrainment (which varies from 0 to 1.0) was directly related to soil exposure (Figure 3.11; Rose 1988) and showed a sharp point of inflexion and increase below 0.1 (10% cover). This also indicated the positive value of even low levels of cover in reducing the entrainment factor which contributes to sediment load.

Wind erosion

In cropping areas much of the wind erosion results from sediment that has been previously entrained by water, transported and deposited. Poorly covered loose sediments are then entrained by wind action, which is another form of fluid flow. Sediments are entrained by a combination of lift, shear force and ballistic impact, whilst gravity, friction and cohesion operate against movement (Lal 1990).

Cover, as for water erosion, is the key protective measure against wind erosion. It may also be evaluated in terms of surface roughness. The second approach is to reduce wind velocity by planting barriers. Alley cropping and agroforestry, as will be further developed in Chapters 9 and 8, clearly provide protection to cultivated areas. The effects of barriers depend on their thickness and porosity, and their height, especially in relation to the length of the barrier. The barrier may extend from a mulch of branches and twigs to short-statured but wind resistant crops to shrubs and trees. The maximum reduction in wind velocity occurred at about twice the height of the shelter belt away from its centre in one study in northern Nigeria, but some reduction in wind velocity was recorded at a distance of 20 times the height of the *Eucalyptus camaldulensis* trees (Lal 1990).

The Universal Soil Loss Equation

Tolerable levels of erosion

One concept on which the determination of a level of soil erosion to be tolerated is based is that of the rate of natural soil formation, which is highest in humid climates. This has the implicit assumption that any reduction of soil depth would represent a loss of potential productivity. The concept is flawed by the quality of the new soil formed by weathering of parent material, which does not have the fertility of topsoil being lost by erosion. A further notion is the level of erosion which occurs under undisturbed natural vegetation, which in the USA is less than 2 t ha^{-1} year^{-1} and more commonly in the range 0.1–0.6 t ha^{-1} year^{-1} (Smith and Stamey 1965). An increased rate of erosion would then represent 'accelerated' erosion. Clearly the significance of this for agricultural production depends on the depth of soil and its nutrient distribution with depth; some deep fertile soils can lose substantial amounts of topsoil without great reduction in crop production. Agricultural advisers suggest limits of 1–5 t ha^{-1} year^{-1} in subhumid tropical situations, and up to 13 t ha^{-1} year^{-1} in more humid areas. The concept of adjusting the duration of the ley on different slope classes in Zimbabwe to restrict erosion to a desired level is discussed in section 10.2, Figure 10.4.

Predicting the rate of soil loss

Once farmers and advisers agree on a local level of tolerable soil loss the next step is to determine whether current land use is meeting this

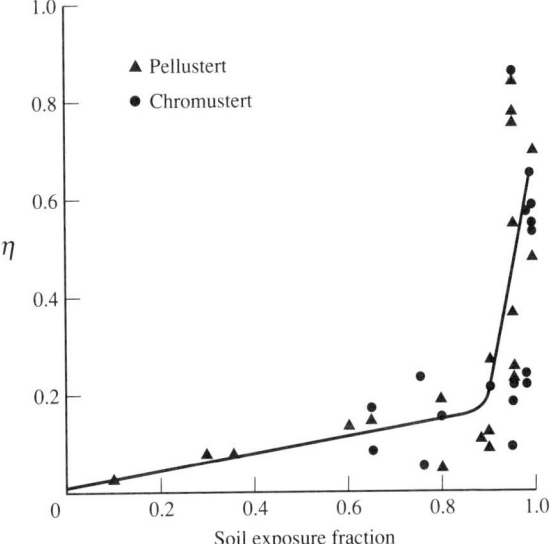

Figure 3.11 Efficiency of entrainment of soil particles (η) and soil exposure fraction on Pellustert and Chromustert clay soils in eastern Darling Downs, Queensland (from Rose 1988).

requirement; if not, what are the feasible adjustments to agricultural practice within the goals and capacities of the farmer which will contain the damage to acceptable limits?

The farmer's estimate of soil loss is poorly guided from research, since there are so many gaps in the data for the tropics and subtropics which need to be employed to describe a local situation. The most widely used concept is that of the Universal Soil Loss Equation, as developed by Wischmeier and Smith (1978). This is expressed as:

$A = RKLSCP$

where the average annual soil loss (A), expressed as t ha^{-1} year^{-1}, is the product of several factors:

- the erosivity of rainfall (R);
- the erodibility of the soil (K);
- the slope length (L) and the gradient of the slope (S);
- the cropping management (C), in so far as this affects erosion;
- the practices of conservation (P) used by the farmer to moderate erosion.

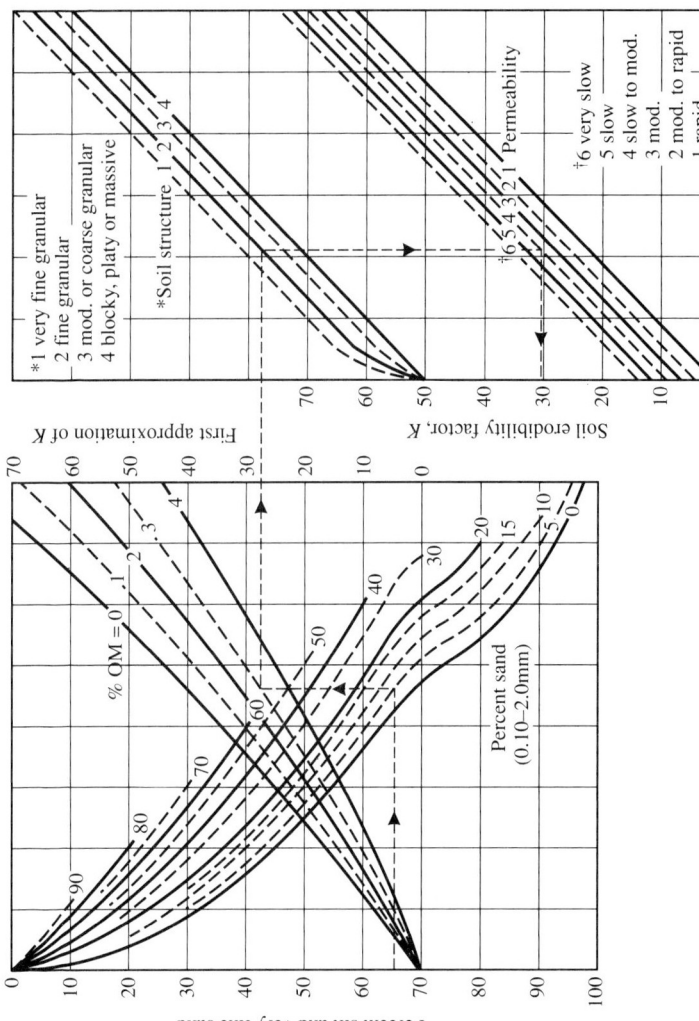

Figure 3.12 Nomogram to determine soil erodibility factor (K) from soil characteristics (from Wischmeier et al. 1971).

Erosivity of rainfall

This question was introduced in Chapter 1 when attention was drawn to the greater erosivity of rainfall in the tropics. Rainfall intensity and the paramount influence of drop size was discussed earlier in this section and exemplified in Figure 3.7. Various indices of erosivity have been developed and expressed in the mapping of isoerodents, or lines representing the same annual erosion index, as illustrated in Figure 1.2. Wischmeier and Smith (1978) used the summation of the number of storms per year and the product of their kinetic energy and rainfall intensity over 30 min to draw an isoerodent map of the USA, and this technique has been extrapolated to tropical areas (for example, Roose 1977 for West Africa, with values from 100 to 1000).

Soil erodibility

An example of a highly erodible ultisol was shown in Figure 1.1, although the usual generalization is that ultisols are less erodible than oxisols or inseptisols, whilst vertisols are especially vulnerable (El-Swaify and Dangler 1982).

The key characteristics determining erodibility are associated with:

- soil texture,
- soil aggregation and stability of aggregates,
- infiltration.

Soil texture reflects two opposing tendencies. The larger soil particles are resistant to detachment and entrainment, and sandy soils are inherently less erodible. On the other hand, very fine particles, which are more readily transported in flowing water, have a greater cohesiveness on the soil surface. The most vulnerable soils are silty, and Wischmeier *et al.* (1971) showed that the fine sand fraction ($<$ 0.1 mm) behaved like silt, so that the sum of silt plus fine sand was positively associated with erodibility.

Soil aggregation provides resistance to detachment, but much depends upon the resistance of the aggregates to dispersion upon wetting. The dispersion ratio is a positive index of erodibility, especially as it relates additionally to crust formation and degree of runoff. The shear strength of the soil, a measure of its cohesiveness and resistance to the shearing forces of water or wind, is involved. The positive effect of OM on aggregation and the stability of aggregation were discussed in Chapter 2, and earlier in this chapter.

The initial acceptance of rainfall (sorptivity) influences the erodibility of storms of short duration, whilst transmissivity or the steady-state rate of infiltration determines runoff from rainfall of longer duration, as discussed in sections 3.2 and 3.3. The latter is often determined by subsoil properties. A pragmatic model which integrates these factors (Figure 3.12) relates soil erodibility to the average soil loss per unit of EI^2 (E, energy; I, intensity of rainfall) on a bare soil on 9% slope of length 22 m (Wischmeier et al. 1971). This soil erodibility nomogram is used by commencing on the left of the diagram with the %silt+fine sand (65% in the example), moving to %sand (5%), to %OM (2.8%) and then proceeding to the right-hand diagram to structure (fine granular, 2) and to permeability (slow to moderate, 4) which gives a soil erodibility factor (K) of 0.31.

In the USA the value of K varies from 0.03 on a gravelly loam to 0.69 on a silt loam. Lal (1990) gives tropical examples in which field determinations of R were often higher than those from Figure 3.12, especially on soils of high OM content. In Thailand the range of K was 0.04 on a sand to 0.49 on a silt loam whilst in Java K varied from 0.03 on an oxisol to 0.22 for an alfisol. In West Africa the majority of soils have a K value less than 0.2 (Lal 1982). Lal (1990) regards Figure 3.12 of most utility for weakly structured or structureless soils, but of lesser applicability on soils high in iron or aluminium oxides, soils of high clay content or high-activity clays (vertisols), or soils with much skeletal material (gravelly and concretionary soils). These three groups constitute a significant proportion of tropical soils.

Slope

Erosion is related to the steepness of slope, length of slope and shape of slope. There are many studies predicting erosion from exponential, polynomial and power functions of grade (S), and Figure 10.4 shows an example from tobacco soils in Zimbabwe. The length of slope (L) is a further variable which controls velocity of flow and which is susceptible to control by the interval between contour hedges or contour banks (Colour Plate 2). These two factors may be united to give an LS factor, described by Wischmeier and Smith (1978). For example a 3% slope with 36 m slope length gives an LS index of 0.3. Runoff velocity is increased on convex slopes, whilst velocity is decreased on concave slopes, sometimes causing deposition.

Crop management

This factor (C) is essentially related to the cover provided by crops in relation to the probable occurrence of erosive rains and prediction may be made for different cropping sequences. The highest values occur where cultivation provides bare soil during the period when the highest expectation of storms of greatest intensity prevails. Many agronomic factors combine to determine this index. Planting density and fertilizer use influence the rate and extent of crop canopy development. Intercropping usually increases canopy. Residue disposal and the use of zero tillage determine the amount of protective cover. Contour tillage and the type of ridging modify the index.

Morgan (1979) gives an example for maize cultivation with contour levels near Kuala Lumpur, Malaysia. Cover is estimated for the different months of the year, varying from 0.001 to 0.7, and weighted for the seasonal erosiveness which is applicable, giving a C factor for the year of 0.208. Other examples may be found in Lal (1990).

Conservation practice

Some of the crop management factors enumerated bear on the conservation practice (P) index. The reference point is a bare fallow cultivated up and down the slope. Hudson (1981) gives an illustration of contour cultivation without other mechanical protection giving P values of 0.6 on gentle slopes of 1–2%, 0.5 on slopes of 2–7% where this practice is most effective, increasing to 0.9 on steep slopes of 18–24%. The use of tied ridges (Pereira et al. 1967) considerably restricts soil loss, and may reduce the above index by at least 50%.

The Universal Soil Loss Equation can be used to guide land use planning. Let us take an example of a farm with rainfall erosivity (R) 300, and erodible soil with K factor of 0.4, a slope of 4% and 100 m slope length, and a farmer planning to grow annual crops with crop management C value of 0.6. Tolerable erosion is set at 12 t ha^{-1} year^{-1}. The first two factors of rainfall erosivity (R) and soil erodibility (K) are local and fixed in character, and give a product of 120 t ha^{-1} year^{-1}. The LS slope factor is 0.65 so that:

$$RKLSC = 300 \times 0.4 \times 0.65 \times 0.6$$
$$= 46.8 \text{ t ha}^{-1} \text{ year}^{-1}$$

Plate 3.1 Contour banks on maize lands in Zimbabwe.

The farmer would reach an acceptable level of control of soil loss by introducing contour banks (Plate 3.1) which restricted slope length to 36 m to give an LS factor of 0.3, and adopting contour cultivation to give a P index of 0.5. Then:

$$\text{Soil loss } (A) = RKLSCP$$
$$= 300 \times 0.4 \times 0.3 \times 0.6 \times 0.5$$
$$= 10.8 \text{ t ha}^{-1} \text{ year}^{-1}.$$

Other options such as mulching, alley cropping, or the introduction of ley pastures might also be considered by the farmer to achieve a similar or better result. There are more exacting models of soil loss than that expressed in the Universal Soil Loss Equation, and the specialist reader will find the review of Lal (1990) a good source of further material.

3.4 Approaches to soil conservation

Most approaches to the control of soil erosion are included in the following:

- Mechanical structures,
- Tillage methods,
- Vegetation barriers,
- Vegetation cover.

Mechanical structures

On sloping land there are two basic concepts of mechanical control. The first, which has been prevalent in Western agriculture, is to reduce the velocity of runoff, and to impound it or conduct it safely in specially constructed channels which withstand erosion. The length of slope which modifies runoff velocity is adjusted by the interval between graded banks. These are constructed either on the contour so that water is impounded and its gradual infiltration into the soil below increased or they are built with a slight gradient which leads to non-erosive velocities in the channel above the bank. Discharge from the banks then flows down waterways with a strong sod cover or with specially prepared surfaces.

The concepts of mechanical engineering, often allied with an obsession about the efficiency of heavy earth-moving equipment, have dominated the culture of many workers in soil conservation. These concepts have proved barren indeed unless allied with input from the biologists. On shallow soils the construction of banks exposes subsoil, which is a poor medium for growing crops, and buries topsoil, leading to a reduction in crop yields. The design and construction of terraces adds to the cost of production; Lal (1982) quotes an instance from Nigeria where the cost of post-clearing land development which incorporated contour banks and waterways was $US430 ha^{-1}. The maintenance of earth or rock structures is an added farm cost. High intensity rains are episodic and difficult to forecast in ways which lead to a structural design that caters for the survival of the structure intact; when a bank ruptures the concentration of flow through the break may cause more local gullying than if the bank were not there. Banks built of sodic soils are subject to tunnel erosion or slippage. Recognition should also be given to the failure of contour banks *per se* to prevent erosion from the inter-bank areas (Braunack and Walker 1985). Figure 3.9 shows the sediment discharge from a contour bank system at Greenmount, Queensland (Freebairn and Wockner 1986b), but the alarming figures for high intensity storms represent only 10–20% of the total erosion occurring and the balance of soil loss is moved off the inter-bank area to accumulate in the bank channel.

The second concept is to construct terraces with a level cross-section which may be safely cropped. These may be bench terraces for rain-grown crops, or they may be edged with low mounds or bunds to contain natural runoff or irrigation water. Some terraced systems in Asia represent the labour of generations of farmers and have been continued in place for centuries, safeguarding the landscape through the continual pastoral care of individuals or communities.

Tillage methods

Tillage merits a monograph in its own right and these comments merely highlight some of the more significant ideas which influence conservation practice. The first is that tillage which produces a bare soil surface should be minimized and agricultural systems devised which do not rely on intensive tillage. Some workers emphasize the need for bare fallowing to reduce evapotranspiration from weeds and to store moisture for subsequent crop use. This objective needs to be balanced by the perception that systems based on long fallowing may provide better water availability to crops, but fewer crops are grown and total crop yields are reduced in the long term (Berndt and White 1976).

The second idea is that tillage should be directed to greater rainfall acceptance. Continuous cereal cropping can lead to structural degradation, as evidenced by increased bulk density, greater dispersion and slaking, reduced aggregate stability and decreased hydraulic conductivity limiting soil water storage (Cook *et al.* 1992). Inversion of the upper soil is avoided in most modern systems, as mentioned in section 10.3. The development of a plough pan is illustrated in Figure 10.5. Alternatively, on hard setting soils tillage is necessary to increase rainfall acceptance, as illustrated in Figure 3.3. Tillage may be directed to increasing total porosity and random roughness, which is sensitive to the moisture content at the time of tillage (Allmaras et al. 1967). Vertisols exhibit especially poor trafficability when wet, and the range of moisture availability which is favourable for tillage is narrow.

The third essential idea is that of contour cultivation, the effects of which were discussed in section 3.3. Contour ridging is commonly practised in tropical systems as a means both of reducing runoff and of providing varying levels of oxygen and moisture availability between ridge and furrow. There are many variations; tied ridges (or basin listing) are especially effective in retaining runoff (Pereira et al. 1967). Systems are devised to favour convenience of intercropping or to meet special crop or implement requirements: camber bed systems, broad-bed furrows, or hillock and mound construction (Lal 1990).

Vegetation barriers

The stability of contour banks is enhanced by the binding action of tree roots, and the planting of shrubs and grass strips on the banks is also beneficial. The alternative is to discard mechanical structures altogether and to rely on vegetation barriers in conjunction with the maintenance of soil cover to control erosion. The establishment of vegetation barriers on the contour will also lead to natural terracing if the soil movement which does occur is arrested by the barrier and accumulates above it.

A barrier of widely spaced single stem trees is ineffective in controlling erosion unless ground cover is also established. The farmer's decision about choice of planting material takes into account the integration of the vegetation barrier into farm activities. A field which is regularly grazed will be protected by unpalatable *Vetiveria zizanioides*, which forms a tall, dense grass barrier (National Research Council 1993). The need for light timber or fuel makes agroforestry attractive, or high value fruit crops may be planted, as discussed in Chapter 8. Alley cropping with shrub legumes provides wood, fodder and vegetative cover for the inter-shrub area, and contributes to the N economy of the ecosystem, as elaborated in Chapter 9; Young (1987) provides a useful review. The necessary width of the barrier is partly determined by the rapidity of shrub establishment and the need for early cover; double rows clearly offer better protection than single row planting, but reduce the area available for cropping. An alternative is to plant the fast growing but short-lived *Cajanus cajan* with the more slowly established but long-lived *Leucaena leucocephala*.

An illustration of the benefits of alley cropping (Paningbatan 1990; Craswell and Pushparajah 1991) comes from the early results of a study at Los Banos, Philippines. The site bore a Lifa clay loam (Typic Tropudalf) and had a gradient of 14–19%. Soil loss over a 3 month period in which 1424 mm of rain fell was 127 t ha^{-1} where maize was grown in rows running up and down the slope. On areas where the shrub legume *Desmanthus virgatus* was planted in 1 m wide strips on the contour 5 m apart soil loss was reduced to 41 t ha^{-1}. If 330 kg ha^{-1} of shrub trimmings were placed on the cultivated area soil loss was only 3 t ha^{-1}, or in a zero-tillage system augmented by trimmings 0.2 t ha^{-1} soil loss was measured. Runoff was 347, 183, 75 and 32 mm respectively in the four treatments.

Vegetation cover

The preservation of surface cover is the most effective defence against erosion. In cropping systems the retention of stubble mulch is effective practice (Juo and Lal 1977) if crops are sufficiently productive to provide enough residues, and machinery has been devised which plant and till through heavy mulch cover. The key issue is the proportion of ground covered, which shows a positive relationship with amount of crop residue. The concept of minimum tillage, often allied with herbicide use, has been widely accepted and adapted to particular crops. This may be extended to zero tillage, in which crop planting and management is wholly dependent on either herbicide application or upon the presence of aggressive plant cover restricting the ingress of weeds.

These ideas are expounded in other sections of the book. In this chapter the relationship between cover and infiltration is illustrated in Figure 3.2, whilst the effects of cover on runoff (Figure 3.8), on sediment concentration (Figure 3.9) and on entrainment of soil particles (Figure 3.11) are shown for different situations. The concept of live mulch is discussed in section 10.3 (Figures 10.8 and 10.9 and Tables 10.8 and 10.9).

In pasture situations grazing management, the choice of pasture species (Hong 1978) and the fertilizer input in relation to stocking rate (Humphreys 1991) are key management factors influencing cover. At La Romelia, Caldas, Colombia (CIAT 1991c) the creeping legume *Arachis pintoi* offered the highest cover relative to other legumes. Soil loss over the period March–October was 31.8 t ha^{-1} from bare soil, but only 2.1 t ha^{-1} from *B. decumbens–A. pintoi* pasture, or 0.8 t ha^{-1} from undisturbed *Paspalum* spp. Figure 3.10 illustrates the difference in runoff and soil loss from bunch and sod-forming grasses, in a specific situation where the occurrence of the sod-forming grass reflects a previous history of heavy grazing. Similarly greater runoff is expected from vegetation dominated by ephemeral species than from perennial grasses (Eldridge and Rothon 1992); these may be indicators of differences in soil conditions (Gifford 1978). Much has been made of the influence of trees in controlling erosion, but the protective cover of well-managed grasslands often provides better control of erosion, especially in subhumid situations where litter may be removed by annual fires and frequent coppicing occurs.

3.5 Conclusion

The maintenance of biological cover on the soil surface is feasible in farm practice; it provides protection against the action of raindrop splash and increases rate of infiltration, whilst the tortuosity of flow reduces its velocity, sediment capacity and degree of entrainment. The seriousness of erosion in the tropics and subtropics was outlined in Chapter 1, and there are many studies indicating the sediment load downstream from the farms (Rapp *et al.* 1972; Dunne 1979; Ciesiolka 1987). However the focus needs to be wider than that of the tonnage of soil loss; the loss of nutrients concentrated in the surface horizon is often the factor most detrimental to crop yields.

The adoption of changed farming practices requires that technologists develop with farmers a suite of conservation options which can be recognized by the farmers as feasible modifications that are in tune with farmer goals. These connections are not necessarily available (Chamala and Coughenour 1986). The distance yet to be travelled is indicated by a survey of farmers working steeply sloping land in northern Thailand (Menz 1992). Rill erosion was evident on 75% of fields and sheet erosion on 97% of fields; 23% of farmers did not appear to connect farming methods with soil erosion and 14% of farmers were unaware that methods existed to control it.

This chapter has indicated that soil structure is a significant component of soil fertility which influences the aridity of the cropping cycle and the delivery of nutrients to crop plants. Mechanical tillage rapidly destroys the gains in aggregate stability which conservation practice brings to many soils (Wilkinson 1975a); this requires that the beneficial effects of forages on soil structure and OM status be regularly or continuously incorporated in cropping systems, as outlined in Chapters 8–11.

CHAPTER 4

Efficiency of resource use

4.1 Introduction

There follow four brief chapters dealing with efficiency of resource use, crop protection systems, outputs from animal production and diversification of income. These themes are not developed in sufficient depth to provide the technology necessary for their expression, which would each require a separate monograph; rather some concepts are mentioned which are central to the motivation of farmers who incorporate forage production in their cropping systems.

Greater efficiency of resource use is claimed for agricultural systems which incorporate forages and livestock relative to those based on cropping alone (Chantalakhana 1990). 'Efficiency' is used in many different senses, according to the criteria employed and the purpose for which a farming system is viewed. The central issues are socio-ecological, and relate to the well-being of rural communities as expressed in the attainment of farmer goals or additionally (Conway 1987; Bawden and Ison 1992) other measures of social value involving the satisfaction of basic human needs for esteem and self-actualization, of equity of allocation, and the quality of human relationships. In this book the focus is on the use of physical and biological resources and attention is given to the lower order farming system and its components rather than to the higher order agro-ecosystem for which value judgements are inherent and of great human significance but of lesser precision.

Many people view efficiency in economic terms, and cost–benefit analysis may be seen as a framework for investigating the pluses and minuses of different options (Edwards, G. 1991), whether these are expressed at the farm or at wider levels; in this analysis it is necessary to attach notional cash values to the component inputs and outputs of

112 Efficiency of resource use

the system. The development of such analyses requires that the physical system of production, its long-term trend line, its responsiveness to inputs, its stability under small fluctuations in components of the environment, and its resilience when major perturbations occur are understood and quantified.

In this chapter efficiency of production is considered in relation to the use of the capital resource of land, the output from the environmental growth factors of light, water and nutrients, and the effects on environmental quality associated with chemical residues and stream and atmospheric pollution.

4.2 Integration of land resources

The capital resources of land are used more efficiently in forage–crop enterprises (Plate 4.1) than in either single animal or crop production operations in two general sets of circumstances.

Plate 4.1 *Desmodium intortum* and *Setaria spacelata* var. *sericea* in the drainage areas, Sri Lanka.

Integration of land resources 113

Plate 4.2 Swamp lands of *Panicum repens* and *Cyperus* sp with cattle in south Thailand.

- Sustainability of crop production depends upon the incorporation of legume-based forages in the crop rotation or as companion species with cash or food crops. This is elaborated in Chapters 8–11.
- Utilization of land of different classes and of varying agricultural capability is more effective if the total landscape is integrated into a crop–livestock enterprise (Colour Plate 5) than if separate crop and livestock operations are confined to particular land classes.

Land of high inherent fertility is usually reserved for cropping, and it is traditional in many tropical areas for pasture to grow on waste land (Colour Plate 6), roadsides, communal grazing areas close to settlement or in more distant open woodland, saline areas (Plate 4.2), on nearby sloping land or on areas of skeletal soil. These pasture areas may be inadequate to support a year-round livestock enterprise. The crop and pasture areas provide complementary seasonal grazing. It is also difficult to sustain draught animals on cropping land alone during the cropping season, when pasture lands and woodland grazing are in greatest demand; conversely dry season grazing of crop residues is needed to sustain animals based primarily on restricted areas of native pasture.

Farmers are skilful in devising systems of integrated land use which optimize crop production but also maintain continuity of forage supply to animals and minimize animal stress. Gibson (1987) describes a self-sufficient dairying system in northeast Thailand which successfully combines crop and pasture production in the utilization of planted legume-based pasture, crop products and crop residues.

The integration of draught and beef production with cropping in the Khon Kaen district, northeast Thailand (lat. 16° N, 1260 mm annual rainfall) was illustrated for three villages by De Boer (1973). From mid-April to mid-October some 85% of annual rainfall occurs. During this period about 57% of the cropping area was planted to rain-fed wetland or upland rice, while kenaf, maize with mung bean as a relay crop, and sugar-cane were the other main crops. Communal pasture, woodland and fallow land comprised 14% of the total area, and was mainly used during the cropping season from May to December. These pastures were dominated by short grasses such as *Dactyloctenium aegyptium*, *Brachiaria miliiformis*, *Chrysopogon aciculatus*, *Digitaria abscendens* and *Eragrostis viscosa* close to settlement; in the more distant woodland areas the tall grasses *Arundinaria ciliata* and *A. pusilla* predominated (Robertson and Humphreys 1976). The cattle and buffalo varied in age distribution and sex ratio in each village according to whether the emphasis was on the use of animals for draught or for breeding and sale. They were grazed on the cropping areas during the dry season and were fed supplementary rice straw in the latter part of the dry season. Despite the relatively small area of pasture available, its integration with the use of the cropping area produced an overall annual SR of 1.1 AU (300 kg LW) ha^{-1}. The complementary character of crop and animal production was evident in that the village with the greatest crop production exhibited the highest SR of 1.3 AU ha^{-1}. These levels of SR also indicate a high efficiency in the integrated system.

A second illustration of the complementarity of land resources is taken from the Ethiopian highlands (Plate 4.3) at Debre Birhan (lat. 10° N, 860 mm annual rainfall multitude) where cropping areas are integrated with communal pastures. The number of cattle, small ruminants and equines was positively associated with the area cultivated to food and cash crops (Gryseels and Asamenew 1985).

4.3 Energy transfer and the use of crop residues

An integrated forage–crop system may capture more light energy or transfer more energy to farm products than a single enterprise if:

- the amount of photosynthesis is greater, arising from the provision of more continuous green cover and the minimization of periods when the soil is bare;
- crop residues are used by the grazing animal to produce draught, meat, milk or surplus animals for sale instead of being returned directly to the soil. A production system based on draught animals minimizes the energy subsidy from fossil fuels.

McDowell (1988) makes a comparison between the use of energy and protein from a maize crop which may yield a metabolizable energy (ME) of 127 kJ ha^{-1} and a protein content of 620 kg ha^{-1} from grain and residual shoots. The use of the grain only for human or livestock consumption represents utilization of 39% ME and 20% protein; the incorporation of the bran and stover as animal feed represents utilization of 56% ME and 28% protein.

The interactions between components and subsystems in an integrated forage-crop system need to be quantified if the comparative energetic efficiency of different interventions is to be evaluated and the needs of the animals for continuity of feed supply are to be met. In the western Kenyan highlands, a typical farm system (McDowell 1988) has a close interdependence between crop and livestock subsystems (Figure 4.1). Farms average c. 1 ha in size and farmers use intercropped

Plate 4.3 Reclamation of steep lands in Ethiopia with *Eucalyptus globulus*.

116 Efficiency of resource use

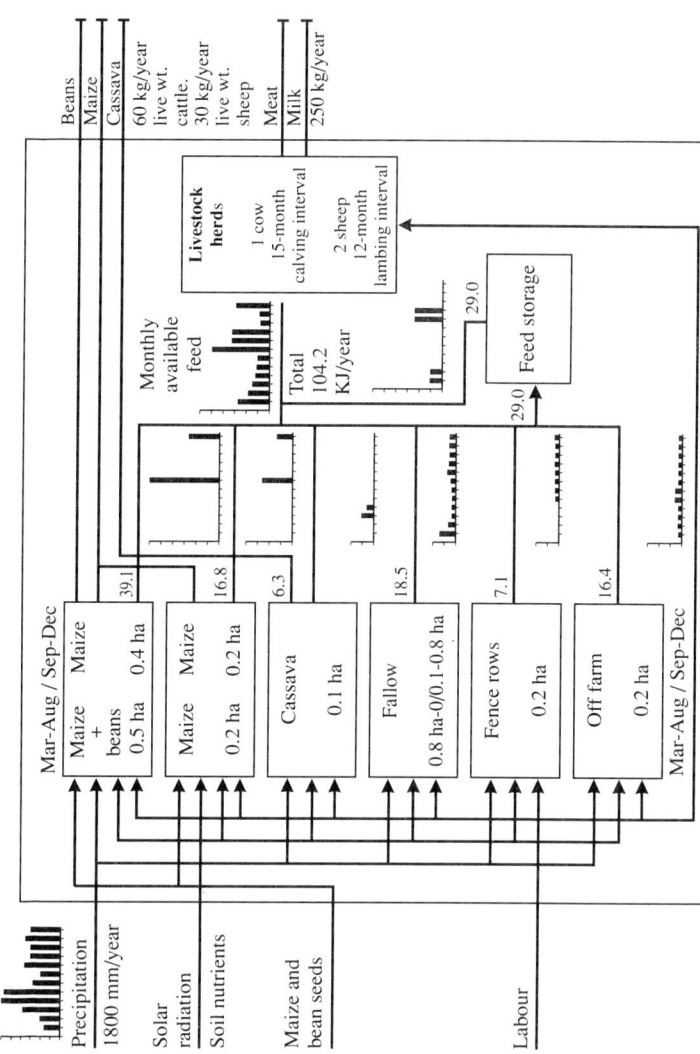

Figure 4.1 Model of low input farming system in western Kenya. Flows between crops and livestock are expressed as kJ DE ha^{-1} year^{-1}, whilst histograms show monthly distribution of rainfall and of feed (from McDowell 1988).

maize and beans, double cropped maize and cassava to provide food and feed. Most cropping is by hand, but livestock of more than one species provide meat and milk when fed on crop products, fence rows and fodder cut from off the farm or obtained by grazing. The net energy output from both subsystems is superior to the output feasible from a single enterprise.

4.4 Utilization of moisture and nutrients
Moisture use

The twin objectives for water management are:

- maximization of output of farm product per unit of rainfall or irrigation water received;
- minimization of environmental damage associated with the characteristics of surface runoff and groundwater.

Pastures and well-managed crops have similar conversion efficiencies of DM produced to moisture used, which are usually in the range 3–40 kg DM ha^{-1} mm^{-1} rainfall, according to environmental conditions. The main advantages of incorporating forages in cropping systems are that pastures will transpire water whilst water will be lost by evaporation during the intermittent periods of bare soil exposure which occur in many cropping programmes; pastures will also reduce runoff, as discussed in Chapter 3. One issue which needs to be addressed is the effect of farm practice on the level of the water table and associated problems of salinity or of nutrient drainage.

The first effect follows the conversion of forest or woodland to crop land or pasture. It is commonly believed that clearing the land will reduce evapotranspiration and thereby increase the amount of flow to groundwater (Williams and Chartres 1991). In monsoon rainfall areas perhaps 70% of the annual rainfall occurs over 3 months and a close sequence of rainfall events may fully recharge the profile and cause water to move deep into the profile beyond the root zone. This may cause leaching of nitrate (Probert and Williams 1986) or if salt is present in the subsoil or regolith the rise in the level of the water table may cause surface salt to appear lower down the slope. At a site near Charters Towers, Queensland (lat. 25° S) water use by deep-rooted *Eucalyptus* spp. woodland was substantially greater than that from native grasses alone where trees had been killed (Figure 4.2), whilst oversowing the native grasses with *Stylosanthes scabra* cv. Seca gave an intermediate level of water use, according to the level of P application.

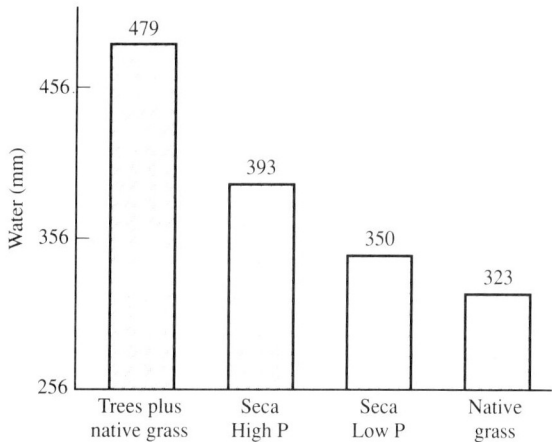

Figure 4.2 Comparative water use by *Eucalyptus* spp. woodland, native grass with trees killed, or oversown with *Stylosanthes scabra* cv. Seca at two levels of P supply (from Williams and Chartres 1991).

A second illustration is the effect of clearing *Acacia harpophylla* (brigalow) woodland near Theodore, Queensland (lat. 25° S), where the hydrological characteristics of three catchments each 12–17 ha in area and of average slope 2.5% were calibrated for 18 years before disturbance (Thorburn et al. 1991). Two of the catchments were then bulldozed and burnt *in situ* in May 1982 (time T_1). One catchment was planted in December 1982 with *Cenchrus ciliaris*, a deep-rooted pasture species, and the second was annually cropped to wheat from 1984. From T_1 to September 1983 (T_2) recharge of groundwater averaged an estimated 70 mm year^{-1} in the bare cropping area, 29 mm year^{-1} where pasture had been sown and which contributed to water use, and only 7 mm year^{-1} in the undisturbed brigalow woodland (Figure 4.3). This period was marked by unusually heavy rain of 450 mm in 5 weeks of April–May 1983, and the results show how groundwater storage can be substantially modified during the early phase of land development.

In the subsequent 4 years (T_3) rainfall averaged 630 mm year^{-1}, and recharge of groundwater was negative or zero for the three systems of land use. This indicates circumstances where land development from forest causes no long-term disturbance to the hydrological cycle and is a reassuring finding. It should be noted that this study was undertaken in a subhumid area where pan evaporation exceeds rainfall on average in all months of the year, where soils are of low permeability and good water storage capacity, and where the trees are shallow-rooted with most roots not extending below 0.8 m depth.

Comparative studies of evergreen forest and of *Imperata cylindrica* grasslands in northern Thailand have shown less soil erosion from grassland than from forest, but greater runoff in the early wet season following annual dry season burning (Gibson 1983). Grassland is therefore advantageous to contiguous crop lands, since planting of wet-rice in the lowlands is often delayed until the paddies fill with runoff water, late planting sometimes occasioning crop failure.

Nutrient flow and nutrient cycling

The main issues which are relevant to the efficiency of nutrient use are:

- retention of the accessibility of nutrients to plants and animals and the minimization of losses to groundwater and to the atmosphere;
- spatial transfer of nutrients to crop lands in order to maximize returns from nutrients;
- cycling of crop and pasture material through animals to increase nutrient availability to plants.

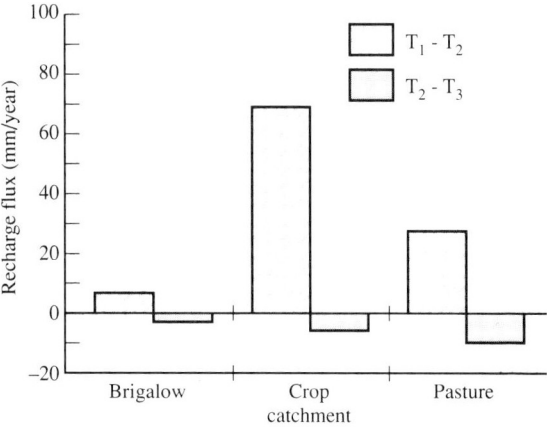

Figure 4.3 Groundwater flux from *Acacia harpophylla* (brigalow) woodland, cropped land, or pasture during early development (T_1–T_2) and during a subsequent phase (T_2–T_3) (from Thorburn et al. 1991).

120 *Efficiency of resource use*

Losses of nutrients

Deep drainage of nutrients occurs in many soils when the profile is completely recharged and follows the movement of moisture into groundwater and stream flow, as discussed in the preceding section. In section 10.3 the effects of zero tillage in promoting deep drainage of nitrate is mentioned (Thompson *et al.* 1987) and Figure 10.10 (section 10.3) illustrates the accumulation of nitrate at depth following a leguminous ley at Katherine, Australia (Jones *et al.* 1991).

This problem is mitigated if vigorous plant cover is maintained which utilizes soil moisture in evapotranspiration to maintain the hydrological balance; the second line of defence is the culture of deep-rooted species. In East Africa the use of tap-rooted legumes such as *Neonotonia wightii* was advocated partly on the basis that cycling of nutrients from the deeper soil layers to the surface soil for subsequent use by crops was beneficial. *Macroptilium atropurpureum* accessed soil layers deeper than those by a companion grass (Sheriff and Ludlow 1984). The 'banking' of nutrients under leguminous covers enhances the production of perennial plantation crops (Agamuthu *et al.* 1981). Subsequently in Chapters 8 and 9 the positive effects of deep-rooted trees and shrubs on the fertility of surface

Plate 4.4 Drying dung for sale.

Utilization of moisture and nutrients 121

Plate 4.5 Donkeys carting dung for sale in Ethiopia (Y. Sedi).

soil are outlined (Kellman 1979; Glover and Beer 1986; Nair 1987; Gutteridge 1992).

Losses of nitrogen to the atmosphere through denitrification or through volatilization of ammonia are substantial in forage systems and were discussed in section 2.4 and illustrated as part of Figures 2.9 and 2.10; the effects of grazing on nutrient cycling are reviewed in Barrow (1987), Floate (1987), Gillingham (1987), Humphreys (1991) and Haynes and Williams (1993). This question is pertinent to the success of animals in effecting spatial transfer of nutrients.

Spatial transfer of nutrients

The sustainability of many cropping subsystems in the tropics and subtropics is dependent upon the transfer of nutrients by animal excreta from other areas on which animals graze or from which forage is cut, removed and fed to animals at another site (Plates 4.4 and 4.5); the grazing subsystem may be in negative nutrient balance. This situation may be rationalized on the basis of greater marginal returns to the farmer from crop lands than from animal products, depending upon relative costs and returns, but the cost to the long-term productivity of the land used for grazing or cut forage should be taken into account. The depletion of N is minimized if forage legumes are grown for grazing, but there is less opportunity for legume N accretion *in situ* in cut-and-remove systems, since the greater part of plant N is located in the shoots.

A simple model (Figure 4.4) of the nutrient cycling in African savanna farming systems (Swift *et al.* 1989) depicts the subsystems of arable croplands, savanna grazing and the household, which also returns nutrients to the total system. The savanna usually contains woodland which contributes fuel, timber and fruit as well as dry season fodder; there is also a transfer of leaf litter from savanna and some farmers spread the enriched soil from termite mounds on arable areas. Previously fallowed areas may also contribute to grazing.

In many systems animals are kept at night and for part of the day in small enclosures (bomas), usually located close to the household for convenience and security (Probert *et al.* 1992). Excreta accumulate in the boma, together with unconsumed crop residues which are trampled into the surface. Usually each year during the late dry season manure is dug out from the boma and carted to crop lands, perhaps using wheelbarrows to convey manure to the nearer fields and using ox-carts for the more distant fields; fields close to the household receive more manure than remote crop lands (Plate 4.6). If the boma is not moved a shallow pit develops which accumulates moisture and provides anaerobic conditions favouring denitrification and also opportunities for leaching of nutrients.

A study of five farms at Mwala near Machakos, Kenya (Probert *et al.* 1992) measured the density and size of manure heaps being spread on crop lands; farmers used surprisingly high rates of application which varied from 36 to 168 t ha^{-1}. In two areas of Zimbabwe Mugwira and Shumba (1968) estimated that rates of application were 14–72 t ha^{-1}, which may be compared with the local recommendation of 37 t ha^{-1} at 4-year intervals. Manure mixed with soil has a low nutrient content relative to that of fertilizer, especially with respect to N. Nevertheless the average input per hectare in the Machakos study averaged 280 kg N, 91 kg P and 44 kg K, indicating remarkably high nutrient applications in this district which might be expected to provide luxury supply. Manure from bomas is of lower quality than manure from commercial feed lots, and the quality of diet and the subsequent method of conservation of the excreta obviously influence nutrient concentration (Tanner and Mugwira 1984). In the Mwale study of Probert *et al.* (1992) the N content of the material spread ranged from 0.17 to 0.63% (Table 4.1) and the high ash concentration indicated much soil contamination, whereas the farmyard manure collected by Ikombo (1984) contained 1.62% N.

The beneficial effects of manure on maize grain yields are illustrated (Figure 4.5) from a field trial at Kampi ya Mawe near Machakos, Kenya (Ikombo 1984). Manure at varying rates was applied before the short rains of 1981, which were disastrously low and led to poor maize yields. The residual effects of manure were strongly expressed in two subsequent crops, and fresh maintenance applications of manure before the short rains of 1982 did not increase yields.

The direct return of nutrients to crop lands by animals camped at high density obviates the labour handling involved in transport of manure and the losses which occur during storage of manure. At Abet, Nigeria (lat. 10° N, 1300 mm annual rainfall) Powell (1986) recorded manure deposition as farmers hired Fulani herdsmen to camp their animals on fields during the dry season; a herd of $c.$ 50 cattle are held on $c.$ 0.04 ha during five consecutive nights before moving to a new position, depositing an average 6900 kg ha^{-1} manure. During the early wet season animals are moved every second or third night to minimize soil compaction, depositing $c.$ 5500 kg ha^{-1} manure. These interventions provided about 41 kg N ha^{-1} and 10 kg P ha^{-1} in the dry season, or 104 kg N ha^{-1} and 15 kg P ha^{-1} in the early wet season, when animals were ingesting higher quality feed. Dry season manuring increased maize grain yields by about 1 t ha^{-1}.

One problem associated with manuring is an increase in the weediness of crops. Conserved manure which has been fermented to produce

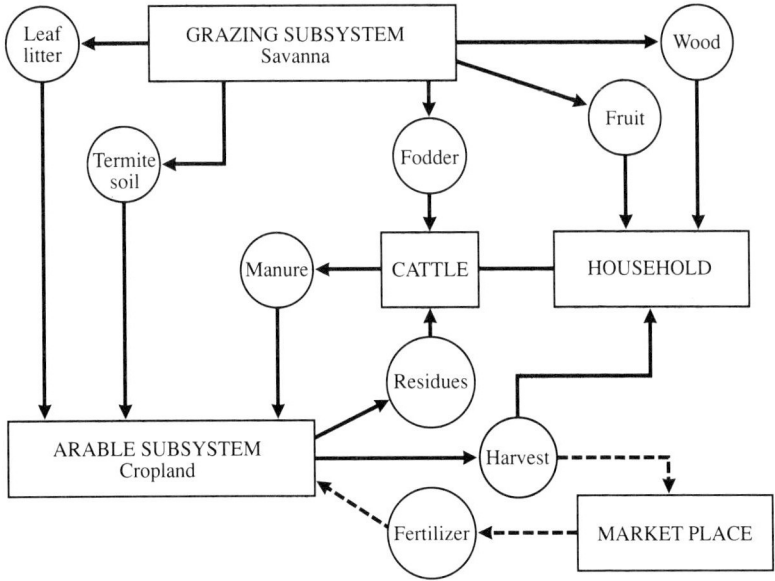

Figure 4.4 Generalized model of nutrient cycling in an African savanna farming system (from Swift *et al.* 1989).

Plate 4.6 Spreading manure from the boma on maize lands in Zimbabwe.

high temperatures in the stack loses considerable N but destroys the viability of seeds of many weed species. In Powell's (1986) Abet grazing study weeds accounted for c. 68% of total N uptake on manured areas, whilst on non-manured areas weeds entrapped only 28% N uptake; the use of manure therefore has a concomitant requirement for good crop husbandry directed to weed control if much of the benefit of manuring to crops is not to be lost to weeds.

Cycling of crop and pasture material through the animal

The processing of crop residues and by-products and of pasture material through the animal provides nutrients to the soil in forms which are more available than from plant material used as surface litter or incorporated directly in soil. Certainly there are losses to the atmosphere and through leaching from excreta, and spatial distribution from grazing is uneven (Jones and Ratcliff 1983). Nevertheless there is good evidence for the beneficial effects of grazing leys on subsequent crop yields, as discussed in section 10.5 and illustrated in Table 10.11. On the other hand the quality and yield of crop residues in the dry season may be too low for their cycling to have much influence on subsequent crop yields; at Abet, Nigeria dry season grazing of cropped fields (as distinct from herding) produced an average of only 111 kg ha^{-1} manure, which would have little impact on crop production.

In alley cropping systems the farmer may have a choice as to whether leaf material from the contour shrub lines is to be placed directly on the cropping area or processed first through ruminants. Catchpole and Blair (1990b) labelled leaf material of *Leucaena leucocephala* and *Gliricidia sepium* with ^{15}N and either applied it directly to soil or

Table 4.1 Nutrient concentration of manure from various African sources (from Probert et al. 1992).

Reference	Ash (%)	C (%)	N (%)	P (%)	K (%)	Ca (%)	Bray 2 P (p.p.m.)	Mineral N (p.p.m.)
Probert et al. (1992)								
Farmer 1	94	4.4	0.63	0.14	0.84	1.24	648	81
Farmer 2	92	5.1	0.55	0.16	1.10	1.94	727	47
Farmer 3	94	1.6	0.17	0.08	0.26	0.58	185	81
Farmer 4	88	3.4	0.33	0.13	0.66	0.96	473	87
Farmer 5	89	4.4	0.50	0.14	0.68	0.84	214	87
Farmer 6	91	3.0	0.35	0.20	0.78	1.47	894	135
Farmer 7 (ex goat boma)	79	5.3	0.62	0.25	1.56	3.09	946	124
Ikombo (1984)			1.62	0.50	1.34	0.26		
Okalebo (unpublished)								
Kimutwa			1.33	0.30	2.11			
Kathonzweni			0.81	0.34	2.44			
P.N. de Leeuw (ILCA) (personal communication)								
Fresh cattle manure	53		1.28	0.45	2.65	1.26		
Old cattle manure	81		0.49	0.31	1.65	0.85		
Small stock manure	74		0.59	0.57	0.57	1.76		
Mugwira & Shumba (1986)								
Chiota communal area			0.98	0.13	0.99	0.48		
Svosve communal area			1.05	0.19	1.47	0.58		
Mokwunye (1980): range for various samples from West Africa			0.48 to 1.95	0.06 to 0.57	0.39 to 2.62			

fed it through goats and collected separately the dung and urine produced. The amounts of each of these materials were adjusted to give equal applications of N to the soil, and were either applied to the soil surface or incorporated with the soil. The recovery of ^{15}N in *Panicum maximum* seedlings growing in the treated soil for 10 weeks was greater if the material was incorporated in the soil rather than applied to the surface, except in the case of *L. leucocephala* leaf (Figure 4.6). The instant availability of urine to grass when incorporated in the soil was evident; the greater losses occurring when urine was applied to the soil surface may have been less if applied to an established grass sward rather than to germinating seedlings. Faecal N was more slowly available than from other materials, especially from *G. sepium* which had a lower N concentration (1.36% relative to 1.76% from *L. leucocephala*).

This study indicates the value of urine in supplying N to plants. In cut-and-remove systems of feeding some urine is absorbed by the plant residues rejected by the animal, which accumulate on the surface of the penned area, or by the soil below. Catchpole and Blair (1990b) suggest collection of urine may be worthwhile, as occurs in dairy slurry systems. Improved efficiency of nutrient recycling can result from better conservation of excreta and from attention to less heterogeneous disposal of material on crop lands. The complementary use of manures with fertilizer or with legume N might also target better the nutrient additions needed to optimize crop yields. The luxury applications of nutrients reported by Probert *et al.* (1992) may in part arise from a recognition of the low N content of boma manure; the combined use of

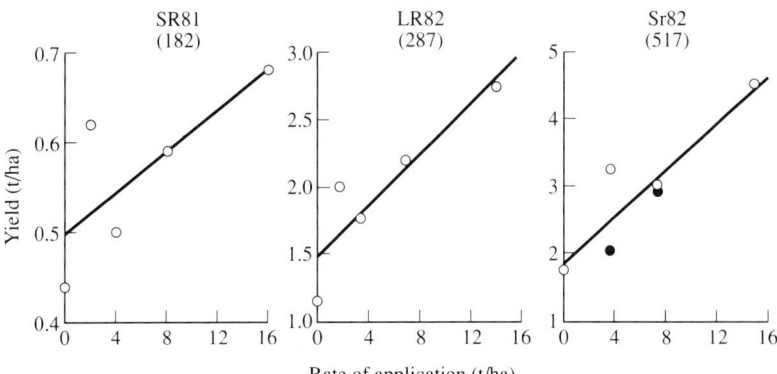

Figure 4.5 Effects of manure on maize yields at Kampi ya Mawe near Machakos, Kenya (SR short rains, LR long rains; rainfall in parentheses; closed symbols indicate effects of fresh applications in 1982) (from Probert *et al.* 1992, citing Ikombo 1984).

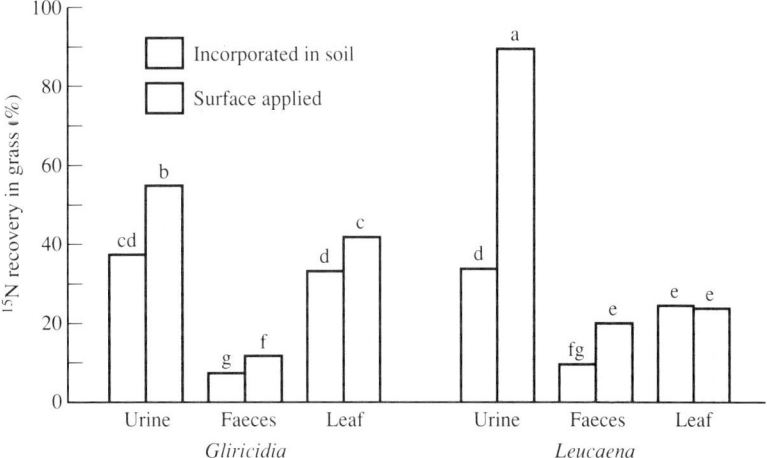

Figure 4.6 ¹⁵N recovery in *Panicum maximum* from equal amounts of N applied as leaf of *Gliricidia sepium* or *Leucaena leucocephala* or as dung or urine derived from these feeds. Treatments showing the same letters do not differ at $P < 0.05$. (From Catchpole and Blair 1990b).

manure to supply P, K and other nutrients and of fertilizer N or legume incorporation to meet the main plant N requirement may enhance the efficiency of the contribution from animals.

4.5 The modification of interference between components of the system

Studies of farming systems recognize the interrelationships that exist among the farm systems, and between the constituent elements and the farm environment. These interrelationships are often complementary and reinforcing; systems evolve in practice from the pragmatic perceptions of farmers seeking to maximize the use of the resources available to them. These perceptions are also influenced by local culture. Farmers in the highlands of northern Thailand devote family labour to the cultivation of subsistence rice, despite the higher returns per unit of labour from, for example, expanding higher value production of fruit crops, since the provision of staple rice for the family has a central cultural impetus; alternatively the reallocation of labour may follow the provision of better information to farmers.

The system which evolves will also reflect the seasonality of particular tasks on the farm, and the scope for the introduction of innovations depends upon the incidence of gaps in occupation or of availability of surplus labour. Spreading of manure in Machakos, Kenya may interfere with early planting of maize unless sufficient labour is allocated in the dry season (Probert et al. 1992); in northeast Thailand the planting of relay forage crops using stored moisture in the rice paddy may compete with the threshing and storage of the rice harvest (Shelton 1980). Conversely the planting of an early short-season legume crop at the beginning of the wet season is feasible in a system where the main wet-rice crop is planted in mid-season; the overseeding of roadsides and waste areas with pasture legumes in the late dry season does not conflict with other operations.

The simplest example of true complementarity between subsystems is the use of animal draught power for crop production, which is discussed further in Chapter 6. The area of land which can be prepared in a timely fashion for crop production depends upon the physical condition of draught ruminants at the end of the dry season; in lowland areas the main source of feed for the animals may be crop residues and crop by-products (Humphreys 1986). The two subsystems become competitive when the requirement for draught is met and when livestock production surplus to this need that produces income from animal sales or meat or milk for farm consumption requires land resources which might be devoted to food or cash crop production; alternatively the introduction of crops such as cotton whose residues are less valuable than maize or sorghum for animal production may restrict the animal enterprise and its surplus off-take (McDowell 1988).

Another example of true complementarity is the use of a legume-based forage system in rotation with crops, where legume N may be the most feasible source of N to maintain crop production. Well-managed biological N fixation augments animal production from pasture, and the level of subsequent crop production depends upon the success of the previous pasture legume, many examples of which are given in Chapter 10.

True complementarity rarely exists in intercropping systems, although it is common for the combined yield of crops to exceed the yield from monocropping on the same area. Legume intercrops with an independent source of N offer less competition for the environmental growth factors of light, water and nutrients than forage grasses and may even be skilfully managed to augment the yield of the cash or food crop (Rika et al. 1981). These questions are subsequently developed in Chapters 8, 9 and 11 for plantation agriculture, alley cropping and annual intercropping respectively. The efficiency of resource use is improved if utilizable forage is produced in place of weeds in multiple cropping, and the balance between the resources used by the forage or crop components may be manipulated according to the value the farmer places on each output. Competition between upland rice and undersown *Stylosanthes guianensis* may be controlled by attention to the density of the forage legume, the timing of its sowing in relation to the planting of rice and the choice of rice variety (Shelton and Humphreys 1975a,b,c).

The system illustrated in Figure 4.4 requires that sufficient land be retained in savanna to provide a spatial transfer of nutrients which sustains production on land devoted to cropping. As mentioned elsewhere, Swift *et al.* (1989) suggest that in Zimbabwe at least 14 ha of savanna are needed to support 2 t maize grain yield from each hectare of crop land. At Machakos, Kenya Probert *et al.* (1992) describe a more favourable system in which the average farm situation is that 5.1 ha of grazing area and 3.4 ha of crop area with 8.7 livestock units might provide a potential annual rate of manure application of 2.5 t ha^{-1} to crop land, containing 50 kg N and 9 kg P.

A combination of crop and animal subsystems provides the farmer with more flexibility of response to changing economic and environmental conditions than is available to farmers dependent upon a single enterprise, as discussed in Chapter 7. Changes in the relative prices of grain and of animal products generate responses in the balance of enterprises. The ley system in northeast Thailand described by Gibson (1987) which sustains annual cropping has in part been overtaken by the success of a burgeoning dairy industry which provides a better incentive for using long-term perennial pastures. Conversely better prices for grain products cause farmers to extend the duration of the cropping phase and to reduce the animal enterprise.

4.6 Control of environmental pollution

The evaluation of farming systems needs to take account of the different forms of environmental pollution which may poison the soil, restricting future production, which may detract from the quality of human life or may adversely affect other organisms. The incorporation of well-managed pastures in the farming system can:

- ameliorate the adverse effects of global warming
- reduce the activity of chemical residues, and
- improve the quality of surface and groundwater entering streams.

Mismanagement of pastures exacerbates these problems.

Global warming

Cattle are popularly conceived to contribute to global warming, since they and other ruminants emit methane. Grain-fed cattle production systems, in which grain is produced from soils which have lost OM in the process of cultivation, may be regarded as a negative factor in the global CO_2 balance. Meat and milk produced from pasture-fed systems are environmentally 'clean' and deserve greater support by the environmentalist lobby. Methane emitted by cattle, sheep and other ruminants contributes no more than 3% of the 'greenhouse effect'; this is trivial relative to methane emissions from rice fields and to CO_2 generated from diverse sources, urban, industrial and agricultural.

In section 2.3 the positive OM accretion under pastures was discussed, and Figure 2.3 provides a serious basis for revising the previous under-estimations (Lieth 1978; Jordan 1981) of the primary productivity of tropical grasslands. The rate of C fixation in well-managed tropical grassland is similar to that of rain forest (Pearcy 1987). The emission of methane from ruminants needs to be balanced by a consideration of the CO_2 absorbed in growing the feed for these animals and the delivery of C to OM, much of it in a refractile form which may last 2000 years. Annual global emission of CO_2 is thought to be about 1.4 Gt (10^{12} kg) C more than the measured increases in atmospheric CO_2. Scientists are concerned to identify the missing sinks for CO_2. The ocean may account for much of it; an alternative is the vast area of tropical grasslands, which use the efficient C_4 dicarboxylic acid pathway of photosynthesis. It may prove to be a major sink for the missing CO_2, perhaps accounting for 0.6 Gt year^{-1} of C not attributed to it (Jones et al. 1992).

This recognition of the positive role of tropical grasslands in ameliorating global warming needs to be qualified by the knowledge that regular burning of grassland, overgrazing, or the frequent cultivation of short-term leys will increase CO_2 emissions.

Deactivation of chemical residues

OM plays a key role in deactivating many chemicals used in agricultural practice; in so far as pastures augment soil OM levels a concomitant beneficial effect in reducing the activity of residual pesticides, fungicides and herbicides is well documented. Vaughan and Ord (1985) and Stevenson (1982) have listed some studies which illustrate this. The generally low molecular weight chemicals combine with soil OM to reduce their mobility and activity. Pesticides such as DDT, heptachlor and endrin are degraded in the presence of OM (Castro and Yoshida 1974; Kahn 1978), and non-biological decomposition is reported for 3–amino-triazole and aldrin (Crosby 1970; Kaufman 1970). Fungicides such as dazomet which are used in disease control in crops and soil sterilization are decomposed (Vaughan and Ord 1985). The herbicide paraquat is adsorbed by base exchange and deactivated by soil OM (Kearney et al. 1965). The long-term damage occasioned by herbicides has been overstated; for example, Thompson (1991) reported no restriction of plant growth and no reduction in microbial biomass following 12 years of fallow weed control using the herbicides paraquat, diquat and glyphosate in a zero-tillage system.

Pollution of streams

Agricultural activities increase the content of nutrients in streams, and most damage is occasioned by accretions of P and N. These result in accelerated growth of aquatic algae. The algal mat which forms consumes the oxygen in the water, resulting in the phenomenon known as eutrophication; fish and other aquatic animals die. Particular algae, such as *Anabaena* spp., form a 'bloom' on the surface of the water, which results in the production of toxins that cause liver damage in humans and may kill stock. These algal growths are especially prevalent in conditions of high temperatures and reduced stream flow, which increases nutrient concentration. P appears to be the limiting nutrient for freshwater algal growth, whereas N may be the principal pollutant in marine conditions (Ford 1991).

Nitrate is of especial concern in determining the quality of water for human consumption; in Europe the designated upper limit is 11.3 μg l^{-1} (Parsons 1984). Reference is made in section 10.3 to an instance where zero tillage, which has resulted in the retention of earthworm tunnels and channels left from decayed roots, promotes loss of nitrate to deep drainage (Thompson 1991). Farmers often have to balance the positive effects of one agricultural practice such as zero tillage on soil conservation against the negative effects on stream pollution and against an increased demand for fertilizer N and its accompanying use of fossil fuel. Similarly N produced from biological fixation is as susceptible to losses to deep drainage as is fertilizer N, although in the latter case excessive levels of fertilizer N have exacerbated this problem. In areas of high N application as much as 5–30% is lost through deep drainage and leaching (Parsons 1984; Williams 1991).

Biological pollution of streams is a further problem which has been shown to result from stock concentration and the associated production of surplus animal wastes from dairies and feed lots. The incidence of *Escherchia coli* in streams is associated with this factor (Ford 1991) and an increased level of ammonia in drinking water is evident following rainfall events on catchments dominated by intensive agriculture.

The incorporation of pastures in the farm system ameliorates these effects if appropriate farm practices are adopted. Increased levels of OM deactivate agricultural chemicals. Maintenance of green cover restricts surface runoff and increases evapotranspiration, thereby decreasing deep drainage, which is especially great on sandy soils with a low adsorptive capacity. The planting of deep-rooted pasture species, shrub legumes and fodder trees also reduces leaching losses. The enrichment of runoff with P is positively associated with the level of surface P application (Sharpley 1980), so that a policy of fertilizer application in synchrony with opportunity for rapid plant uptake and with level of plant demand is less damaging to the environment. Luxury application of fertilizer to tea estates pollutes streams, and Colour Plate 6 and Plate 4.1 illustrate the planting to pasture of poorly drained bottom lands in Sri Lanka, which absorb some of the surplus nutrients.

The concluding comment in this chapter concerns the need for continuing dialogue about the balance between the special needs of the farmer to sustain the farm enterprise and the influence of farming activities on long-term community needs. The well-being of the total community requires the conservation of non-renewable resources and the efficient use of renewable resources. The first requirements are an objectively assessed inventory of resources at the farm and the regional levels, an understanding of the physical and biological processes of production and a choice of technologies available to the farmer which are workable as different levels of intensification are chosen. The challenge to the scientist is to develop in interaction with farmers technologies with in-built economic incentives which result in sustained efficiency of resource use.

CHAPTER 5

Crop protection systems

5.1 Introduction

Monocropping systems require large inputs of agricultural chemicals to control pests, diseases and weeds. The persistence of chemical residues in soil and their content in food rightly occasion concern for the safety of the environment and the ecological security of the farm system. Pests develop progressive immunity to particular chemicals, necessitating either the costly usage of high dose rates or the development of alternative solutions to the problem.

The flexibility of livestock/crop systems enables farmers to use a diversity of environmentally sound approaches to meet particular challenges to their crop protection programme as these emerge. The weed cycle may be broken by the introduction of a ley, whilst the presence of well-managed pastures with good cover will resist weed invasion; alternatively a cropping phase may destroy the weeds in degenerated pastures. The removal of alternative or crop hosts to pests and disease in a ley system, the active suppression of nematodes by forage species, the changed environmental conditions associated with intercropping or the disinfestation by fire used in land preparation can all provide crop protection. The skilled management of pastures can also reduce the density of disease, predation or weeds.

This brief account of approaches to crop protection is limited to those aspects which relate to the contrast between cropping and forage/cropping systems, which affect the motivation of farmers to incorporate livestock in a mixed farm enterprise.

5.2 Weeds

Ley pastures

An acute problem associated with monocropping is the control of weeds whose life cycle matches exactly the life cycle of the crop. Thus fields which are continuously cropped to wheat in south Queensland are invaded by wild oats (*Avena ludoviciana* and *A. fatua*). Wild oats is effectively controlled if perennial pastures of *Medicago sativa* and *Bromus unioloides* are sown to break the life cycle of the weed. When wheat cropping is resumed after pasture the wild oat population again builds up. At Jondaryan Littler (1984) recorded the following densities (plants m^{-2}) in successive wheat crops in the 6 years following pasture: year 1, 0.008; year 2, 0.035; year 3, 0.17; year 4, 0.32; year 5, 0.62; and year 6, 0.74. Concurrently fields growing wheat every year had 1.16 plants m^{-2}. In these circumstances the build-up of wild oats would be contained to manageable levels by limiting the duration of wheat cropping to 4 years. Other benefits for this rotation are illustrated in Table 10.1, section 10.2.

The duration of the ley is partly indicated by the longevity of the bank of weed seeds in the soil, and their loss during land preparation for pastures. The use of forage species well adapted to the environment which are grazed and fertilized in ways that maintain ground cover will minimize weed seed germination and the creation of 'gaps' which allow weeds to invade. This phenomenon depends partly on the change in the red/far-red ratio of sunlight when filtered through pasture, which inhibits the germination of weed seeds (Taylorson and Borthwick 1969); on open soil the predominance of red light in the phytochrome system promotes seed germination.

Traditional systems of repeated tillage which allow weed seeds to germinate and weed plants to be destroyed are hazardous for the environment and expensive of energy.

Chemical herbicides are increasingly accepted in the agriculture of the developing world. Local systems of weed control need to be developed which cater for the specific weeds involved, the cultural conditions and the local climatic environment. At Katherine, Australia native grass pasture is killed with glyphosate (360 g ai l^{-1}) before seeding pasture legumes (Jones *et al.* 1991). Subsequently mid-wet season grazing, mowing or herbicide use may be needed to control weeds; glyphosate may be used with a rope wick applicator to selectively reduce taller weeds and grasses, or Fusilade or Sertin at 1 l ha^{-1} would increase the legume/grass balance. When planting the cereal crop after a legume ley glyphosate may be mixed with 2,4–D to reduce the legume.

Post-emergence weed control may be achieved with a mixture of 720 g ai l^{-1} Metachlor for dicotyledonous weeds and 500 g ai l^{-1} Atrazine for grass weeds, which may be sufficient until crop canopy closure obviates further weed problems.

When pasture degradation with weeds does occur the farmer has the option of returning the pasture land to cropping, as an alternative to other forms of weed control. This is especially indicated for woody weeds; for example gilgai lands of southern Queensland in which *Acacia harpophylla* suckers regenerate in pastures may be destroyed by conventional methods (Scanlan 1984) or by a single cereal crop followed by post-harvest spraying (Johnson and Back 1977); ploughing to a depth of 10 cm is as effective in killing this plant as ploughing at 20 cm depth (Coaldrake 1967).

Intercropping

Systems of intercropping forage or pulse legumes with cereal crops reduce the level of weed infestation (Altieri and Liebman 1986) and this topic has been modelled by Vandermeer (1989). The level of reduction depends upon the nature of the intercrop especially with respect to its rate of early growth and the completeness of canopy cover in suppressing the 'invasion' of weeds. At Morogoro, Tanzania (lat. 6° S, 850 mm annual rainfall, 520 m altitude) weeds are a real constraint to crop production. In a study of differing weeding procedures it was found that the best yield of sorghum was 3.1 t ha^{-1} relative to 0.54 t ha^{-1} from an unweeded crop; comparable yields were 1.0 and 0.23 t ha^{-1} for cowpea and 1.2 and 0.28 t ha^{-1} for green gram respectively. In later work at Morogoro (Mugabe *et al.* 1980) the effects of the pulses green gram or soybean grown as intercrops in maize or sorghum in reducing weed infestation were relatively minor, but in bulrush millet a substantial decrease in the effect of weeds was noted in the intercrop (Table 5.1).

5.3 Pests

Ley pastures

The introduction of a pasture phase in a cropping system may reduce an infestation of insect pests through interrupting the insect life cycle, especially if the pasture species do not provide alternative hosts for the insect. Additionally some pests may be actively suppressed by the pasture species chosen. Nematodes may be controlled by chemical treatment; an alternative technique is a grass ley. This has been used with success in tobacco production in Zimbabwe. Grass roots vary

Table 5.1 Total grain yield (kg ha^{-1}) of bulrush millet plus or minus intercrops and the yield reduction due to weeds at Morogoro, Tanzania (from Mugabe et al. 1980).

Crop combination	Yield under weed-free conditions	Yield under non-weeded conditions	% reduction in yield due to weeds
Bulrush millet + soybean	2550	1375	46.1
Bulrush millet + green gram	2756	1355	50.8
Bulrush millet monoculture	2610	1122	57.0

both in the ease with which nematodes enter them and in their capacity to suppress nematode populations in the soil (Shepherd and Coombs 1979). Particular cultivars of *Chloris gayana* such as Katambora are effective, and *Panicum maximum* is also used. A grass ley of 3 years duration gives good nematode control if weeds which favour nematodes do not invade the sward; a ley duration of 4 years gives more complete control. Recently *C. gayana* cv. Nemcat (CPI 125663) has been released in Queensland. It has a similar yield to cv. Katambora, but is finer-stemmed, more stoloniferous and is resistant to all the major nematode species which afflict tobacco crops. In southern Queensland pastures of *P. maximum* var. *trichoglume–M. atropurpureum* cv. Siratro suppress nematodes in rotation with pineapple production (Colbran 1969). A high level of field resistance in *M. atropurpureum* to the root-knot nematodes *Meloidogyne avenaria, M. hapla, M. incognita* and *M. javanica* is evident (Hutton et al. 1972), and there is a possibility that Siratro roots have a chemotoxic effect on nematodes. Great differences in nematode suppression according to the preceding crop prior to planting pineapples was found (Table 5.2); the effect of fumigating with EDB is also shown in this table, and subsequent infestation with *Meloidogyne* 18 months after planting was prevented in the *P. maximum* var. *trichoglume*–Siratro pretreatment plots even in the absence of fumigation. Colbran (1969) also noted reduced infestation after *P. maximum* cv. Hamil and cv. Coloniao and *Setaria sphacelata* var. *sericea* cv. Nandi, whilst *Lablab purpureus, Melinis minutiflora, Paspalum plicatulum* cv. Hartley, *P. commersonii, S. sphacelata* var. *sericea* cv. Kazungula and *Sorghum sudanense* were ineffective.

Banana production in the wet tropics of Queensland is limited by the burrowing nematode (*Radopholus similis*) which has been recorded on 96% of banana farms (Hall et al. 1993). This pest can cause yield losses of 30% in ratoon crops and reduce the economic life of the perennial banana crop. Fallowing is not a feasible method of control since common weed species are hosts to the burrowing nematode

Table 5.2 Effect of previous crop on larval density of *Meloidogyne* (no. kg^{-1} soil) (From Colbran 1969.)

Previous crop	Pre-plant	Post-plant Fumigated	Post-plant Not fumigated
Panicum maximum var. *trichoglume/ Macroptilium atropurpureum*	13	106	0
Chloris gayana/M. atropurpureum	471	1 041	1 914
Vigna sinensis/Secale cereale	7 770	884	1 730
Cajanus cajan	14 190	504	2 160
Volunteer species	777	354	1 795

and the mainly steep terrain of banana farms makes cultivated land vulnerable to erosion; fumigation is also difficult to apply and may in any event cost $US1200 year^{-1}. An alternative is to rotate banana production with cattle production, which requires the establishment of a rapidly summer growing, sod-forming grass that suppresses nematodes. Following extensive testing *Digitaria milanjiana* cv. Jarra has been released as a plant which meets these criteria, and which can also be readily eradicated by herbicides (glyphosate and Gramoxone) when banana planting is resumed.

Intercropping

Intercropping of plant species of differing host reaction to insect pests may reduce the incidence of the latter since a less susceptible biomass is present and the intercrop offers a barrier to insect movement. There may also be some modification of the microclimate which alters its suitability for the multiplication of its pest or of its natural predators.

This question was explored in a study (Havel *et al.* 1980) at Morogoro, Tanzania where cowpea crops were grown with or without maize crops at differing densities and in differing ratios. There was a constant row width of 75 cm for maize and 37.5 cm for cowpeas, so the latter was sown in paired rows. Within-row spacing varied from 7.5 to 60 cm for maize and 5 to 40 cm for cowpea respectively. Both crops were infested by a large number of insect pests. The leaf-eating beetle *Ootheca bennigseni* was damaging to both intercropped and pure stand cowpeas; the maize provided a subsidiary food source and beetles commonly sheltered in maize funnels. Similarly the heteropteran bugs *Acanthomia* spp. and *Nezara viridula* occurred on both cowpeas and maize, and did more damage to the former.

On the other hand intercropping was beneficial in reducing the incidence of two serious pests (Table 5.3). The density of both pests, the flowering thrip (*Taeniothrips sjostedi*) and the pod borer (*Maruca testalis*), increased with increasing plant density but the presence of maize plants significantly reduced both the numbers of pests and the damage they caused to cowpea. The whole field of plant associations reducing pest damage and the planting of species inimical to the presence of pests (such as the neem tree) is receiving more attention as farmers and the community become disillusioned about the efficacy and costs of chemical control and its environmental implications.

5.4 Disease

There is a commonality of approaches to the control of pests and of fungal, bacterial and viral diseases. The principles of denying alternative hosts to diseases, and the value of fire in disinfestation (Lenné 1981) apply (Plate 5.1). There is also a need to avoid introducing alternative hosts for diseases of crops through the use of contiguous ley pastures. For example, a serious bacterial disease of French bean (*Phaseolus vulgaris*) is halo blight (*Psuedomonas phaseolicola*). *M. atropurpureum*

Table 5.3 Density of thrips (*Taeniothrips sjostedti*) and pod borer (*Maruca testulalis*) on cowpea grown with or without maize at various plant densities at Morogoro, Tanzania (from Havel *et al.* 1980).

Pest and density	Cropping system	Plant population (ratio to 'optimum')			
		0.5	1	2	4
Thrips (no./flower)	Cowpea	10.7	10.3	12.1	13.6
	2/3 cowpea, 1/3 maize	6.2	7.5	6.9	8.4
	1/3 cowpea 2/3 maize	1.7	3.8	4.7	3.3
Pod borer (no./10 pods)	Cowpea	4.3	5.0	6.7	6.3
	2/3 cowpea, 1/3 maize	2.3	1.7	3.0	4.0
	1/3 cowpea, 2/3 maize	1.7	2.7	2.0	2.3

Plate 5.1 *Stylosanthes guianensis* cv. Endeavour infected with *Colletotrichum gloeosporioides* (D.F. Cameron).

cv. Siratro is susceptible to this disease, which is exhibited both in leaf lesions and by seed transmission (Moffet 1973); the pasture legume *Neonotonia wightii* poses a similar threat to crops of French bean.

Crop management directed to soil protection may also have unfortunate effects on crop protection, and the practice of stubble mulching needs to take this into account. In Zimbabwe the retention of maize stover as a surface mulch leads to an increased incidence of stalk borer and of the leaf blight *Helminthosporum turcicum* (Rattray 1961). Maize yields averaged 10% higher if trash was ploughed in. In subtropical Queensland wheat production is reduced by the disease yellow spot (*Pyrenophera tritici-repentis*), and this problem is exacerbated by stubble retention (Rees and Platz 1979). Fortunately the spectrum of organisms exhibits secondary changes in response to changed agricultural practice; the new disease and pest problems which arise are often subsequently reduced by the appearance of new predators which decrease the activity detrimental to crop production.

Intercropping again offers prospects of disease control through the presence of barriers which make the transfer of inoculum more difficult than in monocropped situations. Reduced wind velocity associated with tall intercrops such as maize, sorghum or millet may decrease the speed at which pathogen propagules spread. The changed air circulation and reduced humidity of the intercropped microenvironment may reduce the incidence of powdery mildew (*Erysiphae polygoni*) and transfer of inoculum by rain splash may also be decreased. At Morogoro, Tanzania Keswani and Mreta (1980) recorded reduced incidence of powdery mildew on green gram if intercropped with sorghum or millet, and yield was linearly and negatively related to an index of disease severity.

Crop protection is a central preoccupation of the farmer. The continual evolution of new organisms or the arrival of pests across geographical barriers promote new challenges. The spread of the psyllid *Heteropsylla cubana* from the Caribbean across the Pacific Ocean to Asia and Australia (Bray and Sands 1987) has reduced the production of *Leucaena leucocephala* and led to the use of other shrub legumes and the breeding of psyllid-resistant *Leucaena*. This short discussion focuses only on the aspects which bear on the use of livestock–crop integration. Farm systems involving forages can be developed which enhance food security, reduce farm costs for crop protection, and which do not damage the environment.

CHAPTER 6

Outputs from animal production in farming systems

6.1 Introduction

Livestock contribute to the success of mixed farming involving cropping in different ways, according to the objectives of the farmer, the intensity and specialization of farm production, and the environmental constraints which are locally imposed on cropping and on the forage supply.

Herbivorous animals in developing countries account for $c.$ 66% of the world's cattle, $c.$ 64% of the sheep and goats, $c.$ 80% of the equines and almost all the buffaloes and camels (Jones 1988). They represent an increasing focus for agriculture in most tropical regions where the per capita demand for milk and meat is increasing with associated economic growth; in the regions such as sub-Saharan Africa where population growth has continued to outstrip the growth in food supply animal products are a basic component of food security.

The product output per AU is less in the tropics and subtropics than in temperate agriculture. This does not represent a lower ratio of outputs to inputs, however these may be defined, since animal production in Europe, North America and Japan is supported by a large transfer of economic resources from other sectors of their economies, is in negative balance with respect to energy outputs and inputs (Wilkins 1982), and utilizes production systems which cause severe pollution problems to their communities. In assessing levels of production from tropical systems, due regard might be given to:

- the role of animal traction, which reduces other animal performance indicators (Plate 6.1);
- the climatic stresses on animal physiology and health;

142 *Outputs from animal production in farming systems*

Plate 6.1 Draught ruminants in Burma.

- the availability of capital;
- the characteristics of the feed supply.

As mentioned in Chapter 4, the conversion of crop residues and of pasture on waste land to outputs of food, traction and manure represents a seizure of opportunity which is not available in other options for productivity. The capacity of the ruminant to utilize the long-chain structural carbohydrates present in crop residues and in the pasture grasses with the C_4 photosynthetic pathway which dominate the tropics is unique. Although pigs and poultry are minor users of forage, they are not considered in this book, since monogastric nutrition must be based on other feed sources, and since their consumption of grain and vegetable products may be seen to be competitive with human dietary needs. For related reasons emphasis is given to systems based on grazing forages and crop residues and feeding of collected forage and crop by-products rather than to grain feeding, since ruminants are inefficient converters of grain, and alternative systems are available in which the protein/energy ratio of ruminant diets can be adjusted to give good levels of animal performance (Leng 1990).

There are two essential features of the forage supply in the tropics which modify husbandry practice. The greater photosynthetic efficiency of the C_4 grasses leads to a higher growth potential than is exhibited by C_3 legumes and temperate grasses; not only is photosynthetically active radiation (PAR) at high radiation levels converted more efficiently (see Figure 8.3), but the resources of soil moisture and nutrients are more fully exploited to produce biomass. Higher yields are associated with a greater content of indigestible or poorly digestible structural carbohydrate and a lower ratio of leaf lamina to stem. Good levels of animal production then require selective grazing (or selective consumption) of the more nutritious fractions of the forage: young leaf and seed. Alternatively or additionally, higher quality leguminous forages of greater N concentration or crop by-products of high protein status (such as cottonseed meal) are required to supplement a grass-based diet. This problem is more acute in humid areas, where pasture growth rates are high, than in subhumid areas where growth rates are slower and where continuity of forage supply is the dominant production problem (Tothill et al. 1989).

The second feature, which is related to the first, is that animal diets are more prone in the tropics to deficiencies of essential minerals than in temperate regions. This arises in part from the efficient utilization in growth of minerals taken up by C_4 grasses which leads to high utilization quotients (the inverse of nutrient concentration); some plant improvement programmes have concentrated on adaptation to low levels of soil mineral supply rather than to the maintenance of high nutrient concentration. It is also associated with the highly leached condition and low CEC of most tropical soils. For example, a wide survey of Latin American forages led McDowell et al. (1983) to conclude that elements were deficient or borderline in supply in a high proportion of cases: Zn 75%, P 73%, Na 60%, Cu 47%, Co 43% and Mg 35%. These deficiencies may often be overcome from local sources of minerals, if these are carefully sought out.

The characteristics of the feed supply determine in a primary way the type of livestock system which evolves, and this in turn reflects the species of grazing animals best adapted to the physical and biological environment, and whose utility matches the economic and social objectives of that society, as overlaid by tradition and culture. With decreasing aridity the sequence is from camels to goats and sheep to cattle, with sheep poorly represented in humid zones and buffaloes wholly confined to humid areas. The resistance of goats to dehydration, their predilection for browsing and their wide-ranging grazing confer a special adaptation to subhumid farm conditions containing shrubs (Devendra and McLeroy 1982). Larger animals are preferred for draught, although equines such as donkeys are well used for carriage (Plate 4.5). Nutrient demands for high levels of milk production are greater than for animal reproduction, which in turn are greater than for animal growth, whilst sheep may be regarded as compulsive wool growers; dry sheep such as wethers may be grown for wool production on grasslands which are of nutritional value too low to support growth or reproduction.

This book is limited in scope to animal production systems that are integrated with crop production. However it might be noted that the expansion of cropping into marginal sub-Saharan grazing lands and the development of sedentary grazing herds in these areas impacts on the grazing resources previously available to Sahelian nomadic herds (Traoré and Breman 1993).

6.2 Entrepreneurial objectives for animal production

The traditional objective for livestock in entrepreneurial farming systems is income from the sale of:

- milk,
- meat,
- wool, hair and skins,
- young animals.

In many tropical farming systems animals are also kept for:

- excreta, either as nutrients for cropping land or as dried dung for fuel;
- draught and carriage, as required in farm operations;
- enhanced family nutrition;
- effective utilization of farm resources;
- diversification of income and accumulation of capital savings;
- cultural needs and farmer status.

Milk

Milk production and the use of milk products such as flavoured, sweetened, condensed and dried milk, ice cream, butter, ghee, and to a lesser extent cheese, have made considerable gains in most tropical countries. This has occurred despite the gloomy forecasts of many economic and agricultural advisers, and the recognition of lactose intolerance in Asian populations. For example, the burgeoning dairy industry in Thailand grew by a factor of ten between 1981 and 1988 to a local production of 69 000 t (C. Manidool, personal communication); this represents a considerable change in dietary preference. Most of this increase has been based on smallholder dairying and mention was made in section 4.2 of self-sufficient dairying in northeast Thailand, which combines crop and pasture production to provide a sequence of legume-based pasture, crop products and crop residues (Gibson 1987). The benefits of this system to crop production are illustrated in Table 10.10 and section 10.4, whilst favourable economic outputs are detailed in section 10.2.

The most intensive ruminant production systems in the tropics and subtropics are dairy enterprises (Colour Plate 7), but the level of intensification grades down through dual purpose meat and milk enterprises in Latin America to subsistence milk production in semi-arid Africa, where the milk offtake may be as low as 126 kg per cow lactation in Sudanese Fulani cattle in central Mali (Wilson 1989a). Intensive dairying is carried out in the subtropics of Brazil, Argentina, southern Africa, southern Japan and northeastern Australia and in the cooler highlands of equatorial zones, as in Mexico, Tanzania, Kenya, Malaysia, India and Sri Lanka. Most of the intensive operations are specialized and have limited integration with the production of crop products for sale; they often incorporate crop production as intermediary to dairy outputs, and the degree of land development therefore provides for flexible response to new market opportunities in crop production as these arise. The intensity of development (Humphreys 1991) is reflected in the replacement of native pastures with planted forages, the use of fertilizer to overcome the constraints of mineral deficiencies on plant growth, the introduction of animals with a high genetic potential to respond to high planes of nutrition, and considerable attention to the maintenance of animal health. Interventions to maintain continuity of forage supply through fodder cropping, fodder conservation, irrigation, grain or tuber production and the purchase of concentrates from off-farm (Jennings and Holmes 1985) sources are commonly employed. Many agricultural advisers seek to increase farm self-sufficiency and economic studies sometimes indicate an over-dependence of production on purchased concentrates (Yazman *et al.* 1982 for Puerto Rico). At Muaklek, Thailand (lat. 15° N, 1090 mm annual rainfall, 220 m altitude) cows grazing leafy *P. maximum/S. hamata* pasture gave better economic returns at 1 kg concentrate per 3 kg milk produced than at higher levels of supplementation (Lekchom *et al.* 1989).

The sustained milk yield from grazed tropical grass–legume pastures appears to have an upper limit of *c.* 12 kg hd^{-1} day^{-1} in the absence of supplementation. On the Atherton Tableland, Queensland cows grazing *P. maximum* var. *trichoglume/N. wightii* at the low SR of 1.3 ha^{-1} produced 4100 l per lactation of 330 days (Cowan *et al.* 1974). Production per hectare is increased by the use of N fertilizer and the use of 200–400 kg N ha^{-1} year^{-1} enables SR to be increased to 2–3 cows ha^{-1} but does not increase production per cow (Lowe and Hamilton 1986). The benefit of incorporating legumes in the pasture is illustrated from a study (Lascano and Avila 1991) at Quilichao, Colombia (lat. 3° N, 1840 mm annual rainfall, 990 m altitude), where Holstein or upgraded Zebu × European crossbred cows grazed *Brachiaria dictyoneura* or *Andropogon gayanus* pastures at 1 cow ha^{-1}. The incorporation of *Centrosema macrocarpum* or *C. acutifolium* (Table 6.1) increased milk production in both wet (September–October, February–April) and dry (July–September) periods by 0.7–2.2 kg hd^{-1} day^{-1} according to treatment.

These levels of production are considerably higher than are attained in many less intensive systems. The early phases of the Thailand project previously mentioned (Gibson 1987) gave average production per cow of 5.4 l day^{-1} with 295-day lactation and a calving interval of 384 days. In the Gambia a study (Agyemang *et al.* 1991a) of four villages (lat. 13–14° N, 800–1000 mm annual rainfall) indicated that the trypanotolerant N'Dama cattle, a multipurpose breed used for milk, meat, manure and traction, provided an average milk offtake of 404 kg per cow lactation of 420 days. Calving interval averaged 641 days and calf weaning weight 88 kg. These data provided a higher productivity index than that obtained for larger Zebu cattle. These few illustrations indicate that dairy production is entrenched in diverse tropical farming systems at levels ranging from 126 to 4100 l cow^{-1}; this may be extended to *c.* 6000 l cow^{-1} if irrigated temperate pastures are incorporated (T. M. Davison, personal communication). Milk production from buffaloes, sheep, goats and camels is also significant in some societies.

Meat

Economic indicators are favourable for the expansion of meat production from ruminants (Hill 1988). Artificial barriers to trade limit the access of tropical producers to many developed country markets, but farmers have shown considerable skill in adapting husbandry methods to meet the special carcass requirements of those external markets which are accessible. For example, Queensland producers, at one time oriented to the production of low quality hamburger beef for export to the USA, have met the more exacting standards of the Japanese export

Table 6.1 Milk production (kg cow^{-1} day^{-1}, 4% butterfat) from cows grazing pastures with or without legumes at Quilichao, Colombia (from Lascano and Avila 1991).

Legume	*B. dictyoneura*		*A. gayanus*	
	Wet	Dry	Wet	Dry
Nil	9.6	6.7	9.5	6.2
C. acutifolium	10.6	8.4	10.2	7.8
C. macrocarpum	11.1	8.9	10.3	7.0

trade for animals finished for slaughter at younger age with the desired fat cover. This has involved a greater integration of crop production with forage, and the greater exposure of beef cows and young animals to planted rather than native pastures.

It is the enhanced domestic market in most tropical countries which provides the main stimulus to meat production from ruminants. This is especially evident in Latin America, where growth in economic demand has been outstripping growth in beef production (Toledo 1985), and where beef consumption per capita is higher than in other regions. Meat consumption was 31.8 kg per capita in 1989, of which 46% was beef, and the price ratio between poultry and beef increased between 1980 and 1989 in Brazil, Colombia and Venezuela from a mean of 0.53 to 0.63, providing a favourable indicator for enhanced beef consumption (CIAT 1991e); this is coupled with high elasticity of demand for even the poorest segments of the population. Annual per capita beef consumption in different countries is estimated as 0.7–2.6 kg in south and southeast Asia, 3.6–9.6 kg in tropical Africa, and 7–38 kg in Latin America (Toledo 1985). The introduction of incentive premiums for beef quality has altered the nature of demand and the patterns of production, as in Thailand, whilst especial aspects of seasonal demand associated with festivals create premiums for goat and sheep meat in many tropical societies.

The rate of growth on a sustained basis from cattle grazing unsupplemented tropical pastures is limited to $c.$ 0.7 kg hd^{-1} day^{-1}. This is temporarily exceeded when factors such as compensatory gain occur following stress, but there are recent instances where new technology has lifted the horizon above 0.7 kg hd^{-1} day^{-1}, as with the grazing of the non-flowering leafy Mott dwarf *Pennisetum purpureum* in Florida (Sollenberger and Jones 1989) or with *L. leucocephala* as a grazing component (Quirk *et al.* 1988). Levels of production in the upper range from well-managed planted tropical pastures are more usually in the range 180–200 kg hd^{-1} year^{-1} (CIAT 1991; Humphreys 1991). Cattle performance from a series of studies in southeast Asia, mainly based on integrated feeding systems using forages and crop residues, showed a net LWG of 73–146 kg hd^{-1} year^{-1} (Humphreys 1986).

Small ruminants have many attractive features for meat production; reference has been made to the browsing capacity of goats (which extends to some breeds of sheep), the capital investment is within reach of smallholder savings, and some scientists argue that the metabolic efficiency of small ruminants is superior to that of the larger species (Devendra and McLeroy 1982). A comparison of goat and cattle production in an intensive system in Jamaica on N-fertilized *Digitaria decumbens* pasture gave 1150 kg LWG ha^{-1} year^{-1} from goats at mean SR of 41 ha^{-1}, whilst cattle produced 910 kg LWG ha^{-1} year^{-1} at 4 ha^{-1}.

Wool, hair and skins

Production of fine wool, angora and cashmere gives high individual animal returns, since these are high priced products, but most production of these commodities in the tropics and subtropics comes from natural savanna and not from forages integrated with cropping. The production of carpet wool is often associated with sheep grazing crop residues and volunteer pastures, and is a significant by-product of animal enterprises usually directed to meat or milk production; ewes of tropical breeds may yield 1.5–2 kg hd^{-1} at each shearing and be shorn twice per year (Plate 6.2). Hair from sheep and goats is of importance in local craft industries whilst skins are of domestic utility and also contribute to human diet.

Sale of young animals and reproduction

In smallholder production systems the rearing and sale of young animals may constitute a major source of farm income. In northeast Thailand farm systems oriented to crop production derive an average 51% of livestock income from cattle and buffalo (Manidool 1987), and the

Plate 6.2 Awassi×local sheep for meat and wool production in the Ethiopian highlands.

major part of this income arises from the sale of young stock. Where buffalo or cattle are kept mainly for draught purposes replacement animals are retained where sufficient land is available to rear them, but the working life of large ruminants is such that up to eight progeny may be delivered from an animal of good fertility (Wilson 1989b), providing opportunity for sale of surplus stock. Similarly in smallholder subsistence dairy operations where in many regions of Africa most of the milk may be consumed by the family, the production and sale of calves or animals in store condition are an important source of cash income. In Thailand special government interventions were made to encourage the production of high quality crossbred dairy heifers for sale to other producers.

Reproductive efficiency is a primary objective of both husbandry practice and of genetic improvement. The quality of forage available is the primary determinant of reproductive performance. In many tropical farming systems the nutrition of the cow is such that calving occurs in alternate years; the stresses of pregnancy and of lactation cause anoestrus and the improved nutrition provided by a subsequent season is necessary for the condition of the cow to improve to the point where conception again becomes feasible. Reproductive efficiency may be judged in terms of the following attributes:

- age of mother at first conception or first parturition,
- conception rate,
- frequency of abortion,
- number of animals born at parturition,
- survival of young animals,
- weight of animals at weaning,
- interval between successive births,
- age at final parturition.

There is a large literature on this topic. Reproduction in small ruminants in Africa is chosen to illustrate the performance attributes which need to be measured to assess the efficiency of any system. Wilson (1989b) has collated the literature from many sources to indicate the variation in age at first parturition in Africa of sheep and goats according to country and management system. The variation from 431 to 782 days at first parturition (Table 6.2) was partly associated with controlled breeding imposed under government station conditions. Litter size (Table 6.3) was influenced by nutritional factors, as also demonstrated by Armbruster *et al.* (1991), and interval between parturition was shorter under natural, uncontrolled breeding conditions than under imposed restraints, since seasonal anoestrus is not evident in these breeds.

The key indicator is the annual production of liveweight (LW) of reared animals per unit mother, which takes into account both successful reproduction and milking capacity. For Sudanese Fulani cattle using millet or rice residues in central Mali, Wilson (1989a) refined this attribute to give three comparative indices:

- kg LW of young produced per breeding female per year (mean 36)
- g LW produced per kg breeding female post-partum LW per year (mean 167)
- g LW produced per kg breeding female $LW^{0.73}$ per year (mean 718), an index corrected for metabolic size of the mother.

The above findings gave lower values for these productivity indices than for goats or sheep under comparable conditions. Further indicators may be developed which consider the energetics of the reproductive system and the output of reared young in relation to the feed intake of the mother.

6.3 Other farmer objectives for animal production

Animal excreta

Throughout this book reference is made to the significance of animals in recycling nutrients to cropping lands and in converting nutrients in crop residues to more available forms, and to the dangers attendant upon the interruption of this cycle. The attention of the reader is

Table 6.2 Age (days) at first parturition of African small ruminants (from Wilson 1989b).

Country	Zone	Management system	Goats	Sheep
Burkina Faso	Semi-arid	Traditional, sedentary	455	446
		Traditional, transhumant	423	455
Côte d'Ivoire	Humid	Traditional, uncontrolled	–	431
		Traditional, controlled	–	480
		Ranching	–	494
Ethiopia	Highlands	Station, uncontrolled	–	473
Kenya	Semi-arid	Traditional	556	549
Mali	Semi-arid	Traditional, rain-fed millet	508	497
		Traditional, irrigated rice	486	431
Mozambique	Semi-arid	Station, controlled	781	782
Nigeria	Humid	Traditional, forest	529	–
		Traditional, savanna	507	–
Rwanda	Highlands	Station, controlled	640	713
Senegal	Subhumid	Station, controlled	–	575
Sudan	Arid (irrigated)	Station, controlled	–	723

Table 6.3 Reproductive traits in African small ruminants (from Wilson 1989b).

Country	Goats			Sheep		
	Litter size (number)	Parturition interval (days)	Annual reproductive rate (young/year)	Litter size (number)	Parturition interval (days)	Annual reproductive rate (young/year)
Burkina Faso	1.21	285	1.51	1.02	268	1.36
Côte d'Ivoire	1.08	312	1.17	1.06	280	1.33
	—	—	—	1.23	230	1.97
	—	—	—	1.18	275	1.56
	—	—	—	1.15	267	1.76
Ethiopia	—	—	—	1.08	262	1.66
Kenya	1.23	306	1.47	1.05	312	1.23
Mali	1.15	298	1.53	1.04	290	1.53
	1.18	297	1.63	1.03	259	1.63
Mozambique	1.62	391	1.51	1.36	360	1.37
Nigeria	1.52	267	2.08	1.26	322	1.43
	1.46	251	2.12	—	—	—
Rwanda	1.75	343	1.86	1.43	406	1.24
Senegal	—	—	—	1.12	307	1.33
Sudan	1.57	238	2.41	1.14	275	1.56
	—	—	—	1.17	437	0.98

directed to sections 4.4, 4.6, 9.2, 10.1, 10.5 and 11.2. The positive effects of incorporating shrub legumes in the diet on the value of excreta are illustrated by Cobbina *et al.* (1989). The sale of dung for fuel (Plates 4.4, 4.5) or its household use is an important output in areas where deforestation has reduced fuel availability, and the cost of this diversion in terms of alternative benefits to crop production is mentioned in section 7.3.

A further alternative is the use of excreta for some secondary form of animal production enterprise. For example, Edwards, P (1991) considers a situation where landholders in Thailand having a mean of 1.7 buffaloes each producing annually 750 kg DM manure containing 1.1% N might direct this to 200 m^2 of farm pond, producing 175 kg fish annually. The cost to cropping land should be included in any evaluation of this system.

Draught and carriage

Draught is the primary function of livestock in many tropical cropping systems and the population of working animals exceeds 330 million, of which cattle and buffaloes are the principal species (Falvey 1986). Useful compendia are available in Copland (1985) and Hoffman *et al.* (1989). The introduction of small tractors has displaced draught animal power in some flooded rice lands, but in upland situations the ineffectiveness of current machines, the high cost of fuel and the variable rainwater supply have operated against this trend (Chantalakhana 1990). Thus in Thailand, a country with a growing economy, the reliance on animal traction on farm land is still 91% in the northeast, 68% in the south, 54% in the north and 24% in the central region (Manidool 1987).

The maintenance of animals in working condition at the beginning of the wet season is a key factor in determining the area of land which may be prepared for cropping in a timely fashion, as mentioned in earlier chapters, and husbandry practice is directed to ensuring the animal has body reserves and forage availability to maintain active work. A generous estimate (Falvey 1986) is that working animals should be fed at *c.* $1.7\times$ maintenance requirement. The influence of work on dietary needs is illustrated (Wanapat 1990) for two breeds of buffalo on a positive plane of nutrition in Thailand. Work reduced the rate of LWG, but supplementation increased it substantially and also increased the rate of ploughing (Table 6.4). Ovarian activity in buffalo cows is also reduced by work, as occurs in cattle (Agyemang *et al.* 1991b).

The power required for traction varies considerably with soil conditions and the type of cultivation implement used. A study (Bakrie 1991) of wet soil tillage operations in West Java (Table 6.5) showed that the

Table 6.4 Effect of supplementation on LWG and rate of ploughing for two types of buffalo in Thailand (from Wanapat 1990).

Treatment	LWG (g hd^{-1} day^{-1})		Rate of ploughing (m s^{-1})	
	Supplemented	Unsupplemented	Supplemented	Unsupplemented
Crossbred Murrah				
Working	226	99	0.80	0.68
Not working	408	281	–	–
Swamp				
Working	221	75	0.83	0.73
Not working	384	254	–	–

Table 6.5 Soil shear strength and traction output by working animals in Subang, West Java (from Bakrie 1991).

Villages/operations	Animals	Soil shear strength (kN m^{-2})	Draught force (N)	Walking speed (km h^{-1})	Power output (HP)
Wanareja					
Ploughing	Buffalo	26.5	612	2.5	0.57
Levelling	Buffalo	29.1	362	3.9	0.53
Tanjungwangi					
Ploughing	Buffalo	28.5	540	1.1	0.22
Ploughing	Cattle	28.1	589	2.0	0.44
Levelling	Cattle	–	478	2.1	0.38
Padamulya					
Ploughing	Cattle	28.0	389	3.1	0.44
Levelling	Cattle	25.1	301	2.7	0.31

draught force was less (60, 85 and 77% in three different villages) for levelling operations than for ploughing, and the speed of walking by buffaloes was greater for levelling than for ploughing; the performance of cattle was not consistent in this respect.

Farmers vary greatly in the amount of preparation they give to crop land and in the efficiency with which this is achieved. In Java rice cultivation is intended to invert the soil and bury surface OM, to soften the soil for ease of transplanting, to eradicate weeds, to reduce soil permeability and so avoid loss of water and nutrients, and to level the land so that irrigation is effective. However the development of plough pans reduces the productivity of subsequent rain-grown crops. A common system for wet rice is an initial ploughing followed by three or four passes of a leveller c. 2 m wide. Puddling of the soil may also be achieved by hoof cultivation using herds of buffalo or cattle, called 'merancah' in Indonesia. A study (Liem et al. 1988) in West Timor compared various techniques of preparation for wet rice using tractors, cattle with levellers or ploughs and levellers or used for trampling, and hand cultivation. The range in hours per hectare spent in preparation (Table 6.6) varied greatly for the same type of operation; least time was spent using a leveller.

The amount of working time provided by draught animals depends not only on local farm needs but upon the opportunity to hire out animals to other producers. A study in East Java (Yusran and Yudi 1991) compared the situation in a village, Martopuro, in which wet-rice cropping predominated and a village, Sudimulyo, in which dry land

Table 6.6 Range and mean hours of land preparation per hectare for rice in West Timor using various methods (from Liem *et al.* 1988).

Method of land preparation	Range	Mean
Tractor use	21–96	66
Leveller	43–58	54
Plough plus leveller	96–137	97
Trampling	39–112	73
Hand labour	69–120	83

cropping was the norm. The data were segregated according to farm size. Cattle were rented out for work more at Martopuro, which enjoys a long wet season, than in Sudimulyo where there is a higher cattle density and the wet season is shorter; farmers with less land also employed their cattle for renting income more than farmers with larger holdings (Table 6.7).

The use of animals for draught and carriage provides the closest interdependence of livestock and crop production of any system discussed in this book. Its persistence, despite the growth of industrialized economies, arises from the inherent efficiency of its resource use in farming systems with relatively low capital input.

Enhanced family nutrition

Milk production from smallholder farms is readily sold in most communities, following the successful introduction of small processing plants.

Table 6.7 Days of cattle work annually in various operations in two land ownership groups in two East Java villages (from Yusran and Yudi 1991).

Operation	Martopuro Village farm size		Sudimulyo Village farm size	
	< 1 ha	> 1 ha	< 1 ha	> 1 ha
Wetland ploughing	54	28	10	3
Levelling	12	11	8	2
Dryland ploughing	< 1	< 1	14	6
Pulling carts	0	0	12	2
Total	66	39	43	13
Days rented out	61	9	13	0

The family always has the immediate option of altering the proportion of milk that enters domestic consumption, that is available to the unweaned stock, or that is sold. In some societies working men receive a larger share of family consumption of milk or blood; alternatively in more benign family life pregnant women and young or sick children will benefit. Increased family meat consumption parallels the introduction of small ruminants, if only from 'fallen' animals. The capacity of the family to exchange animals for grain helps to maintain continuity of good nutrition during hard times, such as drought. The overall increase in farm income from incorporating animals in mixed farming (Chapter 7) will also benefit family health and nutrition.

Effective utilization of farm resources

The enhanced utilization of resources of land and climatic growth factors which results from mixed farming was discussed in Chapter 4.

Diversification of income and accumulation of capital savings

The influence on farm income of more effective use of farm resources through the presence of livestock and the reduction of risk associated with diversified outputs is introduced in Chapter 7. For many farmers a more cogent reason for having livestock is the security they provide for savings. In all societies families seek avenues for savings which hold their value but which may be realized quickly in time of need. Monetary investment is insecure in economies exhibiting high rates of inflation or in unstable societies. Property investment is safe in many societies, but is not necessarily realizable in the short term. Livestock hold their value in real terms, thereby acting as a successful hedge against inflation, but also have the advantage of being readily sold in time of family need. Breeding stock form an additional avenue of capital accumulation through their fecundity. In these senses family security is enhanced through the possession of livestock.

Cultural needs and farmer status

It has been fashionable for some sociologists to give great weight to the enhanced status of farmers owning livestock, especially in African communities. This emphasis is misplaced relative to the greater weight families place on security, especially food security, conferred by animal ownership. However the interchange of livestock in customs associated with marriage or with friendship and patronage provides a significant thrust to acquiring livestock in many societies, whilst the need to possess animals used for ritual slaughter and consumption at funeral feasts following the death of a member of the family may be real, as in some highland societies of northern Thailand.

This brief chapter is intended merely to outline the range of imperatives farmers adopt for the outputs which arise from the incorporation of livestock in different farming systems. These themes are further developed in more concrete ways in Chapters 8–11 which describe the four main types of crop–forage system. A fuller list of these is available as an appendix to Chantalakhana (1990).

CHAPTER 7

The diversification of income and flexibility of production

7.1 Introduction

Farmers attain better stability of income and sometimes higher levels of income through diversification of production, which also creates greater opportunity for flexible responses to changed circumstances and new opportunities. This final introductory chapter deals briefly with an additional motivation of farmers towards mixed livestock–crop systems before the technology of these is elaborated in Chapters 8–11.

The minimization of risk is a central consideration for farmers, especially if food security for their families is a primary goal. The control of uncertainty is certainly a primary factor in the thrust to diversification, although probably the most cogent justifications for diversification have already been dealt with in Chapter 4; the use of environmental, economic and human resources is more efficient in animal/crop enterprises than in single enterprises. It is uncommon to find farming circumstances so stable and a land system so uniform that one particular crop provides optimal land use (Anosike and Coughenour 1990) and no opportunity for the involvement of animal production; the integration of crop and animal enterprises utilizes energy satisfactorily in fixation and transfer whilst efficient moisture use and the effective cycling of nutrients can work well in crop/forage systems. The seasonal labour calendar is better filled in diversified production; for example in east Malaysia Cramb (1989) has shown how a combination of annual and perennial cropping provides more continuous use of human resources than does a single crop production system.

For many small farmers the survival of the family has to be the primary consideration, and this conditions attitudes to the introduction of new technology, which may have inherent riskiness, especially if it has not been validated over a sufficient period in farm production. The opportunity for rising living standards may only be seized as there is a cultural advance from the ethos of basic subsistence. The landraces of traditional crops exhibit low yields but yield is inherently less variable than the yield of some bred material; there is a concomitant failure to respond to improved environmental conditions, as for example is illustrated in the insensitivity of local upland rice varieties to fertilizer in slash-and-burn situations in central Laos (Shelton and Humphreys 1972).

A common set of objectives for farmers in the tropics (Clayton 1983) is:

- an adequate and assured family food supply;
- income to purchase a required level of material needs;
- a certain degree of security reflecting farmer circumstances and psychology;
- observance of socio-cultural customs and obligations;
- a satisfactory amount of leisure.

How does diversification of production assist in meeting these goals?

7.2 The minimization of risk

In his classic work Heady (1952) distinguishes 'risk' from 'uncertainty'. For risk, variability or outcomes which are measurable in an empirical or quantitative manner can be described by a probability distribution. Uncertainty is a subjective anticipation of the future. McCarl *et al.* (1987) have developed stochastic dominance models, i.e. models dominated by the laws of probability, which provide procedures for evaluating decisions under conditions of risk and which are used to identify the best enterprise combination that enables the farm to remain profitable. Diversification can be viewed as a form of insurance which involves the substitution of a small known cost for the possibility of a larger but uncertain loss (Bishop and Toussaint 1958); this contingency of cost does not arise in every situation. Diffusion of production elements towards the creation of a larger number of outputs tends to cancel out gains and losses and provide a higher degree of regularity and predicability in the total returns of the enterprise; the chance of loss of a small fraction of a particular output is of less moment than the chance of losing the larger part of a single enterprise (Knight 1971). Vandermeer (1989) attempts to model an adaptive function for the minimization of risk in intercropping.

A concrete example might be drawn from the evolution of small holder farming systems over the past 50 years in medium altitude districts of western Kenya. Many farmers depended primarily on maize crops with some supplementary bean and cowpea production, and production was relatively variable from year to year. The introduction of Sahiwal cattle increased potential milk production and a gradual change to dairying based on *Pennisetum purpureum* and ancillary cropping in this bimodal rainfall environment functioned well. The sale of surplus calves enhanced income, family nutrition improved through dairy products, and food security was better than under the former cropping system.

Environmental variability

The degree of stability in weather conditions and their predicability will influence the thrust to diversification; farmers living in districts with predictable patterns of the occurrence of wet and dry conditions are able to develop farming patterns which reliably accommodate these changes, whilst in more variable rainfall areas, e.g. subtropical Australia, diversification may be essential to provide the resource bank which enables the farmer to survive the unexpected. In Australia the recognition of the relationship between changes in the 'El Niño' Southern Oscillation index and the subsequent occurrence of droughts and floods may be used to adjust SR or choice of crop to minimize the effects of climatic variability (Russell 1991). Farm products respond differently to the same weather conditions (Bishop and Toussaint 1958), so that a combination of production enterprises reduces the possibility of a total loss of production. Weather conditions will also modify the patterns of incidence of diseases and pests.

Risk may be minimized by choosing crop and stock enterprises which give the most stable performance. On the Darling Downs, Queensland the production of winter cereal crops on clay soils which store summer moisture is more reliable than in many winter rainfall areas of Australia. However, variability of the yield of summer cereal crops is greater than that of winter cereals in the former region, and a combination of animal production utilizing summer forage crops and other feed sources requires more exploration. Husbandry, as reflected in crop density, level of fertilizer input, choice of livestock enterprise and SR, may be directed to minimizing risk. The choice of forage varieties may be directed to stability of performance as well as to high yield. For instance, at Tifton, Georgia, grasses exhibit varying ratios of yield in a wet year to yield in a dry year; the yield ratios were 1.6, 2.0, 3.3, 3.5 and 6.9 for *Cynodon dactylon* cv. Suwannee, *C. dactylon* cv. Coastal, *Digitaria decumbens* cv. Pangola, *Paspalum notatum* cv. Pensacola and common *C. dactylon* respectively (Burton *et al.* 1957). The greater yield stability of cv. Suwannee and cv. Coastal was also linked to greater absolute yields in both years and greater water use efficiency.

Market variability

The ratio of product earnings to prices fixed by farmers has decreased in recent decades in many countries, but this ratio varies greatly according to the mix of products being grown. Farmers dependent upon export products are especially vulnerable; copra, rubber and coffee prices are volatile and government intervention does little to cushion these changes, since domestic consumption represents a small fraction of the output. These crops therefore offer an especial opportunity for integration with cattle production, as discussed in Chapter 8.

It is not only market price variability but market access which alters risk. The pursuit of new markets has altered radically the destination of animal exports, for example with respect to trading in beef between the tropical world and eastern Asia. Government intervention to protect local producers, often on the pretext of some developed environmental or disease hazard, may deny farmers access to export markets they have traditionally enjoyed.

Government policies with respect to subsidies on agricultural inputs or to price support of agricultural commodities are further volatile factors; farmers are disadvantaged when 'cheap food' governments are installed. Diversification of products clearly offers some insurance against adverse changes in the market.

7.3 Flexibility of production

Diversification of the farm enterprise provides a better capacity for farmers to adjust to environmental or market disasters and to seize opportunities to improve their position as these emerge. Grain crops which are droughted and fail to reach maturity may be grazed off if the farmer owns ruminants, and total loss is avoided. The composition of the herd may be varied according to the seasonal level of crop production in systems where crop residues are a major source of feed; in mixed herds the farmer has an opportunity in favourable years to retain young growing stock for longer before sale than in unfavourable years where the priority will lie with the retention of breeding or draught stock. The interchange of livestock products for grain, and vice versa, may be determined by the changing price ratios of these. Opportunist harvesting of pastures for seed may be restricted to highly favourable seasons when forage surplus to animal requirements is present. In some regions where fuel is scarce the price for dried dung may favour its marketing; in time of high crop prices the marginal return from applying dung to cultivated lands to enhance crop yields may be greater. For example, the market price of dried dung in some regions of Ethiopia was \$US47–114 t^{-1} whilst its value for promoting a response in grain production was estimated as from \$US61–91 t^{-1}. The foregone response in grain yield through marketing dung needs to be taken into account in assessing the net position (Pearce and Turner 1990).

In many tropical countries livestock act as a banking system and as a cushion against inflation, as discussed in Chapter 6. If currencies are unstable there is a particular incentive to invest in forms of capital that retain their real value but which can be realized as required. Ruminants are therefore used to maintain the stability of family life, providing cash reserves in times of family sickness or need, or meeting investment costs in new farm enterprises.

Complementary and supplementary products

Two product outputs are complementary if an increase in one product causes an increase in the second product, when the total amount of inputs used on the production of these two products is constant (Doll and Orazem 1984). For example a situation might arise where growing leguminous pasture in rotation with maize substantially increased total maize yield over the same period; the two products would only become competitive if total maize yield were reduced.

Two products outputs are supplementary if the amount of one product can be increased without changing the amount of the second output, and indicates a situation where surplus resources are available. For example, the husbandry of interplanting upland rice with the fodder legume *Stylosanthes guianensis* (stylo) can be managed by delaying stylo planting and controlling the density of stylo, which has its own source of N fixation, so that rice yield is similar with or without stylo, at least in the year of starting such a programme, but farmers planting both species have the supplementary benefit of enhanced fodder yield (Shelton and Humphreys 1975c).

Where crops and forages compete for land, for labour or capital, programmes are available (Doll and Orazem 1984) which estimate the best combination of enterprises from a knowledge of the marginal rate of product substitution, known as the slope of the production possibility or opportunity curve, which reflects the amount by which one output changes in quantity when the other output is increased by one unit along a curve where the input is constant, called the isoresource curve. When the prices of outputs are known this information can be used to determine the maximum revenue from the different allocation of inputs chosen for the two enterprises.

This is illustrated diagrammatically for agroforestry (Figure 7.1; Filius 1982) where the production possibility curve denotes the maximum output of forest and agricultural products to be obtained from a given input of resources. This relationship does not always remain the same over the range of possibilities. Within a certain interval there may be complementarity for one production factor and competition or supplementarity for others. Complementarity may exist over the range AB because of savings in the cost of weeding and over the range ED because of savings in the cost of erosion control. Over the range BD competition for light and nutrients dominates. The optimum combination occurs at point C, where a price line is tangential to the production possibility curve.

7.4 Positive indicators for diversification

Farm size

Small farmers in many tropical countries have mixed livestock/cropping enterprises based on land areas as little as 0.4 ha. It is argued that these households, close to the subsistence line, benefit from the incorporation of livestock through improved resource use, better flexibility of production and reduction of risk. Nevertheless some studies suggest that the level of diversification is positively associated with farm size. In the western Sudan larger producers planted a more diversified set of crops than smaller producers, and the larger producers more often

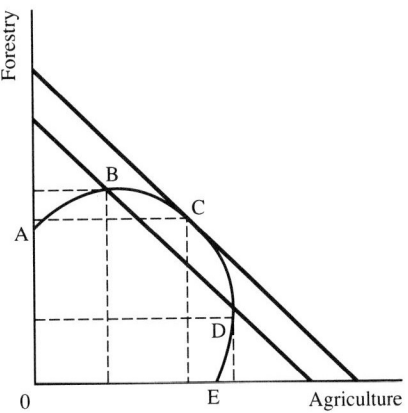

Figure 7.1 Production possibility curve ABCDE for different combinations of forest and agricultural products (from Filius 1982).

were diversified in having livestock (C. M. Coughenour, personal communication). However, large and small producers planted both food (sorghum and millet) and cash (sesame and karkadee) crops. In a more detailed study in Kentucky, USA Anosike and Coughenour (1990) found that farm size was a crucial factor in promoting diversification. This example from Western agriculture may only apply in tropical agriculture for the same reasons in a limited number of situations. It is probably related to:

- economies of scale,
- appropriate capitalization in relation to enterprise size,
- the greater probability of a coarse-grained land system appearing on a large farm.

The first two factors are of less moment in low-input type systems, where the level of mechanization and associated investment are not high. On the other hand on the Queensland Atherton tableland in the 1960s one factor in the failure to promote a combination of dairy farming leys based on the legume *Neonotonia wightii* with maize production for grain was the small size of farms in the area; dual enterprises led to overcapitalization of these mechanized farms.

Management skills

Anosike and Coughenour's (1990) study identified human capital as a factor in diversification, which requires a greater range of technical and managerial skills than single enterprise production. Operators of mixed farms face a more complex and dynamic environment of decision making, with a multiplicity of adjustments to be made over time. The nature of information upon which decisions are to be made is incomplete, and experienced and educated operators are more likely to make judgements which are effective in balancing the gains and losses from diverting labour input and resources to different operations on a day-to-day basis. In some cultures operators develop highly specialized and sophisticated skills directed to efficient production of a single crop, which do not fit them for managing mixed livestock/cropping enterprises. This was especially evident in plantation agriculture associated with estate production of rubber and oil palm; nevertheless in Malaysia there has been a gradual evolution of managerial attitudes which has, for example, led to the incorporation of sheep in plantations, essentially for weed control. It should also be recognized that in traditional mixed farms the skilling of farmers commences in childhood when family labour is used in diverse ways. The availability of off-farm work reduces diversification, and skills are lost from the farm as the urban workforce increases.

Some management advisers have emphasized the need to define the ecological niche a farm occupies in terms of its unique set of scarce resources, for which certain crops or livestock are selected through adaptation or competition (McCarl et al. 1987). What should be avoided is the concept of a static optimum which operates in a particular set of circumstances; the truth is that farmers are continually adapting to changing environmental conditions, the incidence of new pests, diseases and weeds, the availability of new forage and crop varieties, new knowledge about their management, changing prices for inputs and for commodities, and the emergence of new commodities.

In these changing circumstances the continuous process of learning and of adaptation will produce better solutions to optimize the use of resources and to minimize risk if the farmers work flexibly within a livestock/cropping mix than if they are limited to the risks of a single specialized enterprise.

CHAPTER 8

Tree crops with pastures

8.1 Introduction

Farming systems of the tropics

The diverse sorts of farming systems which function in the tropics have been described in numerous studies, for example Ruthenberg (1980), Beets (1982), Pearson and Ison (1987), Fresco and Westphal (1988), Edwards *et al.* (1990) and Humphreys (1991, chapter 2). The systems which involve an interface between crops and forages, and which are the subjects of the next four chapters, may be conveniently grouped as follows.

- Tree crops with pastures (Chapter 8).
- Shrub legumes with annual agricultural crops (Chapter 9). In this case the shrubs are used for fodder, for fuel or light timber, or for spreading on crop lands in order to reduce erosion and maintain soil fertility.
- Pastures with annual crops (Chapter 10); short-term or ley pastures are grown in rotation with annual agricultural crops.
- Annual crops with forage crops (Chapter 11); forages are grown as companion intercrops or are grown in relay for another part of the growing season.

Agroforestry

Agroforestry may be defined (Nair 1989a) as 'a collective name for land-use systems and technologies where woody perennials (trees, shrubs, palms, bamboos etc.) are deliberately used on the same land-management units as agricultural crops and/or animals, in some form of spatial arrangement or temporal sequence. In agroforestry systems there are both ecological and economical interactions between the different components'. It is implied that two or more species (of plants, or of plants and animals), one of which is a woody perennial, are grown for a cycle longer than 1 year, giving rise to two or more outputs and a system more complex than functions in monocropping.

Selection of the most suitable classification of agroforestry systems amongst the many available, for example King (1979), Grainger (1980) and Wood (1990), depends on the purpose for which it is intended. Nair (1985; Table 8.1) categorizes systems firstly according to their structure: the nature and arrangement of components, and their function: the role or output of components. This book is concerned with silvopastoral or agrosilvopastoral systems involving animals and not with systems dealing exclusively with crop products. Secondly, systems may be grouped according to their environmental adaptation within the tropics and subtropics, or to their socio-economic status deriving from the level of technology inputs and their economic orientation.

Regional examples

There are many general reviews of agroforestry the reader may wish to consult (Richards 1977; Thomas 1978; Combe 1983; Torres 1983; Watson 1983; Payne 1985; Young 1988; Nair 1989b; Cameron *et al.* 1991; Shelton and Stür 1991). More specific regional examples of the success or failure of agroforestry systems in field practice are available for different environments and a selection of these follows: Africa (Hofstad 1978 and Baum 1984 for Tanzania; Okafor and Fernandes 1987 for Nigeria; Sturmheit 1990 for Zambia; Cook and Grut 1989 and Kerkhof 1990 for survey project experience); Asia (Wu and Rin 1976 for Taiwan; Gutteridge and Boonklinkajorn 1979, Boonkird *et al.* 1984, Cheva-Isarakul 1986, and Sophanodora and Tudsri 1991 for Thailand; Shekhawat *et al.* 1988 and Jain 1993 for India; Wiersum 1982, Nitis 1985 and Nitis *et al.* 1989 for Indonesia; Moog and Faylon 1991 for Philippines; Monsoon Asia Agroforestry Joint Research Team 1986 for the region); Australia and the Pacific (Cook and Grimes 1977, Vergara and Nair 1985; Shelton *et al.* 1987; Applegate and Nicholson 1988); and America (Weaver 1980, Bishop 1983 for Ecuador; Briscoe 1983, Johnson and Nair 1985, and May *et al.* 1985 for Brazil; Evans 1988 for Paraguay).

This chapter is arranged to consider:

- the rationale for agroforestry,
- tree–pasture relations,
- the management of tree–pasture associations,
- the management of particular crops: coconuts, rubber, oil palm, coffee and other plantation crops, fruit crops and timber in humid and subhumid climates.

Tree crops with pastures

Table 8.1 Major approaches to classification of agroforestry systems and practices (from Nair 1985).

Categorization of systems (based on their structure and function)		Grouping of systems (according to their spread and management)	
Structure (nature and arrangement of components, especially woody ones)	Function (role and/or output of components, especially woody ones)	Agro-ecological/ environmental adaptability	Socio-economic and management level
Nature of components	Arrangement of components		
Agrisilviculture (crops and trees including shrubs/trees)	*In space* (spatial) Mixed dense (e.g. home garden) Mixed space (e.g. most systems of trees in pastures) Strip (width of strip to be more than one tree) Boundary (trees on edges of plots/fields) *In time* (temporal) Coincident Concomitant Overlapping Sequential (separate) Interpolated	*Productive function* Food Fodder Fuelwood Other woods Other products *Protective function* Windbreak Shelterbelt Soil conservation Moisture conservation Soil improvement Shade (for crop, animal, and man)	*Systems in/for* Lowland humid tropics Highland humid tropics (above 1200 m a.s.l; e.g. Anes, India, Malaysia) Lowland subhumid tropics (e.g. savanna zone of Africa, cerrado of South America) Highland subhumid tropics Tropical highlands (e.g. in Kenya, Ethiopia)	*Based on level of technology input* Low input (marginal) Medium input High input *Based on cost/benefit relations* Commercial Intermediate Subsistence
Silvopastoral (pasture/animals and trees)			
Agrosilvopastoral (crops, pasture/animals and trees)			
Others (multipurpose tree lots, apiculture with trees, aquaculture with trees, etc.)			

This chapter draws extensively on Shelton and Stür (1991).

The rationale for agroforestry

The earlier chapters of this book have provided a general rationale for incorporating livestock and forage production in the agricultural systems of the tropics and subtropics. The bases for successful adoption of specific tree–pasture associations will unfold through this chapter and a few introductory comments are made about the benefits and the difficulties of combining the two types of output.

Efficiency of resource use

The rates of removal of the world's forests, which through the 1980s have accelerated to estimated figures of 1.7% in Africa, 1.4% in Asia and 0.9% in Latin America (Serrao et al. 1993), occasion alarm because of the depletion of resources and its impact on the global environment. Some scientists believe it is only through the replacement of natural forest with woody perennial species combined with crop and livestock production that landholders will find the incentives to replace these resources and to stabilize landscapes.

The early phases of establishment of tree crops leave much of the land surface unoccupied by the young woody perennial, which is usually planted at relatively low densities (for example c. 100–200 coconut palms ha^{-1}). This leads to considerable waste of rainfall and incoming radiation, as discussed in section 8.2, unless the palms are intercropped with other crop species or are planted to forage (Colour plate 9, Plate 8.3). Continued production of forage is feasible throughout the life of many plantation species.

The amelioration of the climate in which pastures and animals are produced under trees is a favourable aspect of resource use. Trees provide shade, which is beneficial for animal production. Cows at Atherton, Queensland (Silver 1987) produced an additional 1.4 l hd^{-1} day^{-1} of milk if provided with shade, and the difference in production from animals lacking shade reached 2.2 l hd^{-1} day^{-1} in the hottest week of the study. The air temperature during the daylight hours is cooler in plantations. Wilson and Ludlow (1991) quote examples of reduced temperature above the pasture of 2–3 °C at midday under mature rubber trees, whilst under *Acacia* the daily maximum temperature may be 2–3 °C lower than in the open, and the minimum might be 1.5–2 °C higher in the shaded situation. In the subtropics the incidence of frost may be almost entirely eliminated under a tree canopy, and winter performance of frost-intolerant tropical pasture species is superior to that observed in open pastures (Cook and Grimes 1977; Cameron et al. 1989).

Trees may be specifically planted to provide shade and shelter for livestock, or to reduce wind speed over adjacent pastures or crops. Good shelter belts with a permeability of 40–60% may be expected to reduce wind speed at least 20% for a distance of 25 tree heights leeward, and 5 tree heights windward (Shea 1991). Seguin and Gignoux (1974) provide theoretical models showing the effect of hedges on windspeed. Bird (1991) suggests that shelter belts 250 or 500 m apart reduce windspeed by 55% and 33% and animal maintenance needs by 18% and 10% respectively. In Zimbabwe a windbreak of *Pinus patula* c. 5 m high and planted in three staggered rows reduced mean daily wind by 58, 33, 15 and 9% at points 30, 60, 90 and 120 m to leeward respectively (Payne 1968). Obviously the windiness of the site will determine whether real benefits to stock and pastures result from tree planting.

Maintenance of soil fertility

Control of soil erosion is a significant imperative for the planting of trees and pastures. The destruction of rain forest has less impact on the loss of soil and the acceleration of runoff from catchments if land use involves replacement of forest with plantation crops of rubber, oil palm or coconut, especially if these form double strata with underplanted pasture (Chen 1993). Similar reductions of nutrient losses in surface runoff are also evident if ground cover is maintained.

The effects of forages in plantations on soil fertility are discussed in section 8.2. Positive effects on tree crop yields are feasible if legume-based pastures are planted and appropriate management adopted (Rika et al. 1981), although negative effects may arise from planting high-yielding grasses and neglecting fertilizer needs (Ferdinandez 1972). The addition of OM which occurs under pasture establishes a nutrient bank upon which the trees may draw (Broughton 1977), whilst the tree acts as a nutrient pump which delivers nutrients to companion pastures in an accessible way (Zinke 1962).

Diversified and increased income

As discussed in Chapter 7, the dependence of many farmers in the tropics on export-oriented crops such as copra, palm oil or rubber leads to considerable insecurity of income, since the prices for these products are notoriously unstable on world markets. The addition of animals not only reduces risk but may augment farm income. The establishment of pasture in existing plantations (Colour Plate 9) is inexpensive relative to the costs of developing virgin land for pasture, since land has already been brought into a condition which lends itself to further improvement. The saving in the costs of herbicides and of mechanical weeding through the introduction of sheep into rubber plantations in Malaysia are outlined in section 8.4. Labour costs in mixed plantations may also be reduced, as occurs in Taiwan (Wu and Rin 1976).

Managerial needs

The increased level of management skills required to operate a mixed enterprise rather than an enterprise with a single type of output is the most significant negative factor operating against the introduction of tree–pasture associations, as discussed in Chapter 7. There is not only a different set of skills involved in plantation agriculture or silviculture and in animal production, there are also inherent cultural barriers to the adoption of mixed enterprises, especially if there is a history of animal damage to plantation operations or output. On large plantations this may be partially overcome by specialization within the labour force, but the need to coordinate animal and plantation operational requirements demands good managerial capacity.

8.2 Tree-pasture relations
The shaded environment of the plantation
The light regime of the pasture

Plantations are usually located in humid regions where water stress is a constraint to production that is secondary to radiation as an environmental growth factor. On fertile soils or in intensively managed plantations where optimum fertilizer levels are employed it is the level of radiation and its efficiency of use which determines product output. Radiation receipt is lower in the tropics than in the subtropics, and in many equatorial regions there is an extended seasonal occurrence of low cloud which decreases plant production. Total short wave solar radiation in the humid equatorial tropics varies from low values of c. 11 MJ m^{-2} day^{-1} to maximum values on clear days of c. 22 MJ m^{-2} dat^{-1}, whilst at lat. 22° S in Australia values in excess of 30 MJ m^{-2} day^{-1} often occur for long periods.

Short wave radiation (400–3000 nm wave band) is the most commonly measured meteorological element for which long term records are available, but it is PAR (400–700 nm wave band) which determines the rate of plant growth. Fortunately the availability of integrating pyranometers which record PAR is improving the data that may be used to predict plant–light relations. Under the open sky

PAR approximates to 0.5 short wave radiation. However, since PAR is preferentially absorbed by the green tree canopy, the proportion of short wave radiation striking the pasture below as PAR may be as low as 0.27. Differential absorption of red and far-red light by the tree canopy also alters the ratio of these components (R/FR) which controls the operation of the phytochrome system in plants. The R/FR ratio in full sunlight of 1.2 may be reduced to 0.62 under mature rubber (Wilson and Ludlow 1991). This increases stem elongation of pastures growing in plantations, may decrease tillering of grasses (Deregibus et al. 1985), and may also reduce the germination of light sensitive seeds.

The transmission of light to the pasture depends upon the density and growth of trees, their spatial arrangement, and the architecture of the canopy, and these are also reflected in the total biomass of the trees. A plantation of élite tree germplasm, for example recently released hybrid coconut, planted at a high density in a triangular arrangement on a fertile site, will intercept more PAR than local variety coconuts planted in a square arrangement 10 × 10 m on poor coastal sands. In the Solomon Islands smallholder coconuts showed light transmission of 40–70% (Litscher and Whiteman 1982). Percentage light transmission

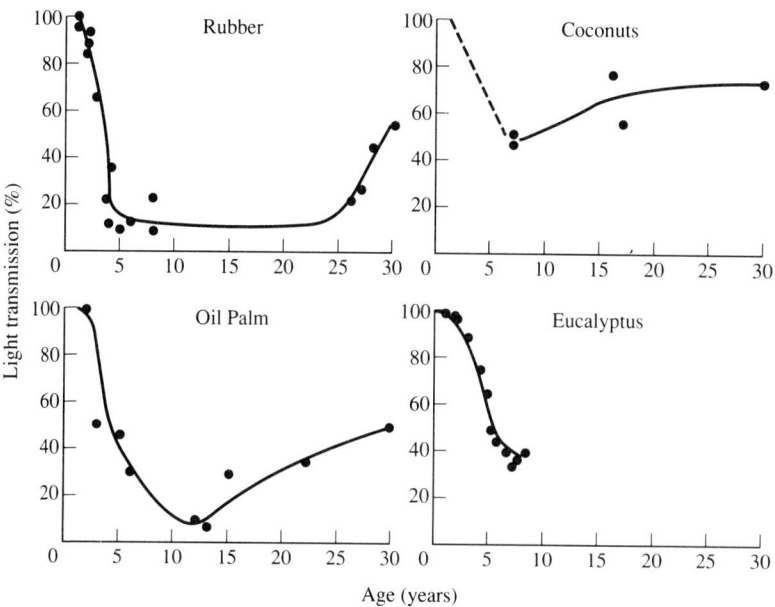

Figure 8.1 Relative light transmission (% of PAR) with age for plantations of rubber, oil palm, coconuts and eucalypts (from Wilson and Ludlow 1991).

(Y) was negatively and linearly related to palm density (X_1, palms ha^{-1}) and palm height (X_2, m):

$$Y = 102.6 - 0.195\,X_1 - 1.47 X_2;$$

age of planting, which varied from 9 to 19 years, and palm height were positively correlated. In later years the reverse applied, and light transmission would be positively associated with age, as shown in Figure 8.1 (Wilson and Ludlow 1991).

The general situation relating age of trees to light transmission for four types of plantation is that the early period after planting provides good opportunity for pasture or crop growth in the first 5 years (Figure 8.1). Thereafter canopy closure in rubber provides a very dark environment. Pasture biomass (DM) decreases from c. 3–5 t ha^{-1} at the immature stage (3–5 years) to c. 0.1–1 t ha^{-1} at age 8–22 years, increasing to c. 2 t ha^{-1} beyond this age (Chen 1993). Good pasture production is feasible throughout the whole life of a coconut planting, with oil palm and eucalypt forest providing intermediate conditions. The amount of transmission to the pasture depends upon the proportion of direct sunlight to diffuse light. The latter arises from the whole sky and is dispersed more widely through the canopy than occurs from the single direct source from the sun on a clear day. Measurements of direct sun therefore underestimate the total light received by the pasture. Irradiance also varies spatially according to the distance from a tree trunk. Wilson and Ludlow (1991) have recalculated data from a Malaysian oil palm plantation 5.5 years old (Figure 8.2) in which light transmission near midday varied from 2.5% near the stem to 73% in the middle of the inter-row. The area under the curve integrates to 47% light transmission overall. Percentage transmission for the whole day tends to be slightly lower (c. 78%) than the estimate obtained from measuring transmission at midday.

Pasture productivity

Pasture growth, which in the final analysis determines SR, depends positively on the amount of PAR falling on the sward, which in turn depends upon the radiation received at the site and the percentage transmission through the tree canopy. This is shown for swards of different grasses established at sites where variation in density of coconut palms on Banika, Solomon Islands (lat. 9° S) gave differing light environments (Table 8.2; Smith and Whiteman 1983). The swards were harvested six times during the year. The data for two 'sun' species, *Brachiaria decumbens* and *B. humidicola*, and for two

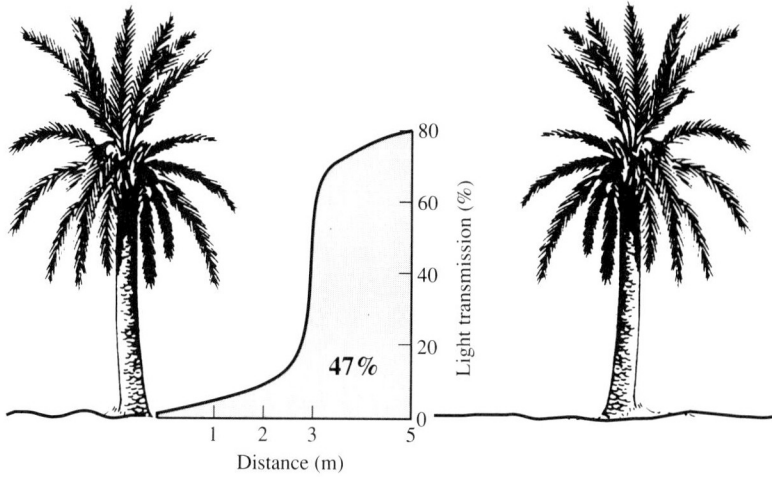

Figure 8.2 Spatial variation in relative light transmission in oil palm measured near midday (from Wilson and Ludlow 1991).

'shade' species, *Paspalum conjugatum* and *Axonopus compressus*, have been selected to illustrate the responses. In full sunlight and at intermediate light levels the 'sun' species were much more productive than the 'shade' species, and the proportionate reduction in growth of the 'sun' species as light transmission decreased was greater than the percentage reduction in transmission. 'Shade' species are susceptible to photoinhibition which reduces their growth at high radiation levels. *P. conjugatum* and *A. compressus*, which are found widely distributed in heavily shaded situations, showed a substantial reduction in yield

Table 8.2 Total shoot yield (t DM ha^{-1} year^{-1}) of grasses growing under coconut palms of differing density in the Solomon Islands (from Smith and Whiteman 1983).

Grass	Light transmission (% PAR)				
	100	70	50	40	20
Brachiaria decumbens	28.2	12.7	10.9	6.1	3.3
B. humidicola	22.8	13.6	11.1	4.7	2.6
Paspalum conjugatum	11.4	5.9	4.3	4.7	2.6
Axonopus compressus	9.3	6.4	6.1	3.7	3.1

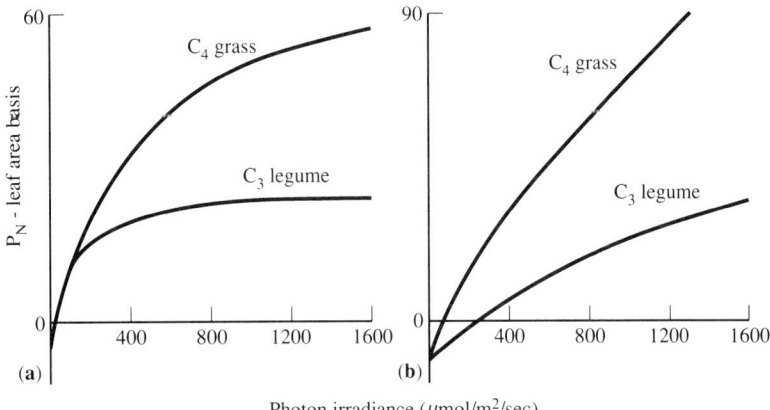

Figure 8.3 Light response curves for net photosynthesis (P_N) of a C_4 grass and a C_3 legume for (a) individual leaves and (b) whole canopy (from Wilson and Ludlow 1991).

in shade, but this was of lesser magnitude than occurred for *B. decumbens* and *B. humidicola*. However, it should be noted that at low light levels of 40 or 20% transmission the yields of all four species were similar. Particular species may show increased growth under shade, especially if N supply is limiting, as discussed later in this chapter.

The rate of net photosynthesis responds differently to changes in the level of PAR (or photon irradiance) depending on whether plants exhibit the C_4 dicarboxylic acid photosynthetic pathway, as occurs in tropical grasses, or the C_3 phosphopyruvic acid pathway, as occurs in tropical legumes and in temperate grasses and legumes. The shape of these responses (Figure 8.3a; Wilson and Ludlow 1991) indicates that the net photosynthesis of the leaves of C_4 plants continues to increase to the highest level of irradiance tested and is superior to that of the C_3 legume except at very low light levels. When this is translated to the photosynthesis of swards (Figure 8.3b), which incorporates other factors relating to leaf disposition, the C_4 grass is believed to maintain a small advantage over the C_3 legume even at low light levels.

There are a number of plant responses which assist the maintenance of growth in shade. The chief of these are:

- Reduced respiration rate,
- Increased ratio of shoot to root,
- Increased specific leaf area.

176 *Tree crops with pastures*

As light level decreases so does respiration (Boardman 1977), so that the compensation point at which C fixation equals C losses through respiration is also reduced. Grasses appear to exhibit rather greater adaptation in this respect than legumes. The partitioning of assimilate to roots is decreased and the most successful species in shade have reduced root systems which may be insufficient to provide adequate anchorage for plants subject to grazing. Wong *et al.* (1985a,b) give for a range of grasses a change of shoot/root ratio of 2.5 to 6.7 as light transmission decreased from full sunlight to 27% transmission; for legumes the change in ratio was 6.5 to 14.3 respectively. Shade also reduces the partitioning of assimilate to seed in *Panicum maximum* (Oliveira and Humphreys 1986). Plants under shade are therefore

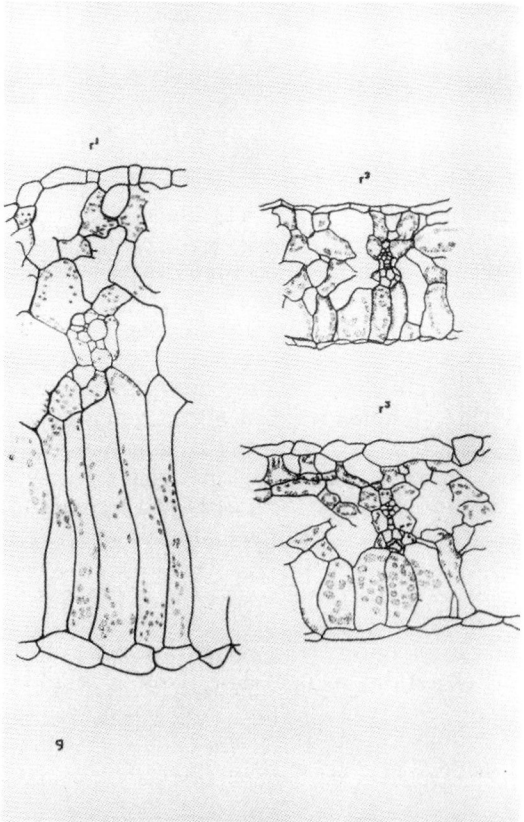

Plate 8.1 Transverse sections of *Macroptilium atropurpureum*.

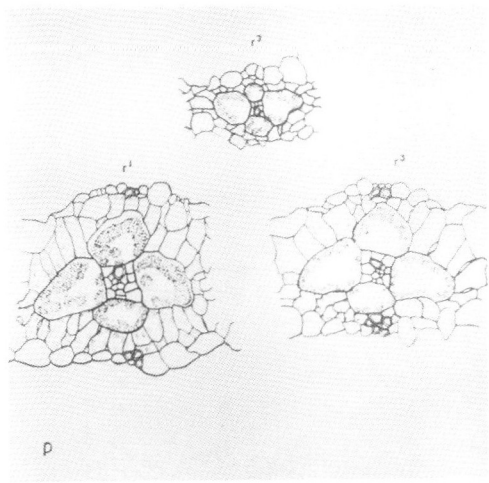

Plate 8.2 *Panicum maximum* var. *trichoglume* that developed under three illuminance regimens: 100% daylight (bottom left), 33% daylight (top) and 11% daylight (bottom right).

effective in maintaining a carbon economy which directs energy to the expansion of the leaf surface.

This is further assisted by a considerable increase in specific leaf area. The leaves of plants are conditioned by the level of illumination in which they have developed, and shaded leaves are thinner than unshaded leaves, and contain fewer, smaller and less densely packed cells. This is illustrated in Plates 8.1 and 8.2 (Ludlow and Wilson 1971), which show transverse sections of the leaves of the C_3 legume *Macroptilium atropurpureum* and the C_4 grass *Panicum maximum* var. *trichoglume* grown under three illuminance regimes: 100, 33 and 11% daylight. These treatments had mean short wave radiation of 15.4, 5.6 and 2.0 MJ m^{-2} day^{-1}. Specific leaf area was 280, 710 and 1000 cm^2 g^{-1} in these treatments for *M. atropurpureum*, and 330, 500 and 930 cm^2 g^{-1} respectively for *P. maximum* var. *trichoglume*. These figures indicate the greater expansion of single leaves which occurs under shade; this confers an advantage for light interception, although photosynthetic capacity is reduced as chlorophyll concentration decreases with shading. Wong *et al.* (1985a,b) found rather lower values for specific leaf area under their conditions than Ludlow and Wilson (1971) report, but a similar radical adaptation in this plant component to shade occurred.

A useful and simple approach to the analysis of growth is to

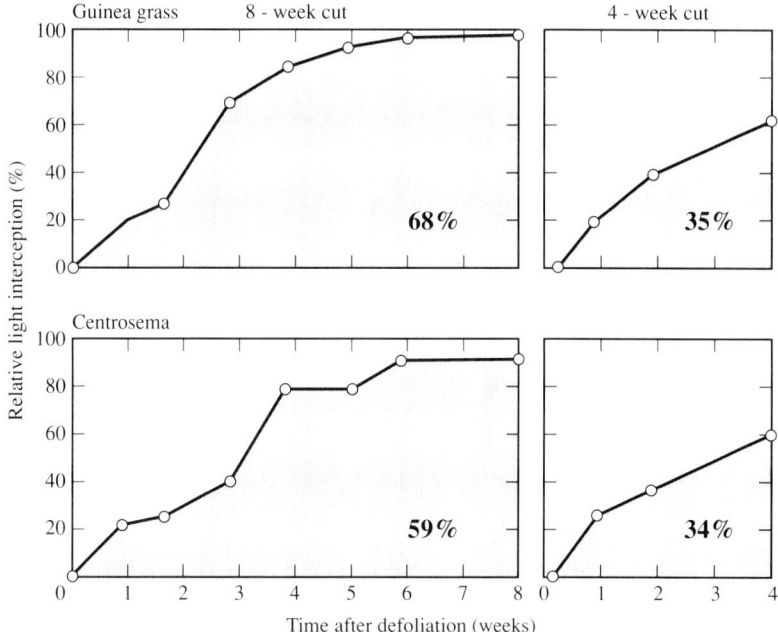

Figure 8.4 Change in relative interception of PAR with time for *Panicum maximum* (Guinea grass) and *Centrosema pubescens* cut at intervals of 8 and 4 weeks (from Sophanodora 1989).

regard it as the product of the amount of light intercepted (J) and the efficiency of carbon fixed per unit of light intercepted (E). There are few reports of total light intercepted, which require continued measurement through each day at sites below the sward. Sophanodora (1989) found that *Panicum maximum* growing under shades of 30% light transmission at Redland Bay, Queensland (lat. 28° S) intercepted 68% of total PAR during 8 weeks after cutting (Figure 8.4) whilst light interception was greatly reduced under more frequent cutting at intervals of 4 weeks. *Centrosema pubescens* was even slower to erect a leaf surface intercepting the available PAR.

Recovery after cutting was substantially slower than occurs in full sunlight (Ludlow and Charles-Edwards 1980) and the lesser interception of radiation and the reduced amount of radiation are the factors which account for the decreased growth of herbage plants in shade. Despite the greater diversion of resources to leaves, the reduced allocation of assimilate to roots and the increased specific leaf

Table 8.3 Estimated radiation use efficiency (g MJ^{-1} of PAR) for herbage species growing under regimes of differing light transmission and nitrogen availability (from Sophanodora 1989).

Species	N supply	Light transmission (% PAR)		
		100	70	30
Brachiaria decumbens	Adequate	2.0	2.3	2.8
	Low	1.5	1.6	2.6
Panicum maximum	Adequate	2.0	2.5	3.4
	Low	1.3	1.4	2.6
Centrosema pubescens	Adequate	1.0	1.0	1.7

area under shade, the expansion of the leaf canopy is insufficient to compensate for the adverse radiation environment. The latter reduces the level of non-structural carbohydrate in the stem base and roots of pasture plants, especially under repeated defoliation. In full sunlight this phenomenon may little affect the rate of leaf expansion after defoliation (Humphreys and Robinson 1966; Beaty *et al.* 1974) but the converse applies in shaded environments. Wong (1993) has shown that the regrowth of *Paspalum malacophyllum* and *P. wettsteinii* in shade is strongly and positively related to the level of storage of non-structural carbohydrate in the stem base, which controls the rate of tillering and the development of the leaf surface.

The other factor in the equation of growth analysis, the efficiency of radiation intercepted (E), is substantially increased in low light environments (Sophanodora 1991). *B. decumbens*, *P. maximum* (Plate 8.3), and *C. pubescens* have been shown in Sophanodora's (1989) study to exhibit higher values of E in 70 and 30% light transmission than in full sunlight (Table 8.3). This table shows the legume *C. pubescens* at a disadvantage relative to the grasses. The efficiency of light use was also enhanced by good N nutrition of the sward. This phenomenon of increased light use efficiency under shade is associated with a number of factors: reduced respiration, reduced allocation of assimilate to roots, absence of photoinhibition, less CO_2 limitation for photosynthesis and increased leaf N concentration, as discussed subsequently.

Wilson and Ludlow (1991) give an illustration of how a knowledge of the light environment in plantations and of pasture response (Plates 8.4, 8.5, 8.6) may be used to estimate the probable shoot DM production of pastures. *P. maximum* is assumed to be growing at low–moderate N level under oil palm with 14% light transmission at the row centre at midday; the oil palm receives 16 MJ m^{-2} day^{-1} short wave radiation.

Plate 8.3 *Panicum maximum* under mature coconuts, Sri Lanka.

1. PAR incident on the pasture will be $16 \times 0.14 \times 0.78$ (factor to convert midday interception to daily interception) $\times$ 0.47 (factor for inter-row spatial variation, Figure 8.2) $\times$ 0.27 (factor to convert short wave radiation to PAR at the pasture surface) = 0.22 MJ m^{-2} day^{-1}.
2. Average daily growth rate of herbage will be 0.22 (PAR incident on the pasture) $\times$ 0.68 (% PAR intercepted by the pasture, Figure 8.4) $\times$ 2.6 (radiation use efficiency, g MJ^{-1}, Table 8.3) = 0.39 g m^{-2} day^{-1} or 3.9 kg ha^{-1} day^{-1}.
3. Yearly production of green leaf DM = 3.9 (kg daily growth) $\times$ 250 (assuming growth limited to 250 days) $\times$ 0.66 (proportion of plantation area actually under pasture) $\times$ 0.4 (ratio of leaf/shoot, Sophanodora 1989) = 258 kg ha^{-1}.

The above figures can be used to estimate carrying capacity, as discussed in section 8.3.

Adaptation of pasture species to shaded sites

The selection of élite pasture species (Plates 8.7–8.10) for particular farming conditions (Humphreys 1987, Part 3) requires a number of desirable plant qualities:

Plate 8.4 Pastures of *B. ruziziensis* fenced for night grazing, south Thailand.

- superior nutritive value;
- growth characteristics, including high leaf/shoot ratio, wide seasonal distribution of growth, appropriate responsiveness to nutrients, and high N fixation in legumes;
- ease of establishment;
- persistence and natural plant replacement under conditions of climatic, edaphic and biotic stress, and compatibility in grass–legume mixtures.

These criteria of merit apply in all farm situations, and are used to determine which species are best adapted to the special conditions on

182 *Tree crops with pastures*

Plate 8.5 Bannur sheep on *B. miliiformis* under coconuts, Sri Lanka.

Plate 8.6 Weedy pastures in a plantation, Solomon Islands.

Tree-pasture relations 183

Plate 8.7 *Pueraria phaseoloides* in immature rubber.

Plate 8.8 Erosion in replanted 1-year rubber, south Thailand, planted to *Pueraria phaseoloides*.

Plate 8.9 Pasture in 5-year rubber.

particular farms where the soils, the climate, the cutting or grazing management and the objectives the farmer has for the pasture will determine which suite of plants are best used. In plantations the response to shade is a primary selection character. Adaptation is judged on differing criteria:

- level of PAR at which saturation occurs,
- high ratio of yield in shade to yield in full sunlight,
- high yield in shaded conditions,
- good persistence in shaded conditions.

The first criterion is not appropriate for C_4 grasses since light saturation is only observed in the tropics if factors other than light limit photosynthesis. The second criterion is appropriate in a physiological sense, since it indicates a high degree of adjustment of plant growth processes to the shaded environment. However, as shown in Table 8.2

Plate 8.10 Deeply shaded pastures in mature rubber.

the grasses which are ecologically adapted to shaded conditions, such as *P. conjugatum* and *A. compressus* do not necessarily yield more highly in shade than selected species such as *B. decumbens* and *B. humidicola* which give high yields in the high light environment of the recently planted plantation, or the medium light environment which follows or which continues in coconut plantations.

Wong (1991) drew on several sources in Asia and the Pacific to summarize the reputation of many grasses and legumes for shade tolerance (Table 8.4). These include species which are not planted, such as *Ottochloa nodosa*, but which are found in the deep shade of mature rubber and oil palm plantations. The species listed as

Table 8.4 Shade tolerance of some widely used tropical forages (from Wong 1991).

Shade tolerance	Grasses	Legumes
High	*Axonopus compressus* *Brachiaria miliiformis* *Ischaemum aristatum* *Ottochloa nodosa* *Paspalum conjugatum* *Stenotaphrum secundatum*	*Calopogonium caeruleum* *Desmodium heterophyllum* *Desmodium ovalifolium* *Flemengia congesta*
Medium	*Brachiaria brizantha* *Brachiaria decumbens* *Brachiaria humidicola* *Digitaria setivalva* *Imperata cylindrica* *Panicum maximum* *Pennisetum purpureum* *Setaria sphacelata*	*Calopogonium mucunoides* *Centrosema pubescens* *Desmodium triflorum* *Pueraria phaseoloides* *Desmodium intortum* *Leucaena leucocephala*
Low	*Brachiaria mutica* *Cynodon plectostachyus* *Digitaria decumbens* *Digitaria pentzii*	*Stylosanthes hamata* *Stylosanthes guianensis* *Zornia diphylla* *Macroptilium atropurpureum*

having low shade tolerance are sometimes grazed on land adjacent to plantations in conjunction with other species actually planted within the plantation area.

An agronomic assessment of the environmental adaptation of 14 of the most frequently occurring forages in Asia and the Pacific tabulates their requirements for soil fertility and management, and their resistance in the stresses which occur (Table 8.5; Stür and Shelton 1991a). *Axonopus compressus* (Plate 8.11) adopts a prostrate habit under heavy grazing, and together with the legumes *Desmodium heterophyllum* and *Mimosa pudica* is found in well-utilized areas (Watson and Whiteman 1981). By contrast, the erect-growing *I. cylindrica* disappears under frequent grazing or cutting; it may be replaced by useful low growing species or broad-leaved weeds may invade the 'gap' created by heavy use. *P. conjugatum* is a grass intermediate in response to grazing and well adapted to acid soils. *M. pudica* bears spines on the stems and is therefore regarded as a weed, but stock eat it well and produce good LWG (Reynolds 1981). *D. heterophyllum* is also a useful component of pastures in plantations throughout the Pacific Islands and Asia, but commercial seed is not available; it is sometimes spread by stem cuttings in conjunction with grasses being vegetatively planted (Humphreys

Table 8.5 Summary of adaptation of frequently occurring forages. High H, medium M, low L. (After Stür and Shelton 1991a.)

	Tolerance to shade	Forage yield	Animal product	Required		Resistance to					Fertilizer response	Potential competition with plantation crops
				Management level	Soil fertility	Grazing	Drought	Soil acidity	Water-logging	Weed invasion		
Naturally occurring												
Axonopus compressus	H	L	M	L	L	H	L	H	M	M	M	L
Paspalum conjugatum	H	L	L	L	L	M	L	H	H	L	L	L
Imperata cylindrica	M	L	L	–	M	L	H	H	–	–	M	M
Mimosa pudica	H	L	M	L	L	H	–	–	–	–	–	L
Desmodium heterophyllum	H	M	H	L	L	H	H	M	H	–	–	L
Cover crops												
Calopogonium mucunoides	M	H	L	M	L	L	L	H	M	L	L	L
Calopogonium caeruleum	H	M	L	L	L	–	–	H	–	L	–	L
Pueraria phaseoloides	M	H	H	H	L	L	M	H	M	L	M	L
Centrosema pubescens	H	M	H	M	L	M	M	M	L	M	M	L
Sown or planted												
Stenotaphrum secundatum	H	M	M	L	M	H	–	–	M	H	–	L
Ischaemum aristatum	H	M	M	M	L	M	–	M	M	H	M	L
Brachiaria decumbens	M	H	H	M	M	M	M	H	M	M	H	M
Brachiaria humidicola	M	H	H	L	M	H	M	H	H	H	H	M
Panicum maximum	M	H	H	H	H	L	M	M	L	L	H	H

1980). The legumes *D. triflorum*, *D. canum* and *Alysicarpus vaginalis* are widespread but contribute little to animal production under shade. The broad-leaved species *Mikania cordata* and *Asystasia* spp. are readily eaten by stock and are discussed further in sections 8.3 and 8.4.

Leguminous cover crops are traditionally established in plantations, and the initial adaptation of legumes to grazed pastures in the humid tropics arose from the testing by the Queensland Department of Primary Industries of cover crop species as pasture plants (Schofield 1941). The twining legumes *Centrosema pubescens* and *Pueraria phaseoloides* have been the most productive of these in silvopastoral terms. In Asian (Eng *et al.* 1978) and Pacific Island (Smith and Whiteman 1985) sites *P. phaseoloides* is readily grazed out, but this does not occur in Brazil and Colombia, where dominance of *P. phaseoloides* (Plate 8.7) and the retention of planted grasses is a problem (Böhnert *et al.* 1985). *Calopogonium mucunoides* is not well accepted by grazing animals (Thomas and Humphreys 1970) which may also lead to legume dominance, although the legume is eaten to some extent in the dry season and is palatable as hay. *C. caeruleum* grows well under shade but is quite inimical to animal production (Middleton and Mellor 1982; Chong *et al.* 1991) since it is poorly eaten in all seasons. *Desmodium*

Plate 8.11 Overgrazed *Axonopus compressus* in deep shade.

intortum is successful in cooler tropical highland situations (Colour Plate 13), whilst *D. ovalifolium* is successful in the humid tropics (Wong *et al.* 1985a). Shrub legumes such as *L. leucocephala* may also have a role.

Of the sown or planted grasses, *Stenotaphrum secundatum* and *Ischaemum aristatum* are mainly grown in the Pacific Islands. *S. secundatum* combines shade and high salinity tolerance with considerable resistance to grazing, since it forms a dense sod which excludes weeds more effectively than many other grasses and provides excellent protection against soil erosion. Animal production is modest unless the grass is combined with legumes (Macfarlane and Shelton 1986), and there is considerable scope for selection of improved varieties. *I. aristatum* is also a stoloniferous grass, and performs best at intermediate light levels. *Brachiaria decumbens* and the closely related *B. brizantha* have persisted in medium shade and exhibit a wide range of adaptation to acid soils of medium to high fertility and good growth into the dry season. However in a high rainfall environment such as the Solomon Islands, selective grazing of grass may be expected to operate year round; the absence of a dry season when selective grazing of the legume might balance competitive relations may lead to the loss of a preferred grass such as *B. decumbens* (Smith and Whiteman 1985) and to legume dominance. *B. humidicola* also has good performance in medium light environments, but is better adapted to low lying situations and to soils of poor internal drainage, or to alkaline soils derived from coralline sediments (Stür and Shelton 1991a); it exhibits problems of legume incompatibility but resists weed invasion well. *Panicum maximum* is widely planted through the whole tropical world on the more fertile soils (Plate 8.3). Its erect habit makes it popular in cut-and-remove systems. It is vulnerable to weed invasion and overgrazing, but if underutilized it is difficult to see coconuts if these are collected from the ground. As mentioned earlier, it is also more competitive with tree crops than *B. miliiformis* (Plate 8.5) in Sri Lanka, and special attention to the pasture fertilizer needs must be given if this undesirable effect is to be avoided (Ferdinandez 1972). *B. miliiformis* is prone to disease attack in the humid tropics (Reynolds 1988).

There are many newly selected grasses and legumes which show promise for silvopastoral situations. Previous pasture research programmes have concentrated on selection of élite material in full sunlight, and recent studies are concentrating on finding superior plant performance for shaded conditions. Amongst the legumes which show high shade tolerance *Arachis pintoi* (Colour plate 14) has wide adaptation in the humid tropics. The variety Amarillo has proven animal production in Colombia, and high voluntary intake and digestibility (Kaligis and Mamonto 1991); the creeping Amarillo is resistant to heavy grazing, compatible with sod-forming grasses such as *B. decumbens* and *B. humidicola*, adapted to acid soils and is capable of seed production in excess of 1 t ha^{-1}. Other perennial *Arachis* fodder species are potentially valuable. Recent studies in southeast Asia and Queensland suggest *Paspalum malacophyllum* and *P. wettsteinii* show promise as grasses for plantations (Wong 1993).

Nutritive value of shaded pastures

The effect of shade on nutritive value is best assessed through its influence on the intake of digestible nutrients. Although there are many laboratory studies of the influence of shade on chemical composition and the components of nutritive value there are few studies of animal response and these results are conflicting. The fertility of the soil, the plant species employed and the ways in which the pastures are managed affect the responses. The generalizations which are well supported are that usually shade:

- increases plant N concentration,
- decreases concentration of non-structural carbohydrate,
- may or may not affect digestibility, intake and cell wall content.

The positive influence of shade on N% may have real benefits to animal production where low levels of N of the order of 1% occur, and this phenomenon of increased N% is widespread (Wong and Wilson 1980; Wilson *et al.* 1986; Samarakoon *et al.* 1990a; Norton *et al.* 1991). This is discussed further in the next section. The energy status of shaded pastures, which is a strong constraint to the individual performance of animals consuming C_4 grass, is less clear and recent studies have not resolved the question.

Samarakoon *et al.* (1990a) shaded long-established pastures of *Pennisetum clandestinum* and *S. secundatum* to 50% light transmission for 5 months. Regrowth after cutting shaded and unshaded pastures was analysed and fed to sheep on two occasions. The contrasting response of the two groups (averaged over two harvests) indicates the difference between a shade species (*S. secundatum*) and a sun species (*P. clandestinum*) (Table 8.6). DM yield and leafiness of *S. secundatum* increased under shade, perhaps associated with better N nutrition and moisture relations, whilst *P. clandestinum* under shade had reduced ratio of leaf and of dead material to total shoot weight, but more stem. Small differences in digestibility were evident, but voluntary intake of *P. clandestinum* was substantially reduced under shade. This was not associated with cell wall content, although lignin% was slightly higher in the second harvest of shaded *P. clandestinum*. Total non-structural carbohydrate (TNC) was low in both species and especially low in shaded conditions. In subsequent pot experiments Samarakoon *et al.* (1990b) confirmed that shade reduced TNC% and

Table 8.6 Effects of shade on plant performance and nutritive value of two tropical grasses (from Samarakoon et al. 1990a).

Plant attribute	S. secundatum		P. clandestinum	
	Full sun	50% light	Full sun	50% light
DM yield (t ha^{-1})	8.8	10.9	8.7	8.4
Leaf (%)	45	57	54	48
DM digestibility (%)	50	53	50	46
Daily voluntary intake (g/kg LW$^{0.75}$)	45	54	49	34
Daily digestible nutrient intake (g/kg LW$^{0.75}$)	19	22	20	13
Non-structural carbohydrate (%)	0.7	0.5	2.1	0.9
Cell wall content (%)	75	71	73	70

increased N%, but reported small and inconsistent effects on leaf DM% and attributes of nutritive value.

Other studies confirm the complexity of the responses to shade. Wilson and Wong (1982) noted decreased digestibility of *P. maximum* var. *trichoglume* under shade. Norton et al. (1991) conducted a feeding experiment with five tropical grasses of varying shade tolerance and found no significant effect of shade on DM digestibility, feed intake or cell wall composition. Higher ammonia and propionic acid and lower acetic acid concentrations occurred in the rumen of sheep eating shaded pastures, and the sheep also maintained a higher N balance. In the present state of knowledge confidence can be expressed in the positive effects of shade on N nutrition, but the intake of energy from shaded pastures will be unaffected or will vary with management or with local morphogenetic effects which increase stem elongation and which might be evident under the changed R/FR light ratio in plantations; the experiments quoted above were carried out under shade cloth, which does not alter the R/FR ratio and would not produce such changes. Tissue constituents such as the ratio of digestible mesophyll to indigestible structural elements are sensitive to stage of development, weather conditions, grazing management and mineral nutrition.

Nitrogen economy of pastures in shade

The concepts of the tree as a nutrient pump, the N fixation of tree species and the influence of pastures in developing an OM pool of nutrients which trees access are significant aspects of the N economy

of tree–pasture associations and these are discussed subsequently. At this juncture further mention is made of the effects of shade on:
- N uptake and concentration in grasses,
- N fixation by forage legumes.

It has long been observed that some grasses have a higher nutrient concentration and grow better close to trees. This is usually attributed to transfer of nutrients to the drip-ring through stock concentration, leaf fall from trees and increased OM accumulation. It is now known that under low fertility conditions shade of itself produces an effect which increases the N% of grass shoots. This was observed in southeast Queensland in *P. maximum* var. *trichoglume* where Wong and Wilson (1980) noted increased N accumulation in shoots of 29–76%, according to shade level and cutting frequency. This effect also occurred under *Eucalyptus grandis* in *Paspalum notatum* (Wilson et al. 1990). Similarly in Hawaii Eriksen and Whitney (1981) recorded increased N yield due to shading in five of six tropical grasses grown under conditions of low N supply.

Reference was made in section 1.3 to the deterioration in growth rate of aged grass pastures, which appears to be associated with a changed distribution of N in the ecosystem, rather than a decrease in the total N present (Robbins et al. 1987, 1989). This deterioration is ameliorated if pastures are shaded, as illustrated by a study at Narayen, southeast Queensland (lat. 26° S, 710 mm annual rainfall) of a pasture of *P. maximum* var. *trichoglume* planted 16 years previously on a grey, brown and red clay soil with a total N content (0–10 cm) of 0.33% (Wilson et al. 1986). Both the growth and the N% of shoots had decreased relative to that of earlier years when shades of 37% light transmission were placed over the pasture.

In the first year of the study dry conditions prevailed over the summer and shade had a small positive effect on DM and N yield (Table 8.7). The second year was also dry but shade had a strong positive effect on growth and N yield, which was further enhanced by irrigation in March and April. The increased N yield mainly arose from better leaf growth, and greater leaf N% in three of five harvests. At the end of the experiment N content present in crown and roots was 205 kg ha^{-1} under shade and 215 kg ha^{-1} in full sunlight, a non-significant difference which could not account for the larger response in shoot N yield recorded in Table 8.7.

How does this effect arise? The phenomenon is a real effect on plant N uptake, since it cannot be accounted for by a diversion of N from roots to shoots. At Narayen it was not due to a difference in root N-fixing activity as assayed by acetylene reduction (Wilson et al. 1986). A subsequent experiment showed that the effect must be associated with plants growing in soil, since it did not occur in shaded plants

Table 8.7 Effects of shade on yield of DM and N in shoots of *Panicum maximum* var. *trichoglume* in southeast Queensland (from Wilson et al. 1986).

Attribute	Treatment	Year 1	Year 2	Total
Shoot yield (t DM ha^{-1})	Full sun	2.6	2.6	5.1
	Shade	2.7	6.7	9.4
Shoot N yield (kg ha^{-1})	Full sun	12	34	46
	Shade	13	85	98

growing in solution culture (Wilson and Wild 1991). It seems that under conditions of low N supply N mineralization is stimulated. Soil microorganisms may operate better at the higher levels of soil moisture operating under shaded pastures (Wilson and Wild 1991) and the more moderate soil temperatures that apply may also favour soil fauna such as earthworms, which may lead to faster rates of litter breakdown and turnover of N than occur in full sunlight. It should be emphasized that under conditions of high levels of N fertilizer application to grasses the benefits ascribed to shade in this section do not occur, nor do they apply to legumes.

The contribution of legumes to N fixation under shaded conditions is a further significant factor in the N economy of mixed pastures. Legumes in young oil palm plantations have been reported to fix 150–200 kg N ha^{-1} year^{-1} (Agamuthu et al. 1980) and there are also favourable accounts of legume effects on levels of soil N under rubber (Watson et al. 1964a). Concurrently it is suggested that shade severely reduces N fixation, so that the full potential of the legume in this respect is not realized. It is known (Othman et al. 1988; Humphreys 1991, figure 4.15) that N fixation in legumes where an effective rhizobial symbiosis exists is directly associated with the rate of delivery of surplus assimilate to the root nodules. It can therefore be expected that the shading which diverts assimilate to shoots rather than roots reduces N fixation.

Eriksen and Whitney (1982) grew six forage legumes near Paia, Hawaii (lat. 21° S, 100 m altitude) in full sunlight or under shade of 70, 45 and 27% light transmission. Annual DM yield was positively related to radiation level, but legumes differed in their production under shade. N yield (Table 8.8) was closely related to DM yield, and the rate of acetylene reduction in soil was positively correlated with radiation intensity. *D. intortum* and *L. leucocephala* produced the highest N yields under light transmission levels of 45 and 27%, whilst *C. pubescens* also showed superior shade tolerance. *S. guianensis* and *M.*

Table 8.8 Annual N yield of forage legumes grown under different levels of light transmission in Hawaii (from Eriksen and Whitney 1982).

Species	Light transmission (%)			
	100	70	45	27
Desmodium intortum cv. Greenleaf	540	528	414	245
Leucaena leucocephala cv. Hawaiian Giant	751	710	456	285
Desmodium canum	340	361	186	128
Centrosema pubescens	461	410	293	205
Macroptilium atropurpureum cv. Siratro	362	265	160	83
Stylosanthes guianensis cv. Schofield	496	321	151	88

atropurpureum exhibited the lowest N yields under shade. These data indicate that appropriate choice of legumes will assist to maintain the N economy of forages grown in plantations.

A second example (Richards and Bevege 1967) shows the differing response of tree species grown for timber when planted with forage legumes on a low fertility lateritic podzolic soil at Beerburrum, southeast Queensland (lat. 27° S, c. 1600 mm annual rainfall). The native conifers *Aracauria cunninghamii* (hoop pine) and *Agathis robusta* (kauri pine) and the exotic pines *Pinus elliottii* var. *elliottii* (slash pine), *P. taeda* (loblolly pine) and *P. caribaea* var. *hondurensis* (Caribbean pine) were planted (1) without fertilizer (C), (2) with a complete fertilizer mixture minus N (F), (3) with fertilized *Lotononis bainesii* (FL), (4) with fertilized *Desmodium uncinatum* (FD), and (5) with fertilized *Macroptilium atropurpureum* (FM). The slow growing native pines responded to fertilizer and grew better in the presence of forage legumes, whilst the growth of exotic species was slightly reduced by the association with forage legumes and did not respond to fertilizer application. The N status of the ecosystems 5 years after planting (Figure 8.5) showed that the exotic pines were more effective than the native species in accumulating N under low fertility conditions; an average of 79% of the N present above ground was in the exotic trees, 20% in the litter and only 1% in the ground vegetation, since the forage legumes were suppressed. By contrast in the native conifer plots only 10% of the total N was in the trees, 75% was in the litter and 15% in ground vegetation. *D. uncinatum* and *M. atropurpureum* grew better than

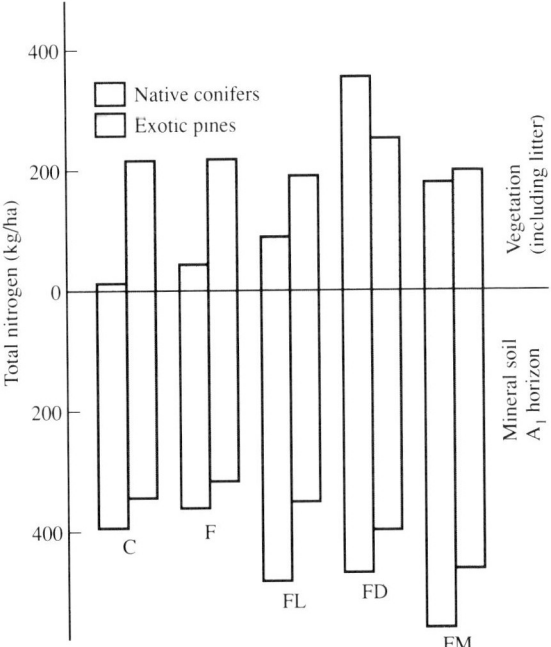

Figure 8.5 Nitrogen content of ecosystems planted to native conifers or exotic pines at Beerburrum, southeast Queensland (C control, F complete fertilizer minus N, FL fertilizer + *Lotononis bainesii*, FD fertilizer + *Desmodium uncinatum*, FM fertilizer + *Macroptilium atropurpureum*) (from Richards and Bevege 1967).

L. bainesii, and this was reflected in the total soil N present in the A_1 horizon (0–10 cm) (Figure 8.5). N content of the dense, mottled subsoil was independent of treatment. N deficiency is clearly a limiting factor to the growth of the native pines *A. cunninghamii* and *A. robusta*, and the planting of adapted forage legumes during the early years of tree establishment can ameliorate this condition.

Interference between trees and pastures

Effects of pastures on tree crops

Scientists often write about competition between plants, but 'interference' has become a more fashionable term to describe the effects of plants upon each other, since it embraces ideas which are not necessarily involved in competition for the resources which control growth. Pastures which are overtopped by trees have no opportunity to compete with the trees for light, but the trees interfere with the light environment in which the pasture grows. The presence of leguminous pasture may augment the supply of N to the companion trees, or may alter the habitat to favour the occurrence of pests or diseases which attack tree crops. Much of the interest of the agriculturalist in tree–crop relations centres on the ways in which plants interfere with the availability of nutrients and of water to each other.

Nutrient relations of trees and crops

There is a spatial concentration of nutrients close to the tree, and a gradient to lower soil fertility in the inter-tree area. This lateral movement of nutrients is well established and may arise from different causes. The resting of stock under the shade of trees brings about a concentration of excreta (see, for example, Sugimoto *et al.* 1987, illustrated in Humphreys, 1991, figure 3.12); this effect is pronounced in grazed pastures with sparse tree cover. It is wasteful of nutrients if the stock concentration under the tree is sufficient to disturb the soil and keep the ground surface bare, whilst the leaching and volatilization losses are great if there is luxury deposition of nutrients, as mentioned in sections 2.4 and 4.4. The concentration of excreta from birds and other animals inhabiting trees is a further factor. Kellman (1979) measured spatial concentration of nutrients under trees in Belize, Central America and attributed this to capture by rainfall.

In most circumstances the processes of greatest influence arise from the growth of tree roots, the active absorption of minerals from the deeper layers of the soil below the tree and laterally from the inter-tree area, the movement of nutrients to leaves and branches, the senescence and detachment of these to the ground surface and the incorporation of their nutrients as surface litter (Young 1987). This is illustrated (Belsky *et al.* 1989) by a study in Tsavo National Park (West), southeast Kenya (lat. 3° S, 400–500 mm annual rainfall, 1000 m altitude) of tropical savanna. Transects extending from the base of *Acacia tortilis* trees, which form a flat-topped canopy, and of *Adonsonia digitata* (baobab) trees were sampled, and the soil analyses of these rhodic ferrasols (Figure 8.6) indicate higher levels of OM (0–10 cm), soil microbial biomass and mineralizable N within 7.5 m of the tree trunks; the data for both species were similar and have been pooled, except for Ca. There was a similar steep decrease in P and K beyond the drip-ring. Soil bulk density (0–20 cm) was greater in the grassland inter-tree area than below the canopy, whilst moisture infiltration was greater in the latter.

A second illustration of the tree as a nutrient pump (Kamara and Haque 1992) arises from measurements under and away from *Faidherbia albida* (syn. *Acacia albida*) trees at Debre Zeit, Ethiopia

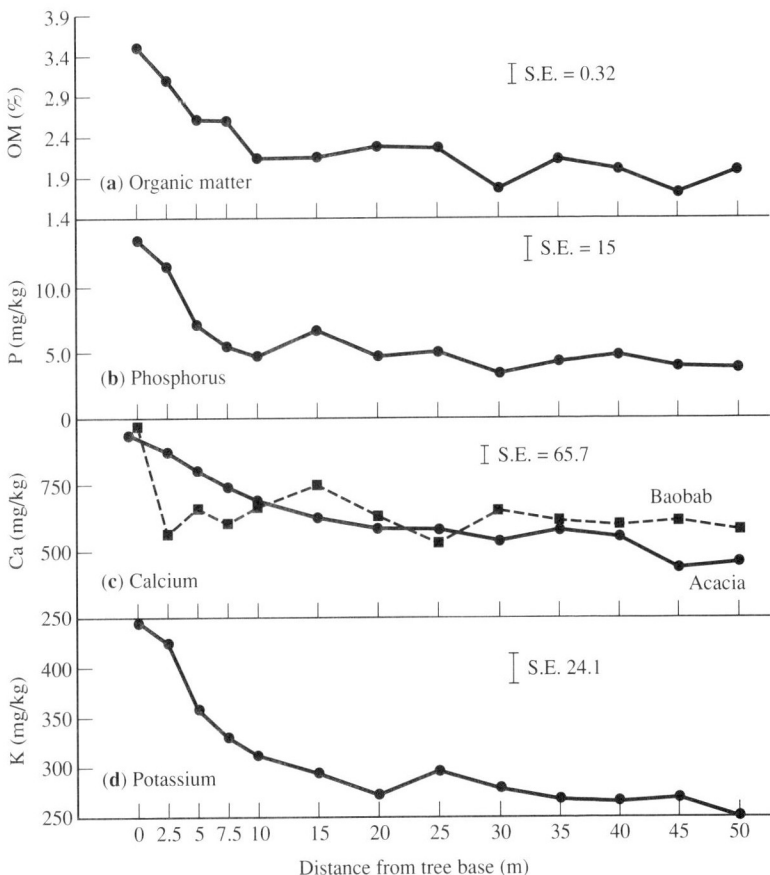

Figure 8.6 Mean concentrations of (a) OM, (b) P, (c) Ca, (d) K as influenced by distance from base of *Acacia tortilis* and *Adansonia digitata* (baobab) (shown separately for Ca) in Kenya (from Belsky *et al.* 1989).

(lat. 9° N, 850 mm annual rainfall, 1850 m altitude), growing on a poorly drained chromic vertisol. Tree density was $c.$ 6.5 ha^{-1}, and trees ranged in height from 3 to 6 m with canopy 26–35 m^2. The soil data (Figure 8.7) when smoothed to give nutrient profiles show profound spatial effects on OM, P, and especially N and K, which operate not only in the surface 0–2 cm layer but at 10–20 cm depth, albeit with lesser amplitude. Litter was concentrated on the west side of the tree, due to the prevalent wind direction. Available water capacity under the tree, measured as moisture storage between field capacity and wilting

198 Tree crops with pastures

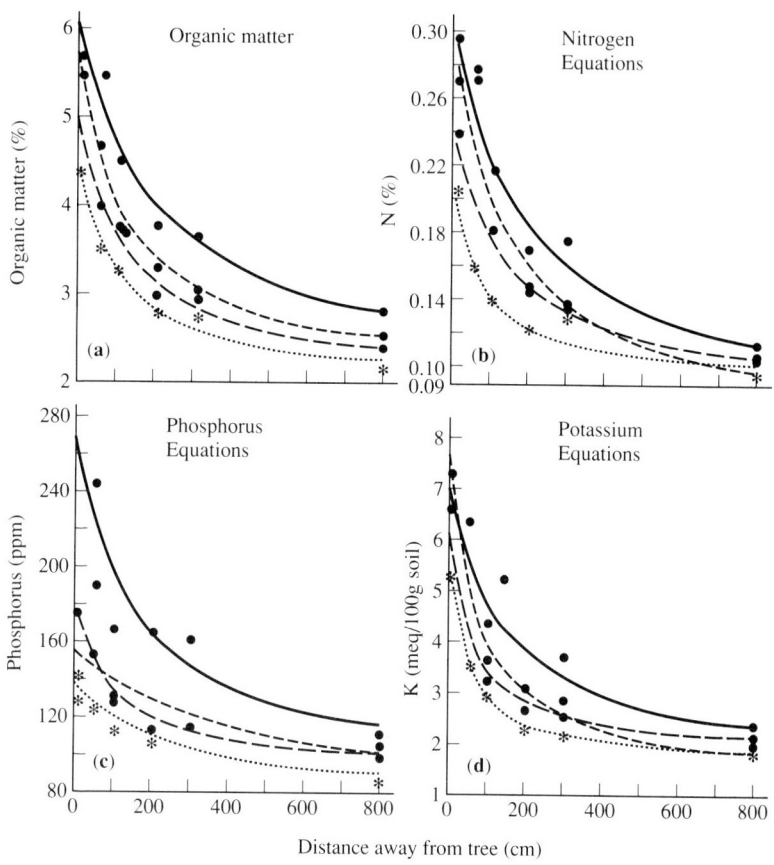

Figure 8.7 Mean concentrations of OM, P, N and K at distances from the trunk of *Faidherbia albida* (syn. *Acacia albida*) at depths of 0–2 (——) 2–5 (- -), 5–10 (— —) and 10–20 cm (---) in Ethopia (from Kamara and Haque 1992).

point, was 1.5–2 times that of sites outside the tree canopy. These measurements help to explain better plant growth which occurs near the trees relative to growth in the inter-tree area (Poschen 1986).

The degree of access of tree roots to nutrients is a controlling factor in tree–pasture relations. Deep-rooted trees with a vertical root distribution concentrated below the trunk offer less competition for nutrients to companion forage species, and may deliver nutrients to the surface soil from the deeper soil layers which would otherwise be inaccessible to herbage plants. By contrast surface rooting trees restrict the availability of nutrients and moisture to companion pastures, and many *Eucalyptus* spp. have this reputation. Alley farming systems have a particular requirement that the shrub N tree species adopted has restricted lateral root development or readily pruned lateral roots which do not affect the nutrients and moisture available to plants growing in the inter-hedgerow area. There is a concomitant advantage in choosing woody perennials with limited root growth in the surface layers within the row. Large woody roots near the surface also add to the difficulty of cultivation.

The study of root development of trees is demanding and there are few pertinent investigations. One such project at Onne, southeast Nigeria (lat. 5° N, 2400 mm annual rainfall) recorded the distribution of fine and coarse roots, the root length and mass in soil layers to 1.2 m depth and 2 m laterally from the centre of the woody perennial (Ruhigwa *et al.* 1992). The soil was a Typic Paleudult and four species of potential utility in agroforestry were grown: *Acioa barteri*, a rosaceous glabrous shrub with slender branches, *Alchornea cordifolia*, a tall shrubby or scrambling member of the Euphorbiaceae, *Gmelina arborea*, a fast growing deciduous tree, and the leguminous *Cassia siamea*. At this site these trees produce similar above-ground yields of c. 13 t DM ha^{-1} year^{-1}. However, their below-ground root growth is radically different. The density of fine root tips (Figure 8.8) has been selected as a good indicator of root activity associated with nutrient and moisture uptake. For this attribute all four species are deep rooted, but *A. cordifolia*, *G. arborea* and *C. simea* have most of their fine roots in the top 20 cm of the soil profile and extending from the base of the tree out to the 2 m lateral width measured; this zone accounted for 73, 74 and 76% respectively of the fine roots counted. By contrast *A. barteri* had only 49% of fine roots in that zone, and Figure 8.8 shows a sharp decrease in root tip density 120 cm from the tree base. Whilst the root density decreased with depth in all species, it is evident that *A. barteri* has many more feeder roots in the 20–80 cm depth than the other plants. These differences were also reflected in the other root attributes measured (Ruhigwa *et al.* 1992) suggesting that *A. barteri* exhibits criteria of merit for use in agroforestry. This was reflected in companion crop yields which favoured *A. barteri*. In the long term a leguminous tree providing N accretion to the system would be preferable to a member of the Rosaceae, if a species with suitable rooting characteristics might be identified. Regrettably *Gliricidia sepium*, *Leucaena leucocephala* and *Flemingia congesta* have high concentrations of roots in the surface soil.

Superior access of tree roots to nutrients increases tree growth. Watson *et al.* (1964b) noted that the above-surface root development of rubber trees was stimulated by leguminous covers but not by grass covers or by *Mikania cordata*, which also reduced root growth of rubber in the 0–7.5 cm layer. The nutrient bank associated with the

OM accumulation under leguminous forages promotes the subsequent growth of rubber as its roots tap into this augmented supply (Broughton 1977), and similar effects are observed in oil palm (Agamuthu *et al.* 1980). Root location varies with crop species, management and age. Netsinghe (1966) concluded from a radioisotope study in Sri Lanka

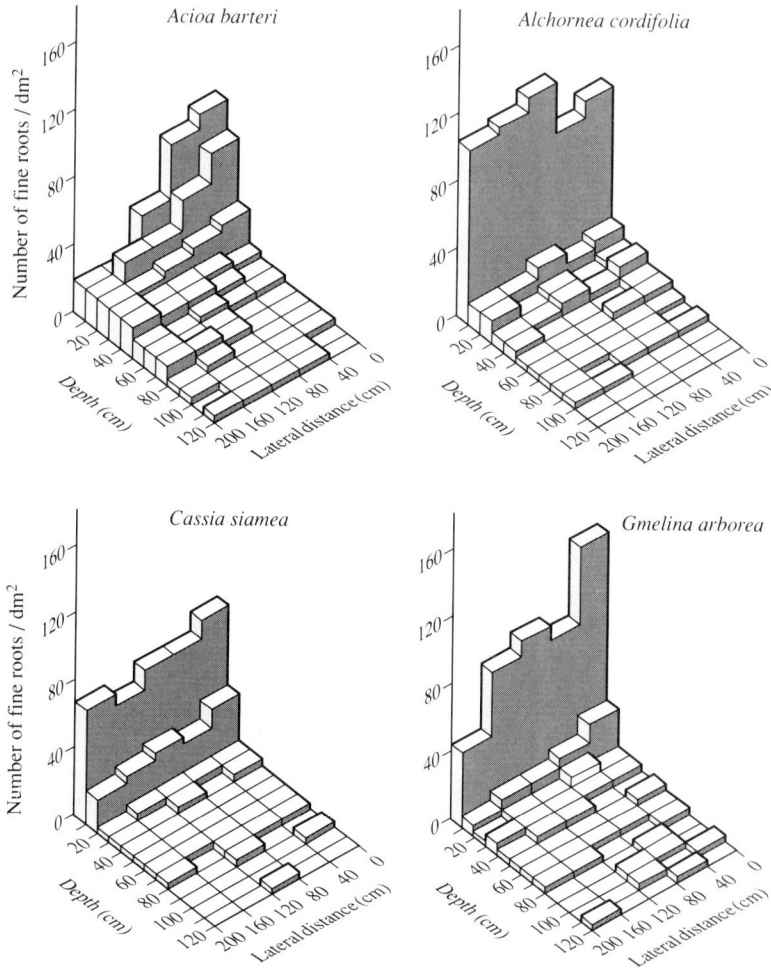

Figure 8.8 Distribution of density of fine root tips (<2 mm diameter) of four species according to soil depth and distance from trunk in Nigeria (from Ruhigwa *et al.* 1992).

of fertilizer placement that the absorptive surfaces of coconuts were concentrated in a radius of only 1.65 m from the bole of the palm. By contrast Steel and Humphreys (1974) found that the roots of mature coconut palms 30 years old and spaced 10 cm apart in a square planting at Kuta, Bali occupied the inter-tree space. A trench dug from 0.7 to 3 m from a tree showed that palm roots were still dense 3 m from the trunk. Soil moisture content was independent of spatial location, and pastures which were blocked according to distance from tree showed no yield variation according to block position. In these palms noon light transmission was 77–80%, indicating a favourable light environment for pasture growth.

Reference was made earlier to the greater competitive capacity of *P. maximum* relative to that of *B. brizantha* or *B. miliiformis* in affecting coconut yield in Sri Lanka (Ferdinandez 1972); the yield of nuts with the latter two species showed a positive trend after 10 years and was twice that of the yield with *P. maximum*. It might be expected that this is associated with the greater yield and nutrient demand of *P. maximum*. However in another Sri Lankan study there was no simple relationship between yield and competitive capacity. Waidyanatha *et al.* (1984) compared the effects of different forages planted with rubber at Paiyagala (7° N, 3100 mm annual rainfall) and used in a cut-and-remove system. In control plots planted to *Pueraria phaseoloides* alone the girth increment of rubber trees 6 years after planting averaged 47 cm, but girth was slightly reduced (Table 8.9) if grasses were planted, or if *Centrosema pubescens* was planted instead of *P. phaseoloides*; the latter yielded more than *C. pubescens*, although both legumes performed poorly when planted with grasses and subjected to regular cutting. *P. phaseoloides* contributed 11% total yield and *C. pubescens* 5%. In this study *B. brizantha* was more competitive with rubber than *B. miliiformis*, with *P. maximum* in an intermediate position. The effect of pasture on rubber growth was only significant in the first 3 years

Table 8.9 Relative girth measurements of rubber in Sri Lanka grown with different forages (*Pueraria phaseoloides* alone = 100) (From Waidyanatha *et al.* 1984).

Grass	Legume		
	None	*Centrosema pubescens*	*Pueraria phaseoloides*
Panicum maximum	84	85	89
Brachiaria brizantha	79	77	81
Brachiaria miliiformis	93	94	98

after planting, and thereafter pasture growth was much reduced as the rubber trees advanced. The cut-and-remove system employed and the low level of N fixation from poorly grown legumes in a mixed sward would have accentuated competition from underplanted grass. It is difficult to find any critical studies which define more clearly the nutrient relations in these associations.

Water relations

Competition for moisture has little impact on plant relations in humid environments of high and well distributed rainfall, as occurs for oil palm plantations, which are often situated in low-lying sites, and for rubber plantations in equatorial regions. Coconut plantations are more usually located on sandy soils of inferior water-holding capacity and in marginal rainfall environments where coconut yield is sensitive to water deficits. Smith (1966) demonstrated a linear negative relationship between yield of copra and the integrated soil water deficit in the 2.5 year period before harvest. Tree performance has been shown to benefit from clean weeding in marginal rainfall environments. It should be recognized that bare fallowing is not a practical alternative in most farm situations, because of the labour and/or fuel costs involved, the hazards of soil erosion and the population pressure on land. The realistic comparison is between the effects of volunteer pasture and planted forages, and these differences are insignificant in so far as water relations are concerned, although other considerations are cogent.

In tropical savanna the management of tree density is a key factor in determining the moisture available for pasture growth (Burrows and Frost 1993). However there is a marked difference in the effects of trees according to the seasonal distribution of rainfall. In the eucalypt savannas of northern Australia trees and herbage limit each other's water supply only when both are actively transpiring. In a monsoon climate competition is restricted to the wet season, and this is less significant if the wet season is reliably wet and dry periods are of short duration during the monsoon. In the subtropics the occurrence of frequent dry spells accentuates competition for water and the removal of trees increases pasture growth from two to seven times that of untreated controls (Burrows et al. 1990). This is illustrated by a model developed and validated from different sources (Scanlan and McKeon 1993) which shows how the negative relationship between tree density and pasture growth is moderated by the depth of soil at the site, which determines its water-holding capacity. When deep soil layers (> 1 m) supply a large proportion of tree transpiration pasture production is less reduced than in shallow soils. At Gayndah, southeast Queensland (lat. 26° S, 730 mm annual rainfall) pasture growth is very low if tree basal area exceeds 15 m^2 ha^{-1} (Figure 8.9), but the degree of reduction in growth over

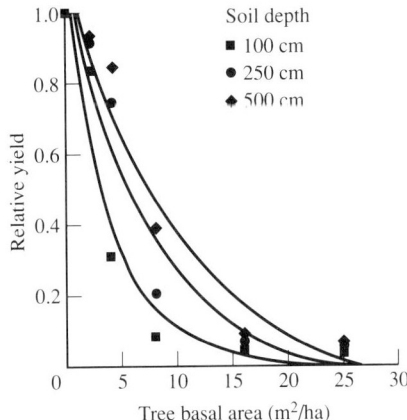

Figure 8.9 Relative yield of pasture (no trees = 1) as influenced by tree basal area and depth of soil at Gayndah, southeast Queensland (from Scanlan and McKeon 1993).

the whole range of tree density is highly sensitive to the shallowness of the soil.

Biotic factors

The influence of the grazing animal on pasture production and on tree crop performance is discussed in section 8.3. Reference is now made to the altered habitat for pests and disease provided by forage ground cover, especially if underutilized. This effect was mentioned in relation to intercropping in sections 5.3 and 5.4. The possibility of common hosts between tree crop and pasture should not be overlooked.

When animals are concentrated in feeding areas in cut-and-remove systems the accumulation of manure has adverse environmental effects, as mentioned in Chapter 4; it may also exacerbate pest problems. The presence of a manure heap favours the breeding of the rhinoceros beetle, *Oryctes rhinoceros*, which is one of the worst pests of coconut palm (Payne 1985).

Allelopathy

Allelopathy is a distinct aspect of plant interrelations. It is the condition when plant roots secrete a toxin, or toxic materials are leached from above-ground organs which inhibit germination and/or growth of other plants. It is one mechanism invoked to explain the spatial isolation of plants or the negative effect of one plant species on the growth performance of another. It is relatively easy to demonstrate in the laboratory the inhibiting effects of some plant extracts on seed germination and growth, but it is less easy to separate the allelopathic effects from the universal application of the effects previously discussed of plant interference with the availability of environmental growth factors between neighbouring plants. This is a controversial question where many non-believers amongst scientists are dismissive of the associative evidence between growth and the occurrence of allelopathic substances, since explanations other than this may have better claims to causation.

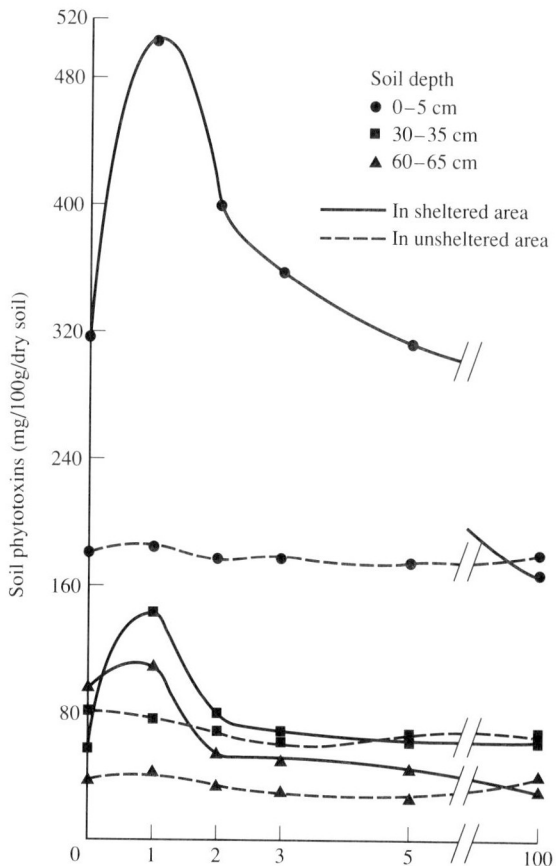

Figure 8.10 Amounts of soil phytotoxins at varying soil depths and distances from a *Eucalyptus tereticornis* shelterbelt or in an unsheltered area at Chandigarh, India (from Singh and Kohli 1992).

One of the early reports of allelopathy in plantation agriculture implicated the weed *Mikania cordata* (Wong 1964). Substances in this plant, which included phenolic and flavenoid constituents, depressed the growth of rubber, tomato and *P. phaseoloides* and inhibited soil nitrification. Many *Eucalyptus* spp. are highly efficient in carbon fixation under conditions of low nutrient supply, and repeated coppicing causes rapid depletion of soil nutrients. This has led to popular belief in 'eucalypt sickness' as a soil condition induced by eucalypts. Certainly surface rooting eucalypts are not ideal species for agroforestry, and Figure 8.9 illustrates their effects on pasture growth in subtropical savannas. There is a concomitant recognition of the occurrence of allelopathic substances in soils located near these trees.

This is illustrated by a study (Singh and Kohli 1992) in Chandigarh, India (lat. 30° N, 280 m altitude) where soils were collected on a transect away from a shelter belt of *E. tereticornis* and from another unsheltered area 100 m distant. Phytotoxins, mainly phenolics and terpenes, were extracted using EDTA and found to inhibit the germination of seeds of *Lens esculentum*. The concentrations of soil phytotoxins (Figure 8.10) were especially high in the 0–5 cm layer within 5 m of the tree line, and at lower depths some increases were observed 1 m from the tree line, whilst values were low in the unsheltered area free of the influence of *E. tereticornis*. It was also noted that arithmetic reductions in the yield of crops of *Cicer arietinum*, *Lens esculentum*, *Triticum aestivum*, *Brassica oleracea*, *B. campestris* and *Trifolium alexandrinum* occurred in a 12 m strip adjacent to the *E. tereticornis* shelter belt.

Many toxic substances are readily deactivated in specific soil conditions or are leached to deeper layers. The field investigation of these phenomena is fraught with many difficulties; the biological importance of allelopathy is open to question and requires critical investigation.

8.3 Management of tree-pasture associations

Pasture management

The principal objectives of pasture management (Humphreys 1991) were summarized in section 1.1, and are recapitulated with a minor modification for the tree–pasture association. The central goal is to effect a synchrony between the forage requirement of the animal and the pasture available. An SR or a cut-and-remove system is chosen which sustains long-term animal production through:

- adequate forage allowance (the forage available per head),
- plant growth and persistence,

- maintenance of cover as protection against erosion,
- desirable botanical composition, especially in relation to legume content and minimization of weeds,

and which promotes yield of tree crop.

Continuity of forage allowance minimizes animal stress and maintains production, and is achieved through:

- seasonal and long-term variation in SR, especially according to the ways forage availability changes with age of plantation;
- providing a sequence of feeds of differing seasonal utility, which may necessitate the planting of shrub legumes;
- modifying the environment in which pastures grow through silvicultural, fertilizer or irrigation practice.

Stocking rate

The choice of SR, or the number and class of animals supported by a unit area of pasture, is the key decision which determines the output of animal product, the longevity of the pasture, the extent of erosion and the profitability or otherwise of the enterprise. Forage intake is not limited on pastures carrying few animals, and the individual animal performance reflects the quality of the pasture on offer. Sometimes underutilized pastures become stemmy or botanical composition changes to disadvantage individual animal production, but usually the highest LWG per head occurs at low SR where the animal has maximum opportunity to improve its diet by selective grazing. High animal performance from C_4 grasses requires that animals eat young green leaf rather than old stem.

Animals compete with each other for the available feed supply as SR increases, and this reduces individual LWG. However the LWG per unit area continues to increase until competition becomes acute and the individual animal suffers a deficiency in both quality and quantity of forage, since more animals are carried on the pasture.

This is illustrated (Chong *et al.* 1991) for sheep grazing pastures under immature (3-year-old) rubber at Sungei Buloh, central Malaysia (2190 mm annual rainfall). Male crossbred lambs (Dorset × Malin) were weaned at 12–14 kg LW and continuously grazed for 6 months at SR of 4–14 ha^{-1}, when they were replaced with a further draft of weaned lambs. Lambs had access to pasture for only 6 h each day, but forage was supplemented with palm kernel cake at 100 g hd^{-1} day^{-1}. Production per head, averaged for two drafts of lambs, was reduced from 106 to 84 g day^{-1} as SR increased from 4 to 14 ha^{-1} (Table 8.10). On the other hand animal production increased from 155 to

Table 8.10 Liveweight gain of lambs grazing pastures under immature rubber at Sungei Buloh, central Malaysia, as influenced by density and age of trees (after Chong et al. 1991).

Stocking rate (lambs ha^{-1})	Liveweight gain	
	Per animal (g hd^{-1} day^{-1})	Per unit area (kg ha^{-1} year^{-1})
4	106	155
6	97	212
8	86	251
10	83	303
14	84	429

429 kg LWG ha^{-1} year^{-1} over the same SRs, a radical difference in total output.

Over some decades of experiments with SR it has been well demonstrated that a linear function relating LWG per head (Y) to SR (X) is a robust one, and $Y = a - bX$ where a is the intercept on the Y axis at zero SR and b is the slope (tan θ) of the regression line. The production per unit area (Z) is the product of individual LWG and the number of animals carried, so

$$Z = XY$$
$$= X(a - bX)$$
$$= aX - bX^2$$

This quadratic function provides that the SR giving maximum LWG per hectare = $a/2b$.

Shelton (1991) has taken the data of Watson and Whiteman (1981) to illustrate this relationship (Figure 8.11). Cattle were continuously stocked under coconuts of 60% light transmission in the Russell Islands group of the Solomon Islands (lat. 9° S, 2900 mm annual rainfall) on pastures with a high legume content. SRs were 1.5, 2.5 and 3.5 ha^{-1}. Figure 8.11 shows there was a linear decrease in LWG per head of 26 kg hd^{-1} year^{-1} for each SR increase of 1 b ha^{-1}, and $Y = 192 - 26X$. This indicated an SR giving maximum LWG per hectare of $192/(2 \times 26) = 3.7$ ha^{-1} and maximum output of 355 kg LWG ha^{-1} at this SR.

It is important to recognize that in field practice this does not represent the 'optimum' SR; the farmer will usually choose an SR less than the value which gives maximum LWG per hectare. This decision will be influenced by the uncertainty of the climate, since variable

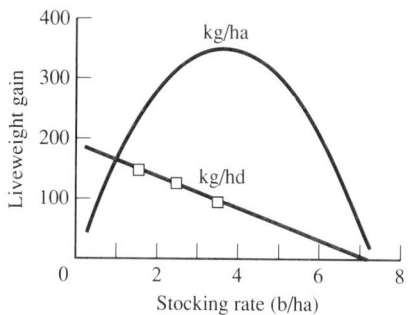

Figure 8.11 Liveweight gain of cattle grazing at different stocking rates under coconuts at Malaita, Solomon Islands (from Shelton 1991).

elements which change the year-to-year rate of forage growth will cause farmers to choose a conservative SR that will ensure animals are well fed in all years. The chosen SR must be one which sustains long-term production and which does not cause weed invasion or soil erosion. Maximizing output per unit area implies accepting half the potential rate of LWG per head, and this may not be the best option if animals and their care are expensive, and if there is a market premium for high rates of LWG and a preference for young animals.

Forage allowance

Units of SR need to take account of the size of the animal and its appetite requirement. The forage allowance, which is preferable to the inverse term 'grazing pressure', is expressed as the amount of forage offered per unit of LW. A more sophisticated measure of LW is the metabolic size of the animal, expressed as $LW^{0.75}$. Many agricultural advisers use the simple arithmetic measure of LW, relating this to the potential feed intake of the animal, which is usually between 2.5 and 3% of LW per day as DM. If lambs average 21 kg LW hd^{-1} and 2 lambs ha^{-1} are being grazed under mature rubber, the feed intake requirement is $21 \times 2 \times 0.025 = 1.05$ kg DM ha^{-1} day^{-1}.

Animals need to be presented with more feed than actually grows to provide opportunity for selection. If the target is 50% pasture utilization, then in the present example forage allowance needs to be $1.05 \times 2 = 2.1$ kg DM ha^{-1} day^{-1} or 770 kg ha^{-1} $year^{-1}$. Under mature rubber trees an SR for weaned lambs of 2 ha^{-1} might show a reasonable synchrony with forage availability (Chong *et al.* 1991). This calculation needs to take account of the level of supplementary feed

Plate 1 Gully erosion near Denbezer, Ethiopia

Plate 2 Contour bank on cropping land in the sandveld of Zimbabwe

Plate 3 Steep slopes in the Thai highlands used predominantly for maize-opium swidden

Plate 4 Population increase in the Thai highlands puts pressure on natural resources

Plate 5 Swamp lands used for grazing and coconuts growing on the upland, south Thailand

Plate 6 Low lands growing pasture in tea production lands, Sri Lanka

Plate 7 Friesian cows for milk production in Ethiopia

Plate 8 Small holder dairy production on pastures and crops in Ethiopia

Plate 9 Three strata system: coconuts, fruit trees and pastures in south Thailand

Plate 10 *Calliandra calothyrsus*

Plate 11 *Albizia chinensis* with sheep

Plate 12 Flowering *Sesbania sesban*

Plate 13 Contour planted young coffee with *Desmodium intortum* cover in north Thailand

Plate 14 Shade tolerant, grazing resistant *Arachis pintoii* in Colombia

Plate 15 Alley cropping maize with *Leucaena leucocephala* in Zimbabwe

Plate 16 *Cajanus cajan* supplementing cattle in Campo Grande, Brazil

(if any) being provided. Sánchez (1991) regarded supplementation at 0.6% of body weight daily as adequate to promote efficient utilization for sheep under rubber in Indonesia. On large estates where regular replanting occurs it is feasible to provide continuity of forage through utilizing different areas of the estate, according to the age of planting, at varying SR. Smallholder agriculture may require cooperative grazing schemes to ensure the same maintenance of SR.

Plant growth and persistence

In section 8.2 pasture growth was analysed in terms of the product of light intercepted by the sward and the efficiency of its conversion to carbohydrate. It follows that defoliation practice should be directed towards maintaining a leaf surface which minimizes light striking bare ground. The rate of leaf removal by the grazing animal and the rapidity with which the leaf surface is restored during and after defoliation are clearly key factors determining the rate of pasture growth. In grazed pastures it is the SR which determines the frequency and the intensity of plant removal, and these in turn influence not only the amount of leaf present but the level of non-structural carbohydrate which is accumulating in the stem bases and roots. In shaded pastures the latter process influences not only plant persistence but also the rapidity with which the leaf surface expands after defoliation (Wong 1993).

In pastures managed under a cut-and-remove system the interval between defoliations is crucial in determining light interception, and Figure 8.4 contrasted the superiority of an 8-week cut over a 4-week cutting interval in light interception by *P. maximum* and by *C. pubescens* swards. In different conditions Wong (1993) noted 90% light interception in 4-weekly cut swards and 55% in 2-weekly cut swards. The DM yield advantage of long cutting intervals needs to be set against the decreasing nutritive value with age since cutting, especially in C_4 grasses. A further factor is the height of cut which influences:

- the proportion of growth which is utilized,
- the residual leaf surface present after cutting,
- the residual buds present below cutting height.

The first factor operates in favour of low cutting height and the last two factors operate in the reverse direction. It has been suggested that sod-forming grasses would have an advantage over erect grasses in maintaining a leaf surface which would maximize light interception (Wilson and Ludlow 1991; Chen 1993). However this generalization has its exceptions, since Wong (1993) found that the erect *P.*

malacophyllum in shade showed superior regrowth and plant persistence under frequent and low cutting management to that of the prostrate *P. wettsteinii*. An overriding consideration was the greater allocation of assimilate in shaded pastures to shoots which was exhibited by *P. malacophyllum* and which promoted more rapid tillering and restoration of the leaf surface than by *P. wettsteinii*, whose carbon economy required the maintenance of a more extensive root system.

Cover and botanical composition

The significance of cover in ameliorating soil erosion on sloping lands has been discussed in Chapter 3, and the level of pasture utilization will modify the degree of bare ground (see Plate 8.11). Botanical composition is also responsive to grazing management. These changes may not affect animal production materially if a good content of pasture legumes is present. Watson and Whiteman (1981) observed a rapid loss of the planted grasses *B. decumbens*, *B. mutica* and *B. humidicola* under grazing in a coconut plantation in the high rainfall Solomon Islands. These were replaced by *Axonopus compressus* and *Mimosa pudica*. Animal production from the pastures was virtually identical with that from naturalized pastures containing *C. pubescens*, *P. phaseoloides* and *C. mucunoides*.

On the other hand a primary purpose of grazing may be the control of edible weeds. This has been a powerful stimulus to sheep production in rubber and oil palm plantations in Malaysia. Cost savings of up to 38% under mature rubber through the introduction of sheep have occurred (Chong *et al*. 1991). Typical responses to selective grazing show a decrease in the palatable *P. conjugatum*, *Borreria latifolia*, *Mikania micrantha* and *Asystasia intrusa* under immature rubber, and in mature rubber decreased content of *Mikania cordata* which may be replaced by *Cyrtococcum oxyphyllum* and *Ottochloa nodosum*. Grazing early in the life of the plantation reduces subsequent weed build-up.

In coconut plantations the presence of weeds is often associated with overgrazing (Reynolds 1988), and the choice of suitable SR and the planting of grasses adapted to shade exclude many common weeds such as *Chromolaena odorata* and *Imperata cylindrica*. In the Solomon Islands (Litscher and Whiteman 1982) the ferns *Sphaerostephanos unitus* and *Nephrolepis hirsutula* together with *Sida acuta* and *S. rhombifolia* become major components of the vegetation under poor grazing management. Woody species such as *Lantana camara*, *Melastoma malabathricum*, *Clidemia hirta* and *Psidium guajava* require special treatment once they have invaded the pasture.

Tree crop yields

Effective grazing management in relation to tree protection requires attention to:

- SR,
- commencement of grazing,
- choice of animal species,
- method of stocking.

The availability of herbage influences the damage occasioned to young trees whose leaves and branches are within reach of the grazing animal. Thus Chen (1991) quoted a case where 4 years after planting oil palm cattle damaged 58% fronds if grazed at 3 b ha^{-1}, but only 22% fronds at 1 b ha^{-1}. Nevertheless, yield of fresh fruit bunches was affected only when frond damage exceeded 57%, and livestock damage to crop production is often overstated in popular culture. On the other hand the detrimental effects of treading on soil structure and infiltration need to be taken into account, especially on low lying sites.

Nutrient cycling under grazing is favoured under high SR; the litter and root components of the vegetation are reduced and the C:N balance increased. In West Bali Rika *et al.* (1981) observed that coconut yield from areas planted to improved pastures was 46% higher if cattle were grazed at 4.8–6.3 yearlings ha^{-1} than if grazed at 2.7–3.6 yearlings ha^{-1}. Cut-and-remove systems of use cause nutrient depletion unless excreta are returned or fertilizer levels increased.

Some agriculturalists recommend exclusion of stock until woody perennials are above grazing height; Ismail (1986) suggests rubber plants should be > 2 m high and 1.5 years old before introduction of sheep. Much depends upon whether herbage supply is adequate. At Pakia, Thailand (lat. 19° N, 1400 mm rainfall, 1500 m altitude), the introduction of cattle either 1 or 2 years after planting *Eucalyptus camaldulensis*, *E. grandis* or *Pinus kesiya* had no significant effect on tree height or survival (Andrews and Kwaengsopha 1982). Damage to bark of established trees may be significant, as reported for cattle in Western Samoa (Pottier 1984) for *Swietenia macrophylla*; in this case damage might be obviated by removing cattle 1 month before the onset of the wet season. In the Solomon Islands growth of *E. deglupta* was reduced by cattle eating bark (Macfarlane and Whiteman 1983; Stür and Shelton 1991b). Natural repellents such as fluffed-up sheep wool or kapok tree fibres reduce the browsing of goats on *L. leucocephala* at Machakos, Kenya (Von Carlowitz and Wolf 1991).

Goats are more difficult to manage in plantation agriculture than are other ruminants. In rubber plantations goats displace latex collection cups whereas sheep cause little difficulty. The browsing preferences, climbing habits and bark consumption of goats restrict their utility as grazers; they are grazed under coconuts (Stür and Shelton 1991b) but often fed in cut-and-remove systems in conjunction with other tree crops.

Damage to trees may be contained by appropriate tethering. Animals linked to an anchored wire in the inter-tree area can graze extensive areas without approaching trees. Solar-powered electric fences have been introduced to control sheep in rubber and oil palm plantations in Malaysia. Rotation of grazing may meet farmer objectives in plantations better than continuous grazing, although this may represent a cost to animal production. A survey of 60 grazing experiments in the tropics and subtropics (Humphreys 1991) showed that rotational grazing gave inferior animal performance to continuous grazing in 51% of experiments, superior performance in 17% of experiments, and had no comparative effect in 32% of cases. Nevertheless where coconuts are harvested from the ground, a system of rotating cattle about the estate and harvesting nuts (and brushing weeds) as the cattle are moved to another area is convenient, since the nuts are more visible. Rotational grazing restricts opportunity for selection, and this favours the consumption of less acceptable weeds, if this is accorded priority by the farmer.

Fertilizer practice

In many plantations the trees are clean weeded (either mechanically or with herbicides) within a short radius of the trunk and fertilizer applied within that circle. The mineral requirements of the pasture and the effect of fertilizer application to pasture on tree crop yields should not be overlooked. In Malaysia yields of latex were independent of forage type (grasses alone, natural vegetation, or leguminous cover crop) if high N status were maintained (cumulative latex yield 20.6 t ha^{-1} over 14 years), but not under conditions of low N supply where grass-alone covers reduced latex yield by 19% (Stür and Shelton 1991b). The responsiveness of N fixation of legumes to mineral nutrition was emphasized in section 2.5 and is also discussed in section 10.4. Reduced growth rates in shade carry a concomitant reduction in mineral requirement, so that lower fertilizer levels are indicated than are recommended for pastures grown in full sunlight. Grasses also vary in their efficiency of N utilization at low levels of N supply; *B. decumbens* is superior to *P. maximum* in this respect (Sophanodora 1989; Table 8.3). In Hawaii Eriksen and Whitney (1981) contrasted the performance of six grasses under varying levels of radiation and N supply. *B. miliiformis* showed the best performance under conditions of shade and low N; *P. maximum* and *B. brizantha* were less disadvantaged by shade if N supply was well maintained.

Management of tree crops

Circumstances have been mentioned where complementarity of resource use by tree crops and pastures is mutually beneficial, but the more common condition is one where trees and pastures are competitive, as illustrated in Figure 8.9. Management is then dictated by the relative priorities accorded to the tree crop or to the output from the pasture. When the market price of copra falls below $US200 t^{-1} (1993 prices) the landholder has a strong incentive to increase animal production from pasture under coconuts, since the return from beef may be positive whilst returns from copra do not meet costs of production. The density of palm planting can be adjusted to favour pasture growth. Conversely when fuel wood or timber for other purposes is in keen demand the planting density or thinning policy of forests can be directed to maximizing production from the tree crop.

Planting density of the tree crop

Effects of planting density will be discussed subsequently for particular crops. The general case is illustrated from a study (Cameron *et al.* 1989) at Samford, southeast Queensland (lat. 27° S, 1100 mm annual rainfall) where the fast growing *Eucalyptus grandis* was planted in a wheel arrangement (Nelder 1962) which facilitates the study of competition. Planting density of trees varied from 42 to 3580 ha^{-1}. The pasture was predominantly *Setaria sphacelata* var. *sericea* cv. Nandi, with some *Paspalum notatum*, *Digitaria didactyla* and *Axonopus affinis*, and was fertilized with 138 kg N ha^{-1}, 41 kg P ha^{-1}, and 55 kg K ha^{-1} in the first 15 months after planting, and was slashed once or twice each year. Pasture growth during the main growing season from November to May averaged *c.* 20 kg DM ha^{-1} day^{-1} at low tree densities. The accumulated yield at the end of each growing season (Table 8.11) was little affected by the density of young trees 6 months after planting, but in subsequent years there was a progressively stronger negative effect of tree density on pasture growth, and pasture was virtually non-existent at 3580 stems ha^{-1}. This result is also illustrated in Figure 8.13 for the stage 3.5 years after planting, when pasture yield was unaffected by tree densities less than 158 stems ha^{-1}, but greatly reduced at higher densities.

Tree growth was influenced by density of planting from the early phases of establishment until the end of measurements reported to 4.6 years after planting. Initially this was evident in greater stem diameter at low density, and the height:diameter ratio was positively related to

Table 8.11 Relative and absolute yield of pasture standing biomass as influenced by density and age of trees in subhumid southeast Queensland (from Cameron *et al.* 1989).

Tree density (stems ha^{-1})	Relative yield (%) at tree ages (years)			
	0.5	1.5	2.5	3.5
3580	81	15	1	2
2150	75	42	9	7
1140	75	105	60	22
595	87	102	94	34
305	97	92	115	73
≤158	100	100	100	100
	Pasture yield (kg ha^{-1})			
≤158	5800	6185	4395	6700

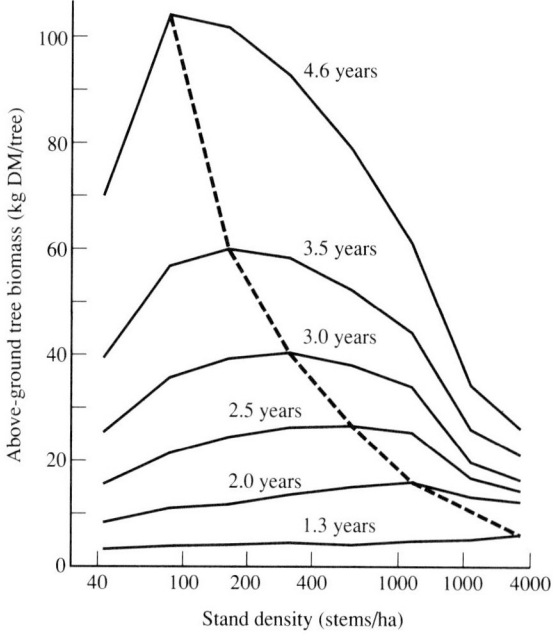

Figure 8.12 Effect of density of *E. grandis* on biomass of individual trees in southeast Queensland (from Cameron *et al.* 1989).

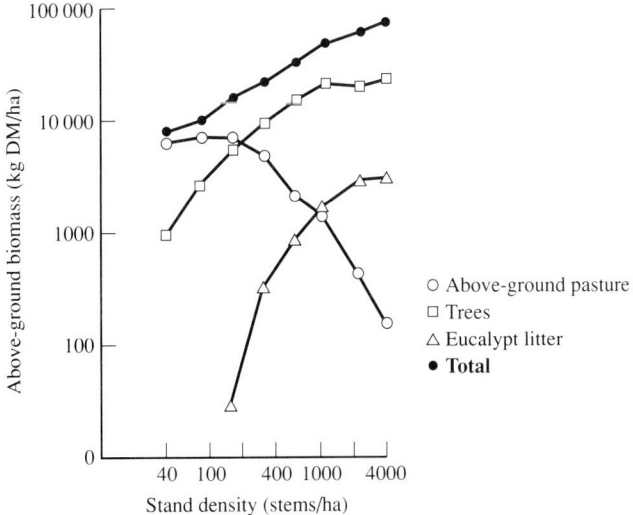

Figure 8.13 Biomass of above-ground pasture, trees, eucalypt litter and total at 3.5 years after planting *E. grandis* at differing densities (from Cameron *et al.* 1989).

tree density throughout the study. By 1.5 years tree height ranged from 3.8 m at 42 stems ha^{-1} to 6.2 m at 3580 stems ha^{-1}. Individual tree biomass was maximal at successively lower densities as the trees aged (Figure 8.12); initially there was a shelter effect which favoured growth of trees at the highest densities, but as competition developed individual tree biomass became greater at the low densities, and at 4.6 years was maximal at 82 stems ha^{-1}. The product of individual tree biomass and stand density (Figure 8.13), which gives the total accumulated biomass of trees, was positively linked to tree density over the whole range in the first 1.5 years, and thereafter was similar at 1140–3580 stems ha^{-1}. Eucalypt litter showed a similar response. Total standing biomass, including pasture, increased with tree density throughout the study. The total growth rate of trees during the growing season at age 2–2.5 years increased to *c.* 80 kg ha^{-1} day^{-1} at the highest density.

This study indicates how different forms and levels of tree production and amounts of pasture production may be manipulated to produce desired results. High stand density (> 1000 stems ha^{-1}) produce maximum wood biomass with narrow stems and little or no pasture. This is unsuited for agroforestry, but is appropriate where the objective is to lower the water table, or to produce charcoal or biomass. Trees established at medium densities of 300–600 stems ha^{-1} could provide round timber for pulp wood, poles and sawn logs, whilst reasonable pasture growth might be attained if early thinning, perhaps earlier than is normal silvicultural practice, were executed, and a final density of 20–50 stems ha^{-1} achieved. At lower densities (< 100 stems ha^{-1}) extensive branching and rapid tapering of the crown would limit the recovery of logs for timber in some tree species, but would provide stock shelter and good pasture growth in this rainfall environment; in drier regions of the subtropics (Figure 8.9) low tree densities would still restrict pasture growth.

The labour and expense of pruning and thinning are compensated if the outputs are saleable at reasonable prices; the option needs to be considered of planting at low initial densities and gaining benefits through greater pasture growth, weed control, and higher SR.

8.4 Management of particular crops
Coconut

Pastures under coconuts (*Cocos nucifera*) are the best-established silvopastoral association (Plucknett 1979), and one of the most feasible, because of the comparatively favourable light regimes (Figure 8.1) for pastures. This may be intensified to a three-strata system of coconuts–fruit crops–pasture, as illustrated in Colour Plate 9, or of coconuts–vanilla–shrub legume. Coconuts are grown in humid coastal tropical areas of Latin America and Africa (Figure 8.14) but over 90% of the world's coconut area of c. 9 Mha (Reynolds 1988) is located in Asia and Oceania, where the Philippines and Indonesia are the main producing nations. This area constitutes an immense potential source of increased animal production.

Recommendations for palm density (Reynolds 1988) vary from 200 palms ha^{-1} (7 × 7 m spacing) for tall varieties to 250 ha^{-1} for hybrids to 285–300 ha^{-1} for dwarf varieties, but many plantings have been made at wide spacings of 100 ha^{-1}, and these provide opportunity for higher SR.

In West Bali Rika *et al.* (1981) found that LWG per head of Bali yearling cattle grazing *B. decumbens–C. pubescens* pastures decreased 8 kg year^{-1} for each increase in SR of 1 ha^{-1}, indicating a greater resilience to change in SR than was evident in the pastures referred to in Figure 8.11. Pastures grazed at 5 yearlings ha^{-1} (800 kg biomass average throughout) should provide long-term production of c. 550 kg LWG ha^{-1} year^{-1}. This level of production is greater

Management of particular crops 217

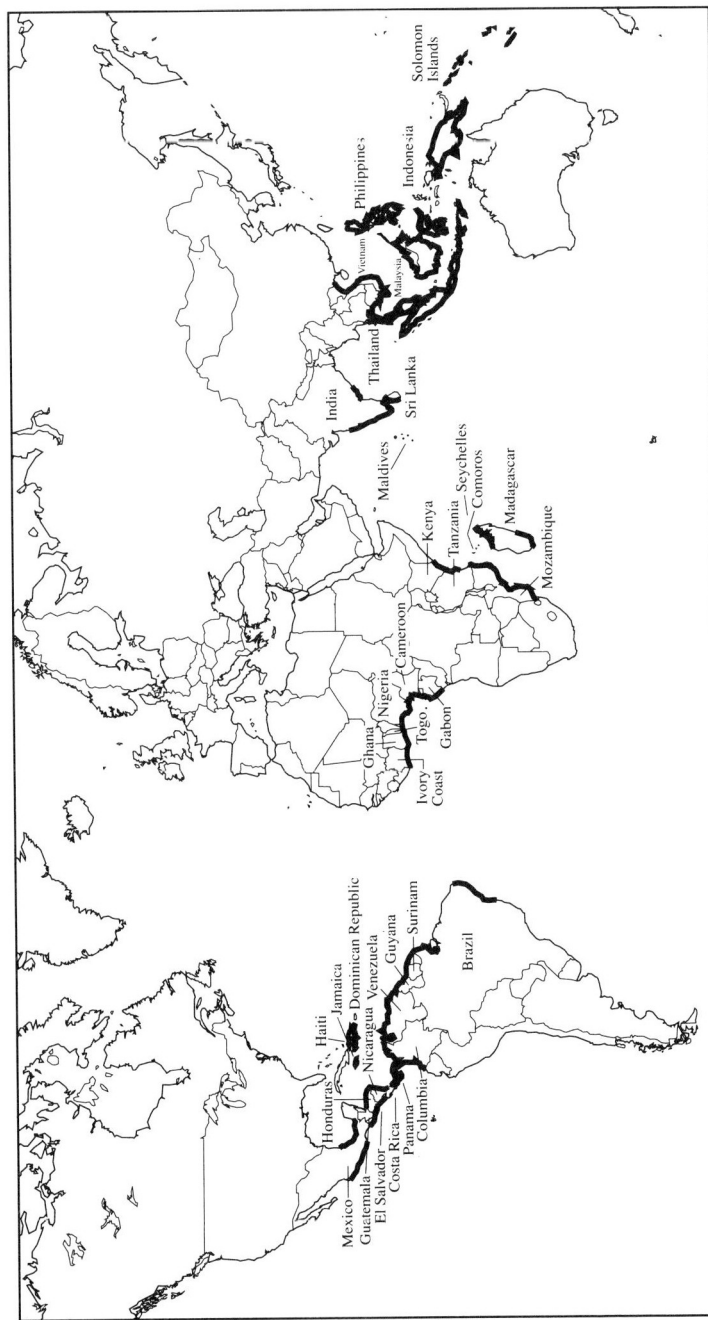

Figure 8.14 Coconut-producing areas of the world (from Feynolds 1988).

than that reported from legume-based pastures in south Thailand, where maximum LWG was 138 kg ha^{-1} year^{-1} at an SR of 2.4 b ha^{-1} (Manidool 1983), or in the Solomon Islands where maximum LWG was 312–355 kg ha^{-1} year^{-1} at an SR of 3.6–3.7 b ha^{-1} (Watson and Whiteman 1981; Smith and Whiteman 1985). In Western Samoa Reynolds (1981) reported cattle gaining 250–400 kg LW ha^{-1} year^{-1} under coconuts of 50–80% light transmission. In Sri Lanka *B. miliiformis–P. phaseoloides* pastures grazed with dairy heifers at 6 ha^{-1} produced 112 kg LWG hd^{-1} year^{-1} when heifers were supplemented daily with 5 kg fresh *Leucaena* or *Gliricidia* leaf (Liyanage 1991). These studies indicate variable production and a widely established truth that incorporation of a vigorous legume in the pastures and management which excludes unpalatable weeds lead to satisfactory levels of ruminant production from pastures under coconuts.

Rubber

Rubber (*Hevea brasiliensis*) plantations are concentrated in the lowland humid tropics on acid soils; c. 80% of production (total c. 4.6 Mt year^{-1}, 1987) emanates from southeast Asia, where Malaysia, Indonesia and Thailand are the main producers (Shelton and Stür 1991). Smallholder production predominates, but on both large and small holdings there is a definite trend to the greater incorporation of small ruminant production. This initially arose to foster cheaper weed control, as mentioned in section 8.3, but as better production techniques have evolved and animal performance has improved, more landholders have recognized the benefits of diversified and increased income.

Reference was made earlier to the change in the nature of forage resources as the radiation regime decreased with age of planting. This is illustrated in more detail from a survey (Chee and Ahmad Faiz 1991) of the forages present in five rubber estates in central Malaysia which did not raise sheep. Planting density varied from 310 to 560 stems ha^{-1}. The mean botanical composition (Table 8.12) showed the dominance of the legume cover *P. phaseoloides* in 1–2 year rubber, and its displacement by *C. caeruleum* at 3–5 years; under grazing it would be expected that *C. caeruleum*, if planted, would assume greater dominance because of its rejection by sheep (Chong *et al.* 1991). In 3–5 year rubber the grass *Ottochloa nodosa* is prominent, together with the palatable weeds *Asystasia intrusa* and *Mikania micrantha*. In mature (6–10 year) rubber, ferns (especially *Nephrolepis biserrata* and *Dicranopteris lincaris*) became the dominant group, and the young shoots of these are eaten by sheep. Forage availability and light transmission follow the expected trends with

Table 8.12 Botanical composition, forage on offer and light transmission in Malaysian rubber plantations of varying age (from Chee and Ahmad Faiz 1991).

Attribute	Age of plantation (years)		
	1–2	3–5	6–10
Botanical composition (%)			
Cyrtococcum oxyphyllum	1	1	12
Ottochloa nodosa	7	28	20
Paspalum conjugatum	6	2	5
Calopogonium caeruleum	0	25	9
Pueraria phaseoloides	79	12	1
Asystasia intrusa	1	11	4
Mikania micrantha	1	11	1
Ferns	0	3	41
Other species	5	7	7
Forage on offer (kg DM ha^{-1})	2600	1250	520
PAR (%)	92	21	10

age (Table 8.12) and this survey indicates the considerable potential of unexploited forage resources under rubber.

Growth of supplemented crossbred lambs was reported in Table 8.10 as 84–106 g hd^{-1} day^{-1}, according to SR. These figures are substantially higher than earlier figures (Ismail 1986) of 47 g hd^{-1} day^{-1} under experimental conditions, or 31 g hd^{-1} day^{-1} for smallholder production of Malin lambs. Birth weight of Dorset crossbred lambs may average 2.3 kg, relative to 1.0–1.7 kg for Malin lambs. Mortality in the latter was greater for twin lambing (50%) than for singles (15%). Local demand for lamb meat is high, and import substitution has followed the expansion of the sheep industry in Malaysia.

Oil palm

Oil palm (*Elaeis guineensis*) (Hartley 1988) is confined to the humid tropics and is better adapted to low lying situations than other plantation crops, since it possesses a hydrophytic root system. Some 77% of world production is concentrated in southeast Asia where Malaysia is the main producer nation (1.8 Mha, Chen 1991) but production in Africa, especially West Africa, is significant and contributes c. 17% of world output (Shelton and Stür 1991). Oil palms are usually planted in a triangular pattern which gives a density of 115–205

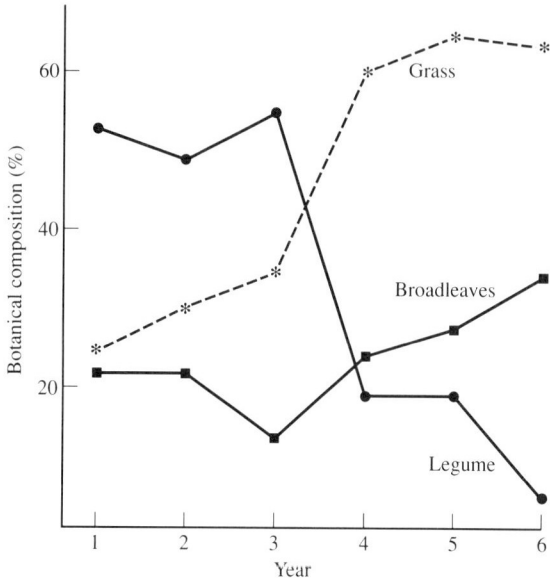

Figure 8.15 Botanical composition of ground vegetation in oil palm plantations of differing age in Malaysia (from Chen 1991).

ha^{-1} and a light transmission profile (Figure 8.1) intermediate between that of rubber and coconut; pastures are in deep shade 8 years after planting.

Forage availability in Malaysian oil palm plantations reaches peak yields of 5.5–9.5 t DM ha^{-1} 3 or 4 years after planting, if the area is not weeded or grazed (Chen 1991). Feed supply then decreases rapidly as canopy closure occurs and there are significant shifts in botanical composition (Figure 8.15). The original legume cover crops disappear by year 6, whilst shade tolerant grasses and broad-leaved weeds, as described for rubber plantations, increase. Studies of the performance of pastures under the closed canopy of the oil palm (Chen and Bong Julita 1983; Chen and Othman 1984) drew attention to the superior yield performance of *B. decumbens* and of *Desmodium ovalifolium*.

Satisfactory levels of cattle production under oil palm have been reported from Malaysia. On unfertilized natural pasture under 2 year oil palm LWG was 332 and 260 g hd^{-1} day^{-1} at 1 and 2 Kedah-Kelantan cattle per hectare respectively, whilst on fertilized *P. maximum* in the open LWG was 307 g hd^{-1} day^{-1} at the higher SR of 6 ha^{-1} (Chen 1991). In another study SR was 3 (years 1 and 2), 2 (years 3 and 4) and 1 (year 5) ha^{-1} as oil palm canopy developed, giving LWGs of 264, 159, 390, 169 and 188 g hd^{-1} day^{-1} respectively in five successive years. In Nigeria N'dama cattle were grazed at 1.2 b ha^{-1} (Payne 1985) under palms spaced at 128 stems ha^{-1}, whilst in Ivory Coast cattle gained 500 g hd^{-1} day^{-1} at 0.5 b ha^{-1}.

The absence of damage to oil palm fruit yield occasioned by grazing at reasonable SR was mentioned in section 8.3, and forage resources under oil palm constitute a valuable resource which is insufficiently utilized.

Other tree crops

Leguminous forage crops are grown with coffee, and Colour Plate 13 illustrates how sloping land bearing young contour-planted coffee in the Thai highlands may be stabilized with *Desmodium intortum* cv. Greenleaf. At Soddo, Ethiopia (lat. 9° N, 1100 mm annual rainfall, 1900 m altitude) coffee berry number per tree over 3 years totalled 5740 and 5120 with and without *D. intortum* respectively, whilst *D. intortum* yielded 3.6–4.0 t DM ha^{-1} year^{-1} cut forage (ILCA 1988*a*). Leguminous shade trees planted over coffee contribute N to the system, and the prunings may also be used for forage. *Leucaena leucocephala* is used in this way in Papua New Guinea (Allen 1985) whilst in Costa Rica *Coffea arabica* is grown under the legume *Erythrina poeppigiana* (Glover and Beer 1986). The timber tree *Cordia alliodora* may be included without detriment to coffee yields (Glover and Beer 1986; Young 1987). Similar systems are employed in Venezuela (Escalante 1985).

Tea plantations are integrated with forage production to a limited degree. Colour Plate 6 and Plate 4.1 show low lying lands in Sri Lanka planted to improved pastures, which benefit from the transport of nutrients in runoff and leachate from the heavily fertilized tea crop growing on the slopes above. Forages are used to rehabilitate soil during the replanting phase (Liyanage 1991), and *Tripsacum laxum* or *Cymbopogon confertiflorus* build up OM and improve soil structure for a period of 18 months before tea is again planted. Leguminous cover crops such as *D. ovalifolium*, *Stylosanthes guianensis* or *Crotalaria* spp. may be sown between the tea hedges and the legume *Gliricidia sepium* planted for shade.

Forages stabilize the inter-tree surfaces in orchards, especially on sloping lands, and provide alternative income. In northern Thailand *Panicum maximum* is grown in lumyai orchards, and utilized with a cut-and-remove system for dairy production, whilst in south Thailand cashew (*Anacardium occidentale*) is combined with pastures (Plate

Plate 8.12 Pasture with cashew trees on sandy soils in south Thailand.

8.12). Cashews may be grown on poor sandy soils which are more attractive to farmers for forage production than for annual crops. In Kade, Ghana sheep utilized various forages under a mango/cashew plantation. Mean daily digestible DM intake of yearling rams was 31, 34 and 39 g kg $LW^{0.75}$ for *P. phaseoloides*, *Brachiaria lata–P. phaseoloides* and *C. pubescens-Asystasia gangetica* pastures respectively (Asiedu *et al.* 1978). Forages are also integrated with macadamia (*Macadamia integrifolia*) and brazil nut (*Bertholletia excelsa*) production (Combe 1983). Sheep are traditionally associated with orchard culture, and low-growing, sod-forming, shade-tolerant legumes such as *Arachis pintoi* (Colour plate 14) offer potential for N fixation, weed control, soil conservation and forage quality in the humid tropics.

Timber production

Humid climates

Where trees are grown for timber in the humid tropics and subtropics pasture–tree relations (Wiersum 1985) have similar processes to those described for plantation agriculture directed to coconut, rubber and oil palm production. The light environment of the pasture is the primary

Management of particular crops   223

determinant of forage growth and the impact of pastures on tree yield is relatively small.

The study (Cameron et al. 1989) at Samford, Queensland (Figures 8.12, 8.13, Table 8.11) was conducted in a regime of 1100 mm average annual rainfall, which is greater than that received by woodland in many subhumid zones, but acute moisture deficiency was an intermittent feature of the environment. By contrast, silvopastoral reafforestation with *Eucalyptus deglupta* on a site such as Kolombangara, Solomon Islands is not constrained by any marked dry season in this lowland humid area (lat. 9° S, 2900 mm annual rainfall) which enjoys year-round rains and cloudy conditions. The area was logged, planted to *E. deglupta*, seeded to improved pastures and grazed with cattle at varying SR (Macfarlane and Whiteman 1983). Light transmission was linearly and negatively related to tree density. As shade developed with increasing tree growth the planted grasses ranked in order of decreasing shade tolerance: *Ischaemum aristatum, Brachiaria decumbens, B. humidicola* and *B. mutica. P. phaseoloides* was successful at low SR, but *C. pubescens* was more resistant to heavier grazing. *Paspalum conjugatum* was the principal invading grass and with good legume composition (50%) gave satisfactory animal production. Early introduction of cattle saved weeding costs.

Mean LWG over the various treatments was 420, 380 and 350 g hd^{-1} day^{-1} in successive years as shade developed, whilst pastures in full sunlight averaged 430 g hd^{-1} day^{-1} over the whole period, suggesting that shading had a negative effect on pasture quality. Macfarlane and Whiteman (1983) suggested the following SR and expected LWG for different levels of light transmission as follows:

- 35% PAR: 0.7 weaners ha^{-1}, 300 g hd^{-1} day^{-1}
- 45% PAR: 1.0 weaners ha^{-1}, 350–380 g hd^{-1} day^{-1}
- 50% PAR: 1.3 weaners ha^{-1}, 400 g hd^{-1} day^{-1}
- 60% PAR: 1.6 weaners ha^{-1}, 400 g hd^{-1} day^{-1}
- 80% PAR: 2.5 weaners ha^{-1}, 400 g hd^{-1} day^{-1}

The commercial project arising from this study suffered logistic difficulties, a problem of matching stock numbers to pasture availability and suboptimal P nutrition, but guidelines were developed with wide applicability for high rainfall silvopastoral practice.

A second example is taken from a site on the Atherton Tableland, Queensland (lat. 17° S, 1600 mm annual rainfall, 760 m altitude) which was burnt, cleared and planted to *Pinus caribaea* var. *hondurensis* at 500 stems ha^{-1} (Applegate and Nicholson 1988; Applegate 1991). The pasture was dominated by *Panicum maximum* with *Melinis minutiflora* and *Axonopus affinis*, and was controlled for a 1 m radius around each tree by spraying with simazine and glyphosate at 6 l ha^{-1} active ingredient. The stand was first thinned at 2.7 years (5 m height) to c. 260 stems ha^{-1}, when half of the thinnings were sold as 'Christmas' trees, and branches were pruned at 3.5 and 5 years. The pastures were grazed by weaner steers and subsequently by horses.

This stand was established at a lower density than the usually adopted density of 750 stems ha^{-1} for *Pinus caribaea* var. *hondurensis*, and trees were pruned earlier than normal. This probably led to better pasture growth, but concomitant timber production was highly satisfactory. At 7.2 years mean basal area was 16.4 m^2 ha^{-1}, diameter at 1.3 m height was 28.2 cm, and mean height was 16.4 m, which compares favourably with other reported levels of production (Applegate 1991). Final harvesting of trees would occur at age 20–25 years, and gross margins over this period, taking into account the value of livestock production, tree thinnings and final timber output, would average $US160–200 ha^{-1} year^{-1}. Additionally the silvopastoral operation provides reclamation of a degraded landscape.

Planting leguminous trees with pasture will augment the long-term N economy of the association. At Mtwapa, Kenya (lat. 4° S, 1260 mm annual rainfall) *Pennisetum purpureum* was established between *L. leucocephala* rows 2 m apart (Jama and Getahun 1991). *L. leucocephala* provided good fuelwood production, but *P. purpureum* was not productive after the first year, due to canopy closure.

The ameliorating effect of trees on frost occurrence on underplanted pasture in the subtropics was mentioned earlier (Cook and Grimes 1977). A native forest site near Gympie, southeast Queensland (lat. 26° S, 1200 mm annual rainfall, 80–120 m altitude) was logged and unwanted trees removed by application of Tordon, providing 18.5 m^3 ha^{-1} of timber. *Eucalyptus maculata* and *E. fibrosa* occurred on the ridges and *E. intermedia*, *E. siderophloia* and *E. tereticornis* occupied the gullies; there was an understorey of *Acacia* spp, *Imperata cylindrica* and *Themeda australia*. The residual forest of 80 trees ha^{-1} (basal area 5 m^2 ha^{-1}) was burnt and aerially sown to a mixture of improved grasses and legumes with Mo superphosphate. Minimum light transmission was 72%, and the most successful species were the legumes *Macrotyloma axillaris* and *D. intortum* (gullies and sandstone-derived soils) and the grasses *M. minutiflora*, *S. sphacelata* var. *sericea* cv. Nandi, *Chloris gayana* and *P. maximum* var. *trichoglume*. Sowing pastures suppressed woody regrowth and tree growth increment was greater with sown leguminous pastures than with unsown natural pasture.

Pastures were conservatively grazed at SRs of 0.28, 0.38 and 0.43 b ha^{-1} in successive years and provided good animal performance. The study (Cook and Grimes 1977; Cook et al. 1984) indicated some disadvantages: difficulties of mustering cattle, fence maintenance (falling branches) and problematical replacement of tree stand, but many advantages:

- low cost of pasture development,
- reduced frost incidence,
- control of soil erosion,
- better retention of moisture,
- control of woody regrowth,
- increased tree growth rate,
- diversification of and increased income.

Subhumid climates

The manipulation of tree–pasture relations is more difficult in subhumid climates, since the occurrence of moisture stress is often severe but erratic in its occurrence and since lower levels of production apply which restrict the inputs available for managerial intervention. Radiation receipt is rarely a restrictive factor on pasture growth; it is the availability of water and of nutrients which are the primary constraints to pasture growth. Trees have a role in providing fuel, timber for farm construction and fencing purposes, shade and shelter for livestock, food for bees, and in some cases fodder and environmental protection. In most subhumid areas livestock (rather than tree) production is the predominant enterprise and the farmers' focus is on directing woodland management to this end. The key objectives (Burrows *et al.* 1988) are:

- control of tree density to maximize pasture growth,
- minimization of the regrowth or colonization of woody species,
- replacement of trees when overclearing occurs,
- avoidance of environmental damage.

Circumstances in which trees provide a stimulus to pasture growth were discussed in section 8.2 (shade increasing mineralization of soil N under conditions of low N supply, and trees acting as a nutrient pump). N-fixing trees may also augment pasture growth. These positive effects are restricted spatially to areas in close proximity to the tree, and the net effect over the whole area is more usually of the type illustrated in Figure 8.9 (Scanlan and McKeon 1993), where reductions in pasture growth occur whose severity depends upon the basal area of trees and the shallowness of the soil. For example, at a site with a soil $>$ 500 cm deep which produces a pasture DM of 3000 kg ha^{-1} year^{-1} if trees are removed, the presence of trees of basal area 10 m^2 ha^{-1} would reduce relative pasture yield to 0.4, or a pasture DM of 1200 kg ha^{-1} year^{-1}. At a site with a shallow soil 100 cm deep which produces a pasture DM of 2200 kg ha^{-1} year^{-1}, the presence of the same tree basal area of 10 m^2 ha^{-1} would reduce relative pasture yield to 0.1, or pasture DM of 220 kg ha^{-1} year^{-1}. The negative effect of tree density on pasture growth will be less in savanna regions with a monsoon climate providing reliably wet and reliably defined dry seasons, since trees and grass compete for water only when both are actively transpiring (Burrows and Frost 1993). Jain (1993) observed at Naila, Rajasthan, that treeless pasture yielded 5.1 t ha^{-1}, whilst pasture growing in association with *Acacia tortilis* and *A. nilotica* produced 3.0 t ha^{-1}. In these circumstances landholders will thin trees to a low density, depending upon their need for the retention and production of tree products.

When trees are removed by clearing, mechanical techniques such as ring-barking, or arboricidal treatment, the control of regrowth, seedling regeneration or the invasion of shrubby species are a focus of attention. Leaving uncleared strips or clumps leads to a more readily managed situation than if scattered trees are left. *Eucalyptus* spp. regenerate from buried lignotubers. The mechanical clearing of *Acacia harpophylla* woodland in Queensland may result in a greater subsequent density of *A. harpophylla* suckers if post-clearing management is ineffective. Fosset and Venamore (1993) killed *A. harpophylla* suckers with an aerial application of 2 kg tebuthiuron; this led to increased LWG in cattle grazing the area of 31 kg hd^{-1} relative to that of cattle grazing on an untreated area. Stock exclosures indicated tebuthiuron increased grass production by 1600 kg DM ha^{-1}.

The pasture which develops following tree clearing or thinning is also vulnerable to invasion by woody shrubs. Pulses of shrub recruitment (Noy-Meir and Walker 1986) follow drought conditions or overgrazing which create 'gaps' in the sward. Many such plants are susceptible to fire; for example in western Queensland invasion of the woody plants *Carissa ovata, Eremophila gilesii, Terminalia oblongata* and *Myoporum desertii* may be checked by a hot fire (Pressland 1982). Once shrub dominance has occurred the accumulation of sufficient grass fuel for a fire is an acute management problem, which is sometimes solved following an abnormal wet season if the farmer seizes this opportunity. It is ironic that removal of woodland may lead to a greater or similar biomass of unwanted woody species if management is ineffective.

When overclearing occurs the landholder has the opportunity to replace the existing trees by natural regeneration, if sufficient 'seed' trees are present. The corner of a paddock may be fenced to exclude stock over an area of 2-5 ha (for example, Kerkhof 1990 for Tanzania), and a period of 5 years or such period as permits trees to exceed browsing height will be needed before the fence is removed. Trees have a finite life and a proper age distribution is required to ensure continuity of production. Alternatively, leguminous or fodder trees may be planted which have the effect of increasing animal production (Mann *et al.* 1989). Mittal (1993) describes a system in Jodhpur, India in which mixed grazing of goats and sheep was augmented with browsing of *Zizyphus nummularia* and lopping of *Prosopis cineraria*.

Environmental damage in woodland is associated with erosion, salting, the loss of species diversity and the destruction of habitat for wildlife. A common guideline is to exclude land of > 20% slope from clearing, and this figure is modified according to the vulnerability of the landscape to slippage. Land within 100 m of recognized watercourses is best not cleared. Clearing may promote a rise in the water table and delivery of salt to the surface lower down the slope on susceptible soils containing salt in the subsoil or regolith, as discussed in section 4.4. Strips of uncleared timber at least 100 m wide which are connected to watercourses and to similar strips on adjacent landholdings provide shade and shelter for domestic stock, but also facilitate the retention of a favourable habitat for wildlife.

8.5 Conclusions

Associations of tree crops with pastures are established farm practice in many parts of the tropics and subtropics (Plates 8.13 and 8.14) and the

Plate 8.13 Two-year-old heifer grazing *Sesbania sesban* with *Brachiaria decumbens*.

Plate 8.14 Bananas intercropped with *Lablab purpureus* in Ethiopia.

expansion and intensification of these is therefore a realistic objective, since much long-term farmer experience is available (Filius 1982). The periodic collapse of world prices for plantation commodities provides a strong incentive for farmers to diversify production, especially since there is a concurrent strong growth in domestic demand for dairy and meat products in almost all of these tropical regions.

There has also been a suite of recent advances in the technology available to intensify production of pastures under tree crops, especially in relation to the availability of pasture legumes and of grazing-resistant grasses. The better scientific understanding of tree–pasture relations is providing realistic guidelines for managing the component elements of silvopastoral systems which lead to better farm income and a more protected environment.

CHAPTER 9

Shrub legumes with annual crops

9.1 Introduction
The rationale for planting shrub legumes

Trees evolved before grasses, and trees provided food for animals before herbaceous pastures supplied the diet of grazing animals. The dinosaur *Diplodocus* and its allies, which were probably the largest animals that ever lived, fed on trees and shrubs. Other animals such as *Stegosaurus* were known as the beaked dinosaurs, and had beaks sheathed in horn carried in front of their tooth-set jaws, which were presumably used to strip the leaves and twigs from trees and shrubs (Gray 1970). The concept of tree foliage as feed is well entrenched in the culture of many tropical societies, although in some communities the high feeding value of woody perennials which are used for other purposes is not recognized, and the concept is alien to many Western cultures. The disappearance of the dinosaurs evokes speculation, but the well-being and survival of modern domestic livestock is often secured by the interest of farmers and scientists in the use of trees and shrubs as sources of high quality forage.

Shrub legumes have many advantages over herbage legumes. The distinction between shrubs and trees is an artificial one; many tall trees are lopped to provide forage and non-N-fixing woody perennials have utility in agricultural systems of the tropics and subtropics. The focus of this chapter is on shrub legumes, since the thrust of much agricultural development is based on the extension of their use (Skerman *et al.* 1988), and their contribution to the N economy of production and its ancillary OM status has primary significance (Gutteridge and Shelton 1993). This book is directed to the relationships of forages and crops; shrub legumes are viewed in this context for their contribution to agrosilvopastoral systems where they are used both to feed animals and to contribute to the sustainability of the yields of annual crops.

The superior features of shrub legumes relative to herbage legumes include:

- legume dominance,
- persistence,
- resistance to grazing,
- seasonal production,
- stock protection,
- alternative uses.

Good performance of these attributes flows on to other characteristics:

- N and OM accretion,
- nutritive value,
- soil conservation.

The greater growth potential in full sunlight of tropical grasses having the C_4 photosynthetic pathway was mentioned in section 8.2 and illustrated in Figure 8.3. Consequently grass dominance in mixed herbaceous tropical pastures has some inevitability, unless grass growth is limited by N deficiency, management interventions (fertilizer policy, stock control) are made which favour the herbaceous legume component, or the legume is not eaten by animals (for example, if *Calopogonium caeruleum* is grown). This situation is completely changed if a shrub legume is planted which has the capacity to overtop and shade the companion grass. Farmers also prefer to plant legumes which persist and do not require frequent replanting. Most improved varieties of perennial herbaceous legumes produce plants which are individually short-lived under grazing. At Mt. Cotton, southeast Queensland, seedlings of the perennial *Lotononis bainesii* had a mean half-life of 4.2 months (Fujita and Humphreys 1992) and the persistence of yield depends upon successful plant replacement through natural regeneration of seedlings or of new bud sites; other examples are given in Humphreys (1991, Chapter 5). By contrast mortality of the shrub legume *Leucaena leucocephala* in southeast Queensland was 13% over 16–20 years (Jones and Harrison 1980) and the capital cost of planting shrubs may be amortized over a long period.

The relatively slow adoption by farmers of herbaceous tropical legumes in farm practice has been partly caused by the promotion of legumes with a twining or scrambling habit (*Centrosema pubescens, Desmodium intortum, Macroptilium atropurpureum* cv. Siratro, *Neonotonia wightii, Pueraria phaseoloides*) whose buds are accessible to the jaws of the grazing animals (Clements 1989) and whose successful persistence requires relatively low levels of utilization which enable the legume to climb and overtop companion grasses, so that their leaves are exposed to the sunlight and their density of residual buds adequate for regrowth. This stricture does not apply to low-growing, free-seeding legumes such as *Stylosanthes hamata*, *S. guianensis* var. *intermedia* (Plate 10.3) or *Arachis pintoi* (Colour Plate 14). Alternatively, tall shrubs have their lower branches and foliage accessible to livestock, but plants persist and grow well if they are managed so that some of the canopy is periodically above browsing height (Colour Plate 11; *Albizia chinensis* with sheep).

The deep-rooted habit of many shrub legumes enables them to access moisture from deeper soil levels, so that some green leaf production occurs during the dry season, or green leaves are retained on the bush, as in the case of *Stylosanthes scabra* cv. Fitzroy. This characteristic is elaborated further in section 9.4 to indicate species differences; for example, *L. leucocephala* exhibits better dry season production than *Gliricidia sepium* (Kang et al. 1990). In the subtropics and tropical highlands the elevation of the shrub canopy above frost height contributes to the availability of green foliage during the cool season. Thus the duration of the seasons when green forage may be offered to livestock is often greater than from herbaceous legumes, and the shrub legume forage may be 'banked' in order to overcome a later discontinuity in the availability of other feeds (Plate 9.1).

Shrub legumes are planted to provide boundary fences to fields or landholdings (Sumberg 1983 for *L. leucocephala*) which give stock control and stock protection. Living fences provide shade and shelter in addition to providing forage. They may also be planted around residential blocks or home gardens to ameliorate the domestic environment. Finally, shrub legumes offer a diversity of potential use, and a flexibility of end-product. The stand may be managed essentially to provide fuelwood or charcoal, light poles for farm and household construction purposes, or canes for trellising horticultural crops. Frequent pruning reduces wood production and increases the proportion of assimilate directed to leafy forage production. Honey and gum production may also result, and shrub legumes are used to shade plantation crops such as cocoa. Hedgerows may also be viewed as sources of biomass.

The successful adaptation of shrub legumes in agricultural practice is then reflected in good levels of DM production and of N fixation and OM accretion, as discussed in section 9.2. In recent years more research attention has been directed to characterizing the nutritive value of shrub legumes (Speedy and Pugliese 1992) especially with respect to their content of condensed tannins and of 'protected' protein (section 9.2). The planting of alleys or hedgerows on the contour impedes the flow of runoff and provides a physical barrier against which soil may accumulate to farm terraces, as mentioned in

232 *Shrub legumes with annual crops*

Plate 9.1 Goat feeding on cut leguminous shrub leaves.

section 3.4. Farmers have adopted shrub legumes for varying reasons, and may not always appreciate the diversity of uses to which they may be put and which need to be tested in their particular farm context. The promotion of particular shrub species outside the boundaries of their environmental adaptation and the advocacy of agricultural systems such as alley farming with insufficient appreciation of the rigorous management required to make them successful have been inimical to the wider adoption of shrub legumes in farm practice.

Sustained yield in annual cropping systems

The yields of annual crops steadily decrease with time in many agricultural systems in the tropics. There are short-term fluctuations in yield, often associated with natural disasters such as drought, but farmers look for the long-term maintenance of yield in systems which exhibit resilience after stresses occur; Figure 1.5 illustrated a sustainable and a non-sustainable system. Continuing depreciation of crop yield occurs if the outputs from the system are exceeding the inputs, and if the natural resource base of the soil, its structure and its nutrients are being depleted. A number of examples were previously given in section 1.2: decreasing yields of sweet potato in

Keravat, Papua New Guinea, of cassava in Thailand, Venezuela and Nigeria, and of winter cereal cropping in southern Queensland. In the next chapter further examples of yield decrease under continuous cropping are shown in Table 10.1 (section 10.2), in Table 10.9 for a variety of crops, and Table 10.10 for cassava and roselle.

These reports from different centres indicate that the problem of yield depreciation is widespread; however it is not an insoluble question and many successful examples of sustained yield occur (Ofori 1973; Sánchez et al. 1982; Sánchez and Benites 1987). The role of shrub legumes in sustaining crop yield is now addressed.

9.2 The objectives of alley farming

Experience with alley farming

Alley farming and other forms of agroforestry

The term 'alley farming' (Kang et al. 1990) denotes a system where shrub legumes are established in rows, pruned to limit the shading of companion crops grown between the hedgerows, and where the prunings are used to mulch the inter-row, to feed livestock and to meet other farm purposes. In some systems the hedges are grazed during the dry season or for an extended fallow period (Atta-Krah 1990). Alley cropping (Kang et al. 1984; Colour Plate 15) refers to a system wholly directed to crop production; shrub legumes may also be established in wide rows with perennial grasses planted in the inter-row area to improve livestock production (Wildin 1993), but the emphasis of this book is on the integration of crops and livestock.

Alley farming has in one sense evolved from shifting cultivation or bush fallow systems in which the natural regeneration of shrubs and forest is periodically slashed and burnt and crops grown on the same land for a limited number of seasons (Nye and Greenland 1960). In alley farming the shrubs and crops are grown concurrently on the same area, and in many societies the loss of crop land to shrub hedgerows can be viewed in this context as small in relation to the greater area of crop land gained.

Shrub legumes are planted in other configurations. Blocks of shrubs may be planted to produce fodder, timber or green manure mulch for crops and these areas may even be rotated with annual crops. More commonly shrub legumes are planted on the boundary of a field, or are placed informally through a cultivated garden or cropping area. Some trees, such as *Faidherbia albida* (syn. *Acacia albida*) in Ethiopia (Poschen 1986) benefit annual crop growth beneath their canopy (Figure 8.7), and surveys of agroforestry projects (Cook and Grut 1989; Kerkhof 1990) indicate the preference of many farmers for informal tree plantings rather than for rigorously managed hedgerows. An intermediate position is to adopt the practice of intermittent pollarding of shrub stands which produces fuelwood and poles, fresh fodder and reduced annual crop yields nearby, as at Hayathnagar in Hyderabad (Hocking and Rao 1990).

Positive effects of alley farming on crop yield

The experience of landholders with alley farming is relatively recent, except where the practice was pioneered in such regions as Timor and Sikka, Flores in Indonesia, Alabang, Rizal in the Philippines, and Mbaise, Imo in Nigeria (Kang et al. 1990). Perspectives are provided by other reviews (Wiersum 1982; Nair et al. 1984; Ssekabembe 1985; Paningbatan 1986; Gutteridge 1988a; Sumberg and Atta-Krah 1988; Anderson and Sinclair 1993; Atta-Krah 1993). There are many examples of improved crop yields where leguminous prunings are brought in from other parts of the farm and applied to the crop area (Onim et al. 1990; Smith 1992a) but a more informative comparison is between the crop yield in the absence of hedgerows and the crop yield between the hedgerows, expressed as output from the total land area of hedgerow and annual crop.

Many positive responses in crop yield have been reported from IITA, Ibadan (lat. 8° N, 1300 mm annual rainfall) and its subsidiary research centres. The production of maize grain on a degraded alfisol alley farmed between *Gliricidia sepium* was -3%, $+12\%$, $+25\%$, $+23\%$ and $+39\%$ in successive crops relative to that of maize grown without hedgerows (Atta-Krah and Sumberg 1988). In a further experiment maize yield (Figure 9.1) was sensitive to the proportion of *G. sepium* prunings applied to the crop area, and maize also responded to the application of fertilizer N even when all prunings were returned; this indicates that the hedgerow legume did not fully meet the N demand of the crop, and perhaps other benefits of the prunings (OM status, provision of other nutrients such as P and K) contributed to crop response at the 80 kg N ha^{-1} level. The low level of yield in the absence of mulch and its elevation in excess of 2 t ha^{-1} with prunings but no N fertilizer are noteworthy. In many experiments at Ibadan low input continuous cropping gives maize yield of c. 1 t ha^{-1} or less, alley cropping without additional N gives sustained yield of the order of 2 t ha^{-1}, and this may be raised to c. 3 t ha^{-1} if augmented with N fertilizer.

Several shrub legumes are successfully used in alley farming. Hedgerows of *Calliandra calothyrsus* (Colour Plate 10) grown at Ibadan on an Oxic Paleustalf produced c. 6 t DM ha^{-1} $year^{-1}$

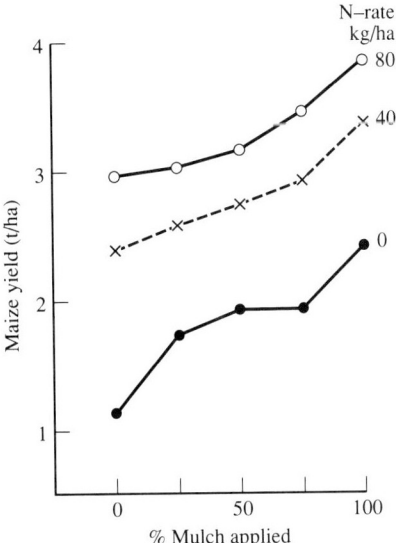

Figure 9.1 Yield of maize grain in 1985 at Ibadan, Nigeria as affected by proportion of prunings of *Gliricidia sepium* hedgerows returned to cropping area and level of N fertilizer (from Atta-Krah and Sumberg 1988).

from four prunings, which contained *c.* 200 kg N (Gichuru and Kang 1989). These increased maize yield from 2 to 3 t ha^{-1}, but decreased yield of cowpea (*Vigna unguiculata*). The standard alley farming legume in many locations has been *Leucaena leucocephala* (Plates 9.2 and 9.3). At Ibadan on the Egbeda soil series (Oxic Paleustalf) the addition of the rosaceous *Acioa barteri*, which has desirable root characteristics (Figure 8.8), to *L. leucocephala* did not benefit maize and cowpea crops (Siaw *et al.* 1991). However *L. leucocephala* hedgerows increased maize yields by 47% in the absence of applied N, or by 24% if 60 kg N were applied, relative to maize yields in the absence of hedgerows. *L. leucocephala* hedgerows also benefited the yield of a number of vegetable crops at Ibadan (Palada *et al.* 1992).

Alley cropping has been successful in other regions. At Mtwapa, Coast Province in Kenya (lat. 4° S, 1300 mm annual rainfall) early studies of alley cropping indicated the powerful effects of shading in reducing crop yield, but when systems of hedgerow use were developed which ameliorated this problem, beneficial effects on crop production emerged. Thus on a low fertility soil (Inceptisol-Oxic Tropepts) *L. leucocephala* hedgerows were repeatedly pruned to single stems 0.5 m high. Weed biomass was substantially reduced,

Plate 9.2 *Leucaena leucocephala* branching well under grazing.

Plate 9.3 Rice straw supplemented with *Leucaena leucocephala* in northeast Thailand.

especially at close hedgerow spacings (Jama *et al.* 1991). Over two seasons maize yield averaged 1.6, 1.8, 2.1 and 2.7 t ha^{-1} respectively in the absence of hedgerows, or when grown between hedgerows 8, 4 and 2 m apart, indicating benefits of 13, 31 and 69% to crop yield.

At Misamfu in northern Zambia (lat. 10° S, 1360 mm annual rainfall) alley farming trials (Matthews *et al.* 1992a) with *Sesbania sesban*, *Albizia falcataria*, *Flemingia Macrophylla* (syn. *F. congesta*), *Gliricidia sepium* (Plate 9.4) and *Cassia spectabilis* were unsuccessful on low fertility acid soils (Orthic Ferrasols). However, when *L. leucocephala* was limed at 2 t ha^{-1} and fritted tree elements applied, application of hedgerow prunings increased maize yield by up to 95%. An economic analysis indicated that at current prices growing *L. leucocephala* and maize without N fertilizer was as profitable as growing maize alone plus 60 kg N ha^{-1}; varying prices of grain, lime and fertilizer would clearly modify this result.

In the Philippine uplands at Claveria, northern Mindanao (lat. 9° N, 2200 mm annual rainfall, 400 m altitude) *G. sepium* and *C. spectabilis* were established on slopes of 18–31% to reduce soil erosion (Maclean *et al.* 1992a,b). The soil was an acidic clay Ultic Haplorthox. Lime was applied at 6 t ha^{-1} to raise soil pH from 4.1 to 4.6. There was some evidence that the non-nodulated *C. spectabilis* was depleting soil nutrient reserves, but *G. sepium* hedgerows grown

Plate 9.4 *Gliricidia sepium.*

at an average of 5 m apart gave large increases in the yields of upland rice and of maize, and reduced the rate of soil loss, although soil accumulated at the lower levels of the terrace.

In Hawaii Rosecrance et al. (1992a) grew several shrub legumes under irrigation on a depleted, kaolinitic, isohyperthermic Vertic Haplustoll (pH 6.1) at Waimanolo. *Sesbania sesban* produced the highest yields of green manure and N, and maize yields in hedgerow systems averaged about 1.8 t ha^{-1}, relative to 0.6 t ha^{-1} in the absence of hedgerows. These responses are discussed later in this section. At Mt. Cotton in southeast Queensland (lat. 28° S, 1400 mm annual rainfall) alley farming of kenaf (*Hibiscus cannabinus*) with *L. leucocephala* led to yield responses equivalent to the application of fertilizer at c. 50 kg N ha^{-1}; shrub legume yield varied from 2 to 4 t ha^{-1}, contributing c. 100 kg N ha^{-1} as prunings (Gutteridge 1988b). Favourable crop responses are also reported to *L. leucocephala* in Colombia (Doorn et al. 1987) and to *G. sepium* in Costa Rica (Kass et al. 1987).

Negative effects of alley farming on crop yield

The illustrations quoted above indicate that alley farming systems have been successfully developed in many ecological zones. The objectives of alley farming are not confined to the maintenance of the yield of annual crops but also embrace the value of other products from the hedgerows such as fuel and forage, soil conservation and efficient and sustained use of natural resources. Nevertheless, farmers are loth to adopt a new agricultural production system which does not result in increased crop yields, and there is now a considerable reservoir of experience throughout the tropics and subtropics of instances where alley farming decreases the production of annual crops. Of 15 comparisons from seven sites in Indonesia, Malaysia, Philippines and Thailand (Sajjapongse 1991) alley farming depressed crop yields in eight cases and increased them in seven cases. Consideration of these and other studies indicates that legume hedgerows reduce or do not benefit the intercrop in the following circumstances:

- high soil fertility,
- marginal rainfall,
- lenient pruning,
- poorly adapted shrub legumes.

The rationale for growing shrub legumes usually relates primarily to the N accretion needed to maintain the productivity of the agricultural system. This will have little force on high fertility soils or in circumstances where cheap (subsidized?) fertilizer N is available.

Thus in Alofua, Apia, Western Samoa (lat. 14° S, c. 3000 mm annual rainfall) hedgerows of *C. callothyrsus* and *G. sepium* were successfully established (Rosecrance et al. 1992a) on a high fertility Tropical Inceptisol and the inter-row areas cropped to taro (*Colocasia esculenta*). The hedges grew well and the prunings decreased the bulk density and increased the water-holding capacity of the soil in this high rainfall environment. No change in soil N, P, K, Ca, Mg or OM was recorded, and yields of taro were reduced in the hedgerows spaced 4 m apart. Crop yield was clearly limited by factors which were not supplied by the legume hedgerows.

Most of the IITA studies are favourable to alley farming, but a long-term experiment of Lal (1989a) using *G. sepium* or *L. leucocephala* hedgerows spaced at 2 and 4 m on an Oxic Paleustalf showed c. 10% decreased yield of maize grain relative to the no-hedgerow treatments, despite better moisture availability, whilst cowpea yield was drastically suppressed by 50–70% of the control values. One hypothesis to explain the maize response is the relatively high fertilizer inputs per hectare: 120 kg N, 26 kg P and 30 kg K. In these circumstances legume N fixation was irrelevant; the cowpea crop had its own nodulation system and as a short crop was probably more susceptible to shading.

There are many reports of decreased crop yields in alley farming systems in subhumid areas, and moisture status is discussed later in this section. At the ICRISAT Centre, Patancheru, India (lat. 18° N, 780 mm annual rainfall, 550 m altitude) *L. leucocephala* hedges pruned four times each year and spaced at intervals from 1.35 to 4.95 m were intercropped with sorghum and pigeonpea (*Cajanus cajan*), or sorghum alone (Rao et al. 1990). Pigeonpea grew moderately well in the first year when the hedgerows were developing but not thereafter. Sorghum was the main component of the intercrop, and the yield of sorghum planted with the first row having a gap of 45 cm to the hedgerow (Table 9.1) was decreased in all hedgerow spacings in all years relative to yield of sorghum without hedgerows. The depression of sorghum yield was negatively related to hedgerow spacing, but even at the widest (4.95 m) spacing the decrease varied from 31 to 49% in different years. There is an interaction between competition for moisture and for light, and at this site reduced availability of moisture to sorghum appeared to be the predominant factor constraining yield.

In the subhumid Mali environment maize yields were 800, 960 and 1160 kg ha^{-1} respectively with *L. leucocephala, G. sepium* or no hedgerows (ILCA 1988b). At subhumid Jhansi, India (lat. 25° N) hedgerows of *L. leucocephala* severely decreased yield of sugar-cane (Gill and Patil 1985).

There are many studies where lenient pruning of the legume hedgerow, which often maximizes N fixation and minimizes labour costs, results in severe shading which reduces crop yields, as elaborated further in section 9.3. High hedgerow density has a similar effect. In another study at Jhansi (Hazra and Tripathi 1986) an *L. leucocephala* canopy reduced the PAR level received by intercropped oats to *c.* 58% of full sunlight and this decreased the oat yield by 50-60%. At Chandigarh in Uttar Pradesh yields of maize, black gram (*V. mungo*) and cluster bean (*Cyamopsis tetragonoloba*) were reduced in *L. leucocephala* hedgerows spaced at 3.75 m which were pruned to 0.75 m height each time they attained a height of 1.75 m (Mittal and Singh 1989); this lenient system of use would have caused deep shading of the crops. In Morogoro, Tanzania yield of maize and beans was depressed by shading from *L. leucocephala* hedgerows unless a drastic pruning regime was adopted (Lulandala and Hall 1987). Shading was also attributed as the cause of yield reduction in maize in Lilongwe, Malawi (Chiyenda and Materechera 1989) grown between *L. leucocephala*, *Cassia siamea* or *Cajanus cajan*; production levels were 85, 70 and 85% respectively of the yield from a no-hedgerow system. A similar effect was noted in The Gambia (Russo 1986).

Other failures of alley farming arise from the lack of shrub legumes well adapted to local soil and climatic conditions. *L. leucocephala* has many admirable features: high levels of N fixation, rapid recovery from coppicing, high acceptability by stock, fast delivery of N upon decomposition, superior drought resistance. The released cultivars are not tolerant of low pH soils or poor drainage, and the widespread devastation of the psyllid *Heteropsylla cubana* (Bray and Woodroffe 1991) has limited its utility. Attempts to grow this legume in unsuited areas has not favoured the adoption of alley farming. Similarly, *Sesbania sesban*, which is ideally suited to areas of poor drainage,

Table 9.1 Yield of grain sorghum (kg ha^{-1}) component of intercrop between *L. leucocephala* hedgerows established at varying spacings at Patancheru, India (from Rao *et al.* 1990).

Width of alley (m)	Year			
	1	2	3	4
1.35	1240	220	30	260
2.25	1500	875	360	830
3.15	1245	1100	1365	1330
4.05	1450	1385	1430	1945
4.95	1780	1680	1470	2380
No alley	2690	2460	3120	3445

does not survive on droughty, upland shallow soils. The failure of species such as *Flemingia macrophylla* (syn. *F. congesta*) in northern Zambia to improve crop yield may be associated with the relatively low N contribution (40 kg N ha^{-1}) and the high C:N ratio of the hedgerow mulch (*c.* 21), which limits its utility as a hedgerow species (Mathews *et al.* 1992b). Other reports of the failure of crop yields in alley farming include Araya Sanchez (1987) in Costa Rica, Rai and Suresh (1988) in Tamil Nadu, India and Peltier and Eyog-Matig (1990) in northern Cameroon.

The balance of different outputs

The objectives of alley farming may be discussed under the following headings:

- increased output of products valued by the farmer. This depends upon:
- maintenance of the soil resource and its fertility,
- effective use of climatic growth factors,
- product quality, especially with respect to nutritive value.

Alley farming provides great flexibility to the landholder and the opportunity to respond quickly to changing markets. The landholder is able to make decisions in the context of a sustainable, resilient system. Long lived perennial shrubs confer stability to the ecosystem and a guarantee of future productivity. Diversity of product output reduces risk and ameliorates the negative effects of single natural, social or market disasters. Landholders are able to manipulate the balance of outputs from annual crops, from fuel and timber, and from livestock. Once hedgerows are established frequent pruning to near ground level minimizes the competition of the shrubs with annual crops, if the latter are the primary focus, and hedgerows may be thinned. Alternatively, high livestock prices may orient the farmer to legume foliage production, or to abandoning cropping temporarily in favour of hedgerow grazing. The response to increasing demand for fuelwood is to lengthen the pruning cycle so that the hedgerows become more woody. Alternatively a demand for staking particular crops or for farming timber can be met by allocating a portion of the hedgerows to production of the desired wood type. Weed control in crop lands can also be favoured by hedgerow management, as illustrated for *Imperata cylindrica* in Nigeria (Aken' Ova and Atta-Krah 1986; Anoka *et al.* 1991) and for other species in Tanzania (Maghembe *et al.* 1986).

The success of these manipulations arises partly from management experience, but is best grounded on an understanding of plant relations in alley farming systems. A review by Anderson and Sinclair (1993)

describes the four types of yield response by two plant components of a system compared with their behaviour as pure stands:
- Competition: yield of both components negative.
- Predation: yield of one component positive, one component negative.
- Commensalism: yield of one component positive, one component neutral.
- Mutualism: yield of both components positive.

Competition arises when both components are using the same resources for growth at the same time; their relative access to these is often susceptible to management and may depend upon whether above- or below-ground competition is involved. 'Predation' is more often a term applied to animal behaviour and perhaps 'dominance' also conveys the concept of one species having enhanced yield and the other species suffering a reduction in activity. Commensalism is a common feature of alley farming where the yield of annual crops is enhanced by the N fixation of the shrub legume, which may itself be independent of soil N availability; this is augmented by the beneficial effects of legume prunings on the supply of other nutrients and the physical fertility of the soil. The occurrence of mutualism is less usual.

Allocation of prunings to crop or livestock production

The farmer may choose to spread the prunings from hedgerows on the land between the hedgerows to protect the soil from erosion and to increase the availability of nutrients to the crops which are planted; alternatively the prunings may be fed to livestock. At some times of the year it may be more appropriate to use prunings when they will most benefit crop production, such as at the beginning of the planting season; it may also be appropriate to feed the prunings to livestock when the crop is less responsive to prunings late in its development, or when there is a discontinuity in the forage supply or a decrease in the quality of forage which would be overcome by feeding foliage of high N content. The principal considerations are therefore:

- the need for control of soil erosion,
- the responsiveness of the crop to additional nutrients,
- the effectiveness of prunings in meeting the nutrient demands of the crop, especially in relation to the timing of demands,
- the degree of return of animal wastes to the crop lands,
- the characteristics of the seasonal feed supply available to ruminants,
- the relative prices of crop and animal products.

The results of some experiments in southwest Nigeria which dealt with crop response to hedgerow prunings (*L. leucocephala* and *G.*

sepium) and with the response of small ruminants to supplementation with prunings were integrated in a study by Jabbar *et al.* (1992) and illustrate an approach which may be taken in different farm situations. In this bimodal rainfall region there are two cropping seasons in each year. Farms differ in their level of maize yield and the responsiveness of the crop to legume prunings.

Table 9.2 incorporates three situations. Where maize yields and crop response to prunings were 'low', maize yields were assumed to be 1.7 t ha^{-1} without prunings and 2.7 t ha^{-1} if all prunings were applied and the latter totalled 7.2 t DM ha^{-1}. A farm designated as having 'high' maize yields produced 2.5 t ha^{-1} without prunings and 4.5 t ha^{-1} if all prunings were applied. An intermediate situation was also modelled.

The animal production system used sheep or goats receiving a low N diet such as cassava peel, in which 1 kg of legume prunings fed might be expected to produce an additional 50 g LWG on a 'low' fertility farm and 60 g LWG on an intermediate or high fertility farm. LWG was valued at $US1.56 kg^{-1} (when 9 Naira=$US1) and maize grain was assumed to be worth $US0.61 kg^{-1}; thus the price ratio of LWG to grain was 2.6.

The economic analysis of additional returns from allocating different proportions of prunings to crop land or to animal forage did not assume that animal wastes were returned to the hedgerow intercrop, which would have materially altered the results in Table 9.2. However,

Table 9.2 Additional returns from small ruminants according to the allocation of hedgerow prunings to fodder or as mulch applied to maize crops for three farm situations in Nigeria (from Jabbar *et al.* 1992).

Per cent prunings applied as mulch						
First season	100	100	100	50	50	50
Second season	100	50	0	50	0	0
Amount prunings (t ha^{-1}) used as:						
Mulch	7.2	5.7	4.2	3.6	2.1	0
Fodder	0	1.5	3.0	3.6	5.1	7.2
Extra returns from small ruminants ($ ha^{-1}) with maize yield and response:						
Low	0	+5	+24	+43	+61	−68
Intermediate	0	+30	+8	−105	−130	−249
High	0	+48	−78	−29	−100	−414

feeding prunings to small ruminants was highly profitable where maize yields were low and the crop was less responsive to the application of prunings, as would occur in dry situations; the best result occurred when 50% of the first season prunings (2.1 t ha^{-1}) were applied to maize land and all the remainder (5.1 t ha^{-1}) was fed to small ruminants. On lands with greater responsiveness to prunings the farmer would lose money by reserving much of the hedgerow material for small ruminants. This analysis is of course highly sensitive to the LWG/grain price ratio, and the profitability of feeding hedgerows to animals would be increased considerably if animal wastes were well conserved and returned to the maize fields, as discussed in sections 4.4 and 10.5.

Allocation of hedgerows to livestock production or to wood

The above example sets a local value on LWG and its increase through legume supplementary feeding. Comparative values are needed for uses related to wood production: stakes required for crop support, fencing materials or other farm construction uses, or fuel. The level of wood production relative to foliage growth can be manipulated by the farmer, according to management of pruning, grazing and hedgerow density, as discussed in section 9.3. At Mtwapa, Coast Province, Kenya relatively close spaced *L. leucocephala* provided leaf manure for maize, but the wood prunings provided a sizeable return as fuel of 24.0 and 17.7 t ha^{-1} in successive years, leading to gross revenue from maize and fuelwood of $US1,070 and $US700 ha^{-1} (Macklin *et al.* 1988).

In the Phu Wiang watershed of northeast Thailand, Wannawony *et al.* (1991) contrasted tree monocropping with agroforestry systems incorporating cassava or mung bean. Tree growth was used for charcoal production, on the basis that 0.7 t wet wood occupies 1 m^3 and produces 100 kg charcoal, valued at $US5 per 100 kg (assuming 25 baht = 1$US). *L. leucocephala*, *Acacia auriculiformis* and *Eucalyptus camaldulensis* produced 3.4, 2.8 and 4.0 t charcoal ha^{-1} respectively as monocrops by their third year, but intercropping with cassava or mungbean substantially reduced charcoal production of *L. leucocephala* and *E. camaldulensis*. The value of cassava more than compensated for this, but annual crop yields decreased with time, suggesting that intercropping might best be confined to the early phase of establishment. In many areas of the tropics the shortage of fuel places a financial premium on management which favours wood production.

Maintenance of the soil resource and its fertility

Soil erosion

Alley farming is especially pertinent on sloping lands. The establishment of living barriers to soil movement and the production from them of material to be used as a protective surface mulch is preferable to the excavation and building of mechanical structures. Young (1989) and Lal (1990) provide excellent examples of the effects of contour hedgerows in increasing infiltration and reducing erosion, whilst Pellek (1992) models the effects of hedgerows on soil loss. Soil erosion from cultivated sloping lands at four of five sites in southeast Asia (Table 9.3) is at unacceptably high levels, but the actual loss from the catchment may be minimized by contour hedgerows if established and maintained in an adequate manner. At the same time it must be recognized that the accumulation of soil against the barriers represents movement from the upper section of the inter-terrace area, and the surface condition of the cultivated area needs to be accepting rainfall if soil movement down the slope is to be avoided.

Nutrient additions from hedgerows

The level of nutrient additions from hedgerow prunings varies greatly according to soil and climate conditions, the legume species chosen, and its density and cutting management. Local equations therefore need to be developed to guide farmers about the expected effects of prunings on their crops. The first example gives an estimate of nutrient yield which may well represent the highest range of potential contribution to cropping, and this estimate may then be revised

Table 9.3 Soil loss (t ha^{-1} year^{-1}) in southeast Asia as affected by farmers' practice and the establishment of contour hedgerows (from Sajjapongse 1991).

Site	Soil type	Slope (%)	Farmers' practice	Legume hedgerow
Indonesia	Oxisol	8–18	50	35
Philippines				
Mabini	Entisol	15–25	97	2
Tanay	Ultisol	25–40	36	2
Thailand				
Chiang Rai	Ultisol	20–50	69	14
Chiang Mai	Alfisol	18–40	6	0.3

downwards, as elaborated further in this discussion. At Apiodoumé near Abidjan, Ivory Coast (lat. 5° N, 2130 mm annual rainfall, 30 m altitude) Budelman (1988a) grew three legumes at varying density in a wheel design (Nelder 1962) on a ferralitic sandy soil of high OM content. Leaf DM production using four prunings annually at 50 cm height was sensitive to shrub density; for example *G. sepium* yielded 5.1, 8.9 and 10.5 t ha^{-1} year^{-1} respectively at densities of 1000, 5000 and 10 000 plants ha^{-1}. The latter density might be regarded as an upper limit (or beyond) for an alley farming system in a humid tropical area. The average nutrient concentration of leaf material (Table 9.4) varied with age of stand and with species (Budelman 1989). N concentration was usually higher in *L. leucocephala* than in *G. sepium*, whilst *Flemingia macrophylla* (syn. *F. congesta*) exhibited the lowest values. The concentrations of the other nutrients K, Ca and Mg were also lower in *F. macrophylla*. The nutrient yields represented by these concentrations are calculated (Table 9.4) for the high density 10 000 plants ha^{-1} situation. These figures indicate massive augmentation of N, and the yield of other nutrients is also substantial; in particular the application of K at 122–235 kg ha^{-1} year^{-1} would cause a substantial spatial transfer from the hedgerow to the intercrop area.

These figures are higher than encountered elsewhere. At Ibadan, Nigeria, Kang *et al.* (1990) quote yields of N, P, Ca and Mg from *L. leucocephala* and *G. sepium* hedgerows at 4 × 0.5 m spacing (half the density of the Table 9.4 plants) less than half the above values, but showing higher K concentration.

Changes in soil properties

Changes in the properties of the soil in the inter-row depend upon the balance between mulch addition, nutrient removal in crops, leaching and nutrient loss in runoff. No change may be detectable if the system is in balance (Rosecrance *et al.* 1992b); in Lal's (1989a) study at Ibadan, Nigeria, the ecosystem was being depleted, although less rapidly with *L. leucocephala* hedgerows than in other circumstances. Atta-Krah and Sumberg (1988) observed positive effects in farmers' fields of alley farming with *G. sepium*. Soil analysis with and without hedgerows averaged: organic C 1.54 and 1.08%; total N 0.211 and 0.145%; available P 9.7 and 5.7 mg g^{-1}; and CEC 68.6 and 45.4 mEq kg^{-1} respectively. Another Nigerian study (Kang *et al.* 1985) shows the situation after 6 years of alternate cropping of maize and cowpeas between *L. leucocephala* hedgerows spaced 4 m apart on a loamy sand Apomu series Psammentic Ustorthent (Table 9.5). Application of *L. leucocephala* prunings maintained higher OM and exchangeable K,

Table 9.4 Nutrient concentration and yield in leaf of three shrub legumes grown near Abidjan, Ivory Coast (from Budelman 1989).

Nutrient	Shrub species		
	L. leucocephala	G. sepium	F. macrophylla
Concentration (%)			
N	2.95–3.67	3.18	2.35–2.83
P	0.22	0.19	0.19
K	1.52	1.68	0.98
Ca	0.92	1.34	0.65
Mg	0.31–0.37	0.40–0.61	0.20
Yield (kg ha^{-1} year^{-1})			
DM	15 400	10 500	12 400
N	455–567	333	292–352
P	34	20–26	24–31
K	235	176	122–174
Ca	142	140	81
Mg	48–57	42–64	25

Ca and Mg levels throughout the profile depth. The application of 80 kg N ha^{-1} lowered surface soil pH slightly, and reduced exchangeable K, Ca and Mg levels relative to the treatment where prunings were retained but no N fertilizer applied. Available P in the soil was independent of treatment. The diversity of these responses indicates the need to monitor carefully the effects of management interventions on soil fertility.

Legume N fixation

It is difficult to measure the level of N fixation in deep rooted shrub legumes and to separate this from the level of soil N tapped by legume roots; the key question is the amount of N delivered by legume shoots for use in crop, animal and wood production. The factors involved were discussed in section 2.5, and reference is again made to the occurrence of non-nodulating woody perennials (Corby 1974, 1990), and to the importance of the soil actinomycetes of the genus *Frankia* and its association with non-leguminous trees such as *Casuarina* (Dreyfus *et al.* 1987). Agriculturalists who give weight to biological N fixation in alley farming will exclude from their recommendations non-nodulating species such as *Cassia spectabilis* and *C. siamea* (Giller and Wilson 1991), however well these might grow when first planted.

Table 9.5 Soil chemical properties after 6 years of alley farming with L.leucocephala and varying management at Ibadan, Nigeria (from Kang et al. 1985).

Fertilizer level (kg N ha^{-1})	Prunings	Soil depth (cm)	pH	Organic C(%)	Exchangeable		
					K	Ca (mEq/100 g)	Mg
0	Removed	0–15	6.0	0.65	0.19	2.90	0.35
		15–30	6.0	0.30	0.12	2.86	0.37
		30–45	6.0	0.12	0.08	2.00	0.31
		45–60	6.1	0.10	0.07	1.63	0.26
0	Retained	0–15	6.0	1.07	0.28	3.45	0.50
		15–30	6.0	0.52	0.22	3.46	0.47
		30–45	6.1	0.22	0.15	2.60	0.39
		45–60	6.1	0.14	0.12	2.16	0.38
80	Retained	0–15	5.8	1.19	0.26	2.80	0.45
		15–30	6.0	0.49	0.17	2.96	0.41
		30–45	6.2	0.19	0.13	2.35	0.35
		45–60	6.2	0.12	0.10	2.33	0.38

The second observation is that in effectively nodulated plants there is a strong and direct relationship between shoot growth, which determines the level of assimilate available to the nodule, and N fixation. The selection of species and their management in terms of density, defoliation and fertilizer application are the primary

Table 9.6 Potential dry matter and mulch-N contributions from L. leucocephala hedgerows in the lowland humid tropics (from Sumberg and Atta-Krah 1988).

Alley width (m)	Total length of leucaena rows per ha (m)	% of land planted to maize	Per ha alley farm		Per ha maize	
			DM (t ha^{-1})	Mulch-N (kg ha^{-1})	DM (t ha^{-1})	Mulch-N (kg ha^{-1})
1.50	6667	50	7.25	300	14.5	600
2.25	4444	66.7	4.83	200	7.2	300
3.00	3333	75	3.62	150	4.8	200
3.75	2667	80	2.90	120	3.6	150
4.50	2222	83.3	2.42	100	2.9	120
5.25	1905	85.7	2.08	86	2.4	98
6.00	1667	87.5	1.81	75	2.1	86

factors determining N output. This is illustrated by two studies. The first attempts to model the level of mulch DM and N simplistically in terms of hedgerow spatial arrangement. The assumption is made from many studies in the areas of the lowland humid tropics where *L. leucocephala* is adapted that the annual N yield of legume prunings averages 45 g m^{-1} hedgerow, and the DM yield 1.08 kg m^{-1} (Sumberg and Atta Krah 1988). On this basis the area occupied by inter-hedge maize crops and the rate of mulch addition may be computed for different hedgerow spacings (Table 9.6). Spacings from 1.5 to 6 m deliver mulch DM from 14.5 to 2.1 t ha^{-1}, and 600 to 86 kg N ha^{-1}. The high N input and weed suppression associated with close hedgerow spacing may then be balanced against the loss of land for maize production, the greater shading of the crop, and the larger labour costs of pruning, so that farmers may choose from this range of options the spacing which delivers the level of N fixation desired and the correct ancillary considerations.

The second illustration concerns the effect of pruning management on delivery of N from the hedgerow (Duguma *et al.* 1988); lenient management in terms of the frequency and intensity (height of cutting) of defoliation maximizes the output of N from prunings (Table 9.7). In this study at Ibadan N yield increased from 30 kg ha^{-1} year^{-1} from monthly cutting at 0.25 m height to 289 kg ha^{-1} year^{-1} from 6-monthly cuts at 1.5 m height, indicating the powerful effect of decisions about pruning on the delivery of N. Associated competitive effects, especially in relation to shading are discussed subsequently. Whilst the main pathway of N transfer to the crop is via the prunings, the below-ground reservoir of N turning over through root and nodule senescence should not be neglected (Sanginga *et al.* 1990).

Table 9.7 N yield (kg ha^{-1} year^{-1}) of prunings from *L. leucocephala* hedgerows at Ibadan, Nigeria according to height and frequency of pruning (from Duguma *et al.* 1988).

Pruning height (m)	Pruning frequency (months)				
	1	2	3	6	Mean
0.25	30	85	153	159	107
0.50	36	134	191	188	137
0.75	50	165	201	200	154
1.0	53	143	217	245	164
1.5	45	146	198	289	169
Mean	43	135	192	216	

N availability in hedgerow residues

Reference is made in section 10.3. to the need to synchronize the timing of soil N supply with the N demand of the planted crop. If rapidly decomposing residues are applied before crop planting rains occur there will be losses of N to the atmosphere and leaching of N below the root zone of the young crop when rains do occur. When the crop is growing most rapidly the application of residues (and their incorporation in the soil) may be mechanically difficult. The agriculturalist therefore needs to consider not only the level of N applied in prunings but the timing of its mineralization and release to crop roots.

Variation in the rate of decomposition can be a useful tool if the farmer is growing hedgerows of differing species and has the flexibility to use these on crops as conditions indicate, and to feed the balance to livestock. On sloping land a slowly decomposing mulch (such as *F. macrophylla*) will provide more sustained surface soil cover to protect the soil against erosion than a rapidly decomposing mulch, as is provided by *L. leucocephala*. Section 2.3 addressed factors controlling the rate of OM decomposition. At this point species effects are illustrated.

Budelman (1988a) at Adiopodoumé, Abidjan, Ivory Coast measured the rate of decomposition of fresh leaf material placed on the soil surface. Leaf DM and N had a half-life (time to 50% disappearance) of 22 days for *G. sepium* and 53 days for *F. macrophylla*; for *L. leucocephala* half-life was 31 days for DM and 39 days for N. IVDOM was 31% in *F. macrophylla* and 64% in the other two species, indicating a feature of the protection of the former. Similar differences between *G. sepium* and *Flemingia* were evident in a study at Ibadan (Yamoah et al. 1986), whilst in Costa Rica Haggar and Beer (1993) considered that only 10% of the N applied in *Erythrina poeppigiana* prunings were taken up by a maize crop in the year of application.

Species differences were also evident in a pot experiment (Gutteridge 1992) in which leaf material, including fine stem to 5 mm diameter of various shrub legumes, was dried at 60° C and applied to the surface of pots at rates of 2.5 and 5 t ha^{-1} equivalent. Since the species differed in N concentration the level of N application varied, as shown in Figure 9.2; N fertilizer was also applied at five rates as NaNO$_3$. Basal nutrients ($-$N) were supplied to give good growth of maize, which was harvested at 56 days. Maize shoot yield increased to the highest level of N application (200 kg N ha^{-1} equivalent) and was less at equivalent fertilizer N levels in the legume treatments. The immediate response to leafy material was greatest in *S. sesban*, and was similar for *G. sepium* and *L. leucocephala*. Maize yield was not

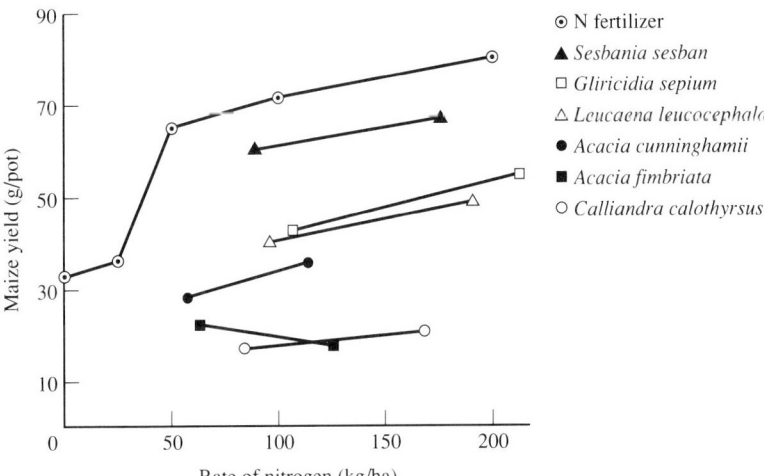

Figure 9.2 Shoot yield of maize as influenced by application of leafy material of shrub legumes (*Sesbania sesban*, *Gliricidia sepium*, *Leucaena leucocephala*, *Acacia cunninghamii*, *A. fimbriata*, *Calliandra calothyrsus*) and N fertilizer (from Gutteridge 1992).

increased by leaf of *A. cunninghamii*, *A. fimbriata* and *C. calothyrsus*. Fertilizer N recovery varied up to 78% for fertilizer N, and up to 31% for *S. sesban*. The low response to and level of N recovery from *C. calothyrsus* and *Acacia* spp. may be associated with the presence of polyphenol compounds such as condensed tannins and/or lignin in the leaf, which protect the protein against decomposition. These three species all have high tannin content. However, it was earlier noted (Gichuru and Kang 1989) that *C. calothyrsus* was successfully used in alley farming with maize at Ibadan, and in the long term the slower breakdown of foliage material may still provide substantial N release from the large bank of OM accumulated.

A second example of diversity of species effects on N nutrition is taken from an irrigated field experiment in Hawaii (Rosecrance *et al.* 1992a) on a depleted Vertic Haplustoll. This study was exploring the use of legumes not denuded by the psyllid *Heteropsylla cubana*. Hedgerows were established with plants spaced 0.5 m in the row and 3 m between rows. Maize was established in rows 0.75 m apart, with the rows closest to the hedge only 0.38 m from the shrubs. Shrubs were cut at 0.5 m height and prunings incorporated into the soil a week before planting maize; a further pruning 5 or 7 weeks after planting was applied to the surface of the hedge inter-row. On this impoverished soil maize yield on each of two occasions responded linearly to legume N application, c. 12 kg additional grain being produced per kg N applied. The response and yield were greater in the middle maize rows than the row next to the trees (Figure 9.3), presumably due to shading. The N input from shrubs increased in the following order: *Cassia siamea*, *C. calothyrsus*, *L. leucocephala* K584, *L. pallida* × *L. diversifolia* KX1 hybrid, *L. pallida* and *S. sesban*. This study also illustrates the activity of scientists in attempting to develop new shrub legumes for alley farming.

Effective use of climatic growth factors

As discussed in Chapter 4, one rationale for the diversification of farming is the more effective capture of environmental resources which may be used in crop and animal production on a sustained basis. Food supplies, farm output and farmer income are expected to be better if radiation is intercepted by plants rather than soil, water is transpired rather than lost as runoff, deep percolation or evaporation from soil, and nutrients are circulated on the farm and retained or replaced there. Alley farming is directed to these ends.

Ong et al. (1991) note that many annual monocrops have unused water stored in the soil at the end of the growing season, and that the introduction of hedgerows increases water use. This is well illustrated (Figure 9.4) by the moisture present at six successive dates over 2 years in the soil profile of areas cropped to sorghum, pigeonpea, a

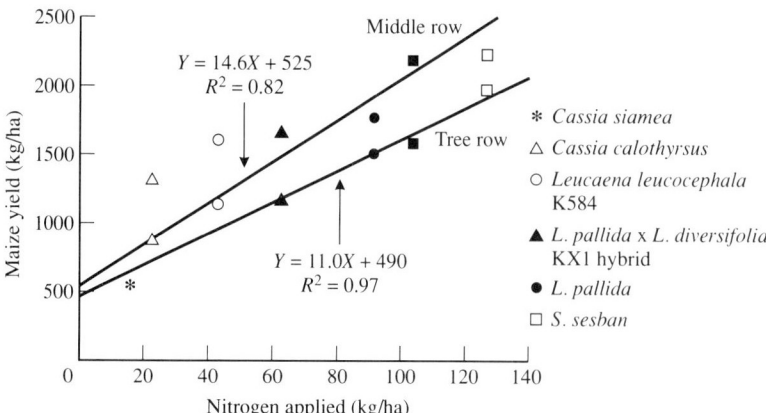

Figure 9.3 Maize yield in relation to row position and level of N applied from various hedgerows of a July crop in Hawaii (from Rosecrance et al. 1992a).

The objectives of alley farming 253

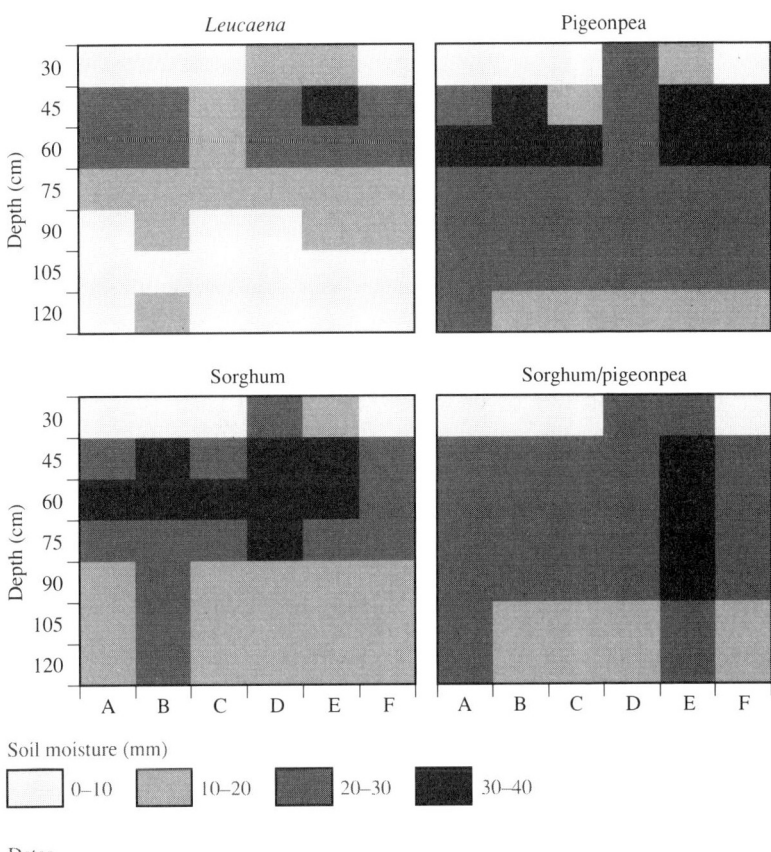

Figure 9.4 Soil moisture content (mm) on a shallow vertisol under three annual cropping systems and *L. leucocephala* during two seasons at Pancheru, India (from Ong *et al.* 1991).

mixture of these, or *L. leucocephala*. The darker areas in the diagram indicate greater moisture availability than the light coloured areas. The exploitation of moisture below 75 cm is incomplete when sorghum is grown; the addition of pigeonpea increases moisture use at depth, but it is the deep rooted *L. leucocephala* which uses all levels of the soil profile, produces a drier environment, and reduces loss from deep percolation. It should also be noted that roots of *L. leucocephala* spread into the inter-hedgerow area. Kang *et al.* (1981) excavated roots of *L. leucocephala* 2 m from the hedge in the 0–20 cm depth zone in an uncultivated situation; even where the inter-row area was cultivated significant root occupation occurred 1 m from the hedge, indicating competition for moisture with inter-row crops would be a feature of the association. The degree of interference between plants expressed in the modification of availability of water is dependent upon density of roots; high root densities decrease the inter-root distances (Gillespie 1989). Jonsson et al. (1988) report a vertical root distribution of several tree species similar to that of maize at Morogoro, Tanzania, but the biomass of fine roots was about twice that of maize. Less competition for water and nutrients is expected if trees and crops develop at different times, especially as expressed in canopy cover, or if the spatial arrangement of plants reduces interference between them.

The combination of hedgerow and annual intercrop intercepts more radiation than an annual crop alone, which provides bare ground during preparation for planting and low leaf area index (LAI) in the early phases of crop growth. Management is directed to minimizing shading of the intercrop, and short crops naturally are more vulnerable to this effect than tall crops. Cowpea grown in a 4 m space between *L. leucocephala* hedgerows received 46 and 85% radiation in the edge and middle rows before hedgerow pruning, relative to 60 and 78% respectively for maize (Kang et al. 1985). Even when hedgerows were regularly pruned to 0.6 m crops received 20–90% incident radiation (Lawson and Kang 1980), and Haggar and Beer (1993) in Costa Rica drew attention to the poor growth of maize next to *Erythrina poeppigiana* hedgerows spaced 6 m apart, which they attributed to the rapid recovery of the hedgerow after pollarding; the lesser reduction of maize growth by *G. sepium* in this study might be associated with its differing shoot architecture of more upright branching habit and with its slower recovery from pollarding. In the northern Rwandan highlands Yamoah (1991) found that pole beans were more shade tolerant between *S. sesban* alleys than potato and dwarf beans, but maize was least shade tolerant.

The most detailed recent study of the microenvironment of hedgerows and intercrops and its effects on growth is that of Corlett et al. (1992a,b) at Pancheru, India (lat. 18° N, 780 mm annual rainfall, 550 m altitude). Double rows of *L. leucocephala* 0.5 m apart were planted at varying inter-row spacings and intercropped with pearl millet (*Pennisetum glaucum*). Hedgerows were pruned to 0.7 m 2 weeks before planting millet and 43 days later. Hedgerows altered the microclimate of the intercrop according to its proximity to the hedge, the hedge shape and the relative size of the two components. Millet between 3 m spaced hedgerows experienced wind speeds 0.5-1.9 m s^{-1} less than occurred for sole millet crops. Leaf and soil temperatures within the alleys were usually warmer at night and colder during the day than for sole millet, according to the degree of shading. The effect of hedgerows on saturation deficit of the atmosphere was probably insufficient to alter the productivity of millet (Corlett *et al.* 1992a).

The most profound effects of hedgerows were on light interception and moisture availability. The hedgerows were planted slightly off-centre from a north–south axis. High levels of light interception occurred in *L. leucocephala* and sole millet (Figure 9.5), but substantial reductions in the radiation received by the intercropped millet occurred, especially on the western side of the hedge. The progress of millet shoot growth in 1986 (Figure 9.6) was similar in intercropped treatments with hedgerow spacing of 2.8 m (five rows) and 3.3 m (six rows), but was only *c.* 60% of growth produced by sole millet. However, the introduction of a polythene barrier to a depth of 50 cm between the hedgerow and the millet led to no reduction in millet growth by the presence of the hedgerow.

This result is analysed in terms of row position and growth attributes (Corlett *et al.* 1992a). In Table 9.8 the five–millet row treatment (2.8 m hedgerow spacing) has the rows nearest the hedge designated E1 and W1 for the east and west sides respectively; these rows have reduced yields relative to the centre (C) and adjacent rows. Similarly in the six–millet row treatment (3.3 m hedgerow spacing) the E1 and W1 rows have greatly decreased yield. Where a root barrier was inserted the positional effect was much reduced. Growth analysis is available for the five–millet row (2.8 m hedgerow spacing). This shows that relative to the situation for sole millet crops, radiation striking the millet canopy was greatly reduced at the edges near hedgerows (61 or 65% of full sunlight) and the actual interception by the canopy of this light was 84 and 74% respectively. The efficiency of light use in terms of DM produced per unit of light intercepted was also lower close to the hedge. The reduction of shoot yield in the intercrop to 54% of the sole crop value was due to all three components, but of these the most influential factor was the reduction in radiation falling on the canopy. Where a root barrier was introduced the millet crop suffered less moisture stress, grew taller and suffered less shading. The simplest view of this interaction may be stated that in these semiarid conditions crops growing between hedgerows experience a shortage of moisture which is expressed as shortness of stature, resulting in shading of the reduced canopy and decreased crop yield.

In humid environments radiation receipt is often the key factor determining crop yield, especially if the nutrient supply is not a gross constraint. The interference by the hedgerow in the radiation receipt of the intercrop is mainly determined by pruning management. It is suggested that many pruning regimes are too lenient to minimize the shading effect. This is illustrated for maize crops grown at Ibadan, Nigeria (Duguma *et al.* 1988) between *L. leucocephala* hedgerows

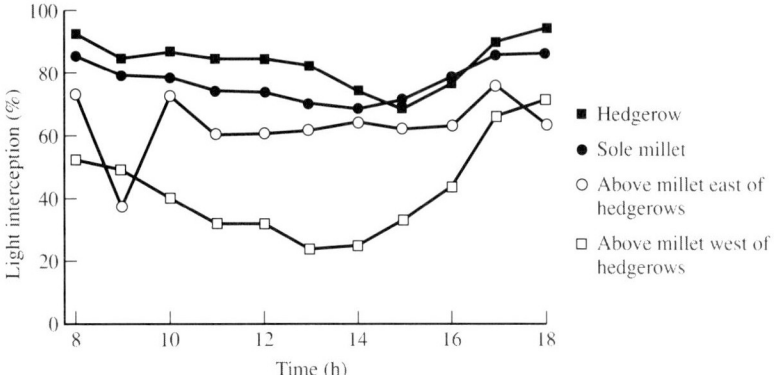

Figure 9.5 Percentage light interception by *L. leucocephala* hedgerows and millet on 25 August (57 days after sowing millet) at Pancheru, India (from Corlett *et al.* 1992a).

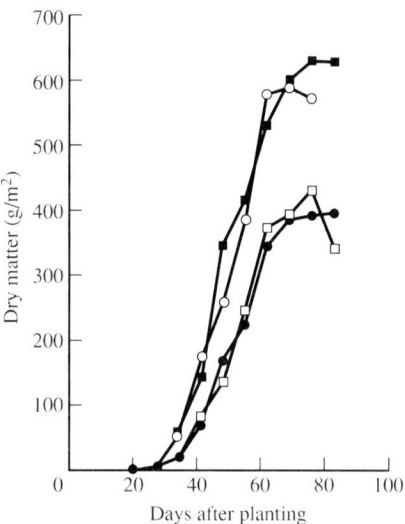

Figure 9.6 Shoot growth of millet grown alone (○), between hedgerows spaced 2.8 m (□) or 3.3 m (●), or 3.3 m with barrier to 50 cm depth between hedgerow and millet (■) (from Corlett *et al.* 1992b).

Table 9.8 Shoot yield and growth attributes of millet grown between *L. leucocephala* hedgerows relative to sole millet crop (= 100). (E1, W1 rows nearest to hedge; C centre row in five-row treatment) (from Corlett et al. 1992b).

Treatment	Variable	Row position							Mean
		E1*	E2	E3	C	W3	W2	W1	
2.8 m row spacing	Shoot yield	32	63	—	58	—	75	42	54
	Radiation receipt	61	81	—	86	—	82	65	76
	Radiation interception	84	86	—	88	—	82	74	83
	Light use efficiency	63	91	—	76	—	108	88	85
3.3 row spacing	Shoot yield	47	74	71	—	69	76	38	63
3.3 row spacing with root barrier	Shoot yield	80	88	88	—	125	98	121	100

*Row closest to hedge.

pruned at differing frequencies and cutting heights. Maize yield (Figure 9.7) was greatest in hedgerows pruned monthly, and was satisfactory with hedgerows pruned every 2 months to 0.5 or 0.25 m height, but was much reduced in other treatments. If Table 9.7 is consulted it will be noted that the N contribution of prunings operated in reverse direction to crop yield, indicating that shade was a more potent factor than N supply under the conditions of this experiment. The pruning combinations of 0.25 and 0.5 m pruning heights cut every 2 months provided 85 and 134 kg N ha^{-1} year^{-1} and may have represented the best field combination.

Alley farming leads to more effective capture of the resources which determine product outputs, but careful estimation of the value of each of these products by the farmer is needed if systems of management are to be devised which provide the desired combination of outputs, since reduced output of particular products is the usual outcome of a complex, mixed system. In humid areas the system can be manipulated to increase annual crop yields; it is a skilful farmer who also achieves significant fodder and wood production from the same area.

Nutritive value of shrub legumes

Concepts of nutritive value

Animal production from shrub legumes is not usually determined by the nutritive value of the shrub assessed as a sole feed, but rather by its contribution to the total diet of the animal from all farm sources. Complementarity is therefore sought with the characteristics of the seasonal feed supply available from herbaceous forage, crop residues and crop by-products, and perhaps from other shrub legumes and trees, in order to meet the demands of livestock for targeted levels of maintenance, growth, milk yield and reproduction. Leaves, pods and fine stem are variously fed and comparisons between the reported levels of nutritive value between species are insecure because of the differences in stage of growth, plant organs included, and climatic and soil conditions. These comparisons also need to take into account the occurrence of anti-nutritive and toxic components.

Nutritive value is a function of the amount of feed ingested and the efficiency with which nutrients are extracted from the feed during digestion (Norton 1993). The level of rumen fill, which is associated with the rate of digestion and the passage of particles through the rumen outlet, controls the amount of intake of fibrous feeds, which is determined initially by the acceptability of the feed to the animal and its physical characteristics. The intake of digestible DM, often expressed as intake of metabolisable energy (ME), needs to be

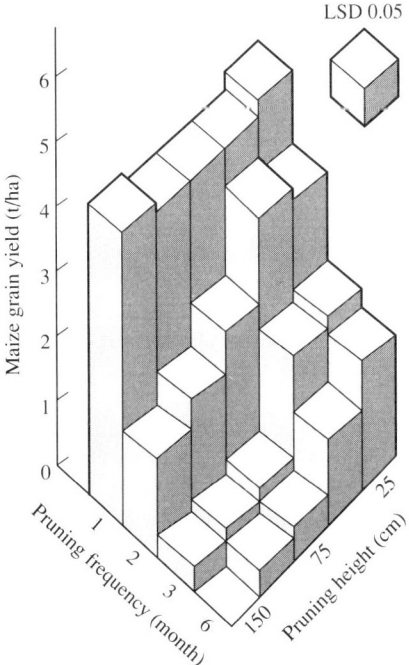

Figure 9.7 Yield of maize grain when intercropped between *L. leucocephala* hedges pruned at varying intensity and frequency at Ibadan, Nigeria (from Duguma et al. 1988).

qualified by information about the composition of absorbed nutrients if a satisfactory prediction of animal performance is to result.

Norton (1993) ranks the nutritive value of feeds according to the following attributes:

- potential for voluntary consumption,
- potential digestibility and capacity to support high rates of fermentative digestion,
- high rates of propionic acid synthesis (glucogenic) relative to total volatile fatty acid (VFA) synthesis expressed as fermentation glucogenic/energy (G/E) ratio,
- capacity to provide nutrients (protein, starch, lipid) which by-pass the rumen and are absorbed from the small intestine, expressed as absorbed P/E (protein/energy) and G/E ratios.

Our knowledge of the last three objectives with respect to shrub legumes is highly imperfect.

Energy supply

The value of studies of chemical composition is enhanced if plant material is separated into a wholly digestible fraction (neutral detergent solubles), partially digestible fraction (neutral detergent fibre or NDF) and indigestible lignin. Variation in the fibre and N fractions of the leaves and edible twigs of *Leucaena* and *Sesbania* accessions grown at Kampi ya Mawe, eastern Kenya is substantial (Wandera *et al.* 1991; Table 9.9). This study partly arose in response to the threat of the psyllid invasion; *L. pulverulenta* has field resistance to the psyllid *Heteropsylla cubana* but has high NDF and lignin content. All accessions showed N concentrations well in excess of the figure of 1.3% N judged as desirable if the microbial population in the rumen is to produce a minimum level of ammonia (70 mg Nl^{-1}) required to maintain activity for digestion.

The DM digestibility is often determined in the laboratory with *in vitro* techniques, which are used to predict actual *in vivo* digestibility by the animal. A further refinement is to measure the rate of digestion in nylon bags suspended in the rumen to give *in sacco* estimates. Norton (1993) has collated some estimates of DM digestibility of the edible fractions of the shrub legumes which show greatest promise for alley farming (Table 9.10). This table, which excludes some legumes of lesser forage value such as *Acacia angustissima*, *Albizia chinensis* (Colour Plate 11) and *Flemingia macrophylla*, shows values for DM digestibility that are high relative to those for C_4 grasses at the same development stage; the *in vivo* digestibility and intake figures were mainly derived from sheep and goats with some entries from cattle.

Variation in voluntary intake is considerable, with *C. calothyrsus* ranking poorly and *L. leucocephala* showing consistently high acceptability. The relationship between DM digestibility and intake is imperfect. The presence of tannins or alkaloids depresses intake in feeds of good digestibility, and animals may reject *G. sepium* on the basis of smell or unfamiliarity. The low particle size and high packing density of small-leafed legumes may give a fast rate of passage out of the rumen which leads to high intake and reduced digestibility. The ME value of shrub legumes is high relative to that of tropical grasses. Devendra (1992) quotes ME (MJ kg^{-1}) values as follows: *L. leucocephala* 12.1, *Calliandra* 12.6, *G. sepium* 12.8, *C. cajan* 13.4, *Sesbania* 13.6, *Erythrina* 14.3. However, the efficiency of LWG per MJ ME increases with diet quality (Leng *et al.* 1992), and at the same level of ME per DM animal performance varies substantially according to the adequacy of minerals and the presence of by-pass protein, which increases the absorbed P/E ratio. Some indication of the relatively high mineral content of selected shrub legumes was displayed in Table 9.4.

Table 9.9 Chemical analysis of leaves and edible twigs (< 6 mm diameter) of *Leucaena* and *Sesbania* accessions grown in eastern Kenya (NDF neutral detergent fibre, ADF acid detergent fibre) (from Wandera *et al.* 1991).

Species	%			
	N	NDF	ADF	lignin
Leucaena leucocephala K8	4.4	26.2	18.5	4.8
L. leucocephala cv. Cunningham	4.4	28.6	19.6	4.8
L. leucocephala cv. Peru	3.8	28.9	19.6	6.4
L. leucocephala CPI 91098	3.6	30.5	21.5	6.2
L. leucocephala CPI 84511	3.5	27.5	19.0	7.2
L. leucocephala CPI 58396	4.0	30.6	21.4	8.1
L. leucocephala (Katumani)	4.1	28.4	20.4	5.8
L. pulverulenta	2.7	39.9	28.3	9.5
Sesbania sesban var. *nubica*	3.9	33.7	27.8	6.0
S. grandiflora	4.5	23.2	19.1	4.5

Table 9.10 has been limited to seven shrub legumes of principal interest. Preston (1992) would add *Erythrina glauca*, *E. edulis*, *E. poeppigiana*, *Acacia mangium* and *Prosopis juliflora* as shrubs of special value to farmers in Colombia, whilst Riveros (1992) characterizes *P. tamarugo* in Chile and *P. cineraria* in Rajasthan, which have superior nutritive value. *Albizia falcataria* has shown promise in northeast Thailand (Gutteridge *et al.* 1984). *Albizia lebbeck* is attracting interest, especially for the high nutritive value of its flowers and pods (Prinsen 1987; Lowry 1989). Wider lists of

Table 9.10 Some values from the literature for the DM digestibility and voluntary intake of selected shrub legumes (from Norton 1993; Mahyuddin *et al.* 1988).

Species	DM digestibility (%)			Voluntary intake (g kg^{-1} LW)
	In vitro	*In sacco*	*In vivo*	
Cajanus cajan	–	53	47–65	22–26
Calliandra calothyrsus	35–48	53–59	60	16–26
Chamaecytisus palmensis	66	–	60–76	23–35
Gliricidia sepium	66–68	68–79	55–56	17–35
Leucaena leucocephala	69	53–82	55–68	27–36
Sesbania grandiflora	67	63	–	–
Sesbania sesban	68	91	–	–

shrub legumes of value are available in Le Hourou (1980), Nair et al. 1984, Skerman et al. (1988), Lefroy et al. (1992), Owina (1992) and Smith (1992b).

Protein

Protein supplementation is seen by many agriculturalists as the primary function of shrub forage legumes (Speedy and Pugliese 1992), and Devendra (1992) recommends that shrub legumes be fed to ruminants as 30–50% of the diet, or 0.9–1.5% LW daily; a lesser role is also available for monogastrics (D'Mello 1992). The value of the protein fed depends partly upon characters intrinsic to the shrub legume and partly on the nature of the other constituents of the total feed.

Highly digestible protein is used as an energy source in the rumen, and this form of protein degradation may be regarded as wasteful of N. Species which do not contain tannins, such as *Albizia lebbeck, A. saman, Enterolobium cyclocarpum* and *Sesbania* spp., exhibited *in sacco* N digestibility of 78, 90, 96 and 96% respectively; these species provide high levels of rumen ammonia, much of which is excreted in the urine.

Condensed tannins (procyanidins or proanthocyanidins) form complexes containing proteins which protect the latter from rumen digestion and augment the by-pass protein available for absorption in the small intestine. This seems to apply to species such as *L. leucocephala* and *G. sepium* which often contain condensed tannins of the order of 2–5% and exhibit moderate protein degradability. On the other hand, high levels of condensed tannins are inimical to intake and protein complexes may be protected from digestion even in the small intestine; *C. calothyrsus* may contain 11% condensed tannin and have *in sacco* N digestibility of only 36% (Ahn *et al.* 1989).

Many farmers will prefer to use shrub legumes for grazing or for cutting and supplementary feeding protein rather than for spreading on the soil in the expectation of higher grain yields (Plate 9.3). The assumption in Table 9.2 was directed to assessing the value of additional income gained from animals fed shrub legumes relative to the income from shrub legume as green manure; a more realistic approach would be to consider the returns from feeding shrub legumes and to add a component of maize response to animal excreta delivered to the cropping fields.

An illustration of the beneficial effects of legume addition in feeding systems is given from the Eastern Province of Zambia (Phiri *et al.* 1992). Maize husk is an abundant crop residue in Zambia, with low N content of *c.* 0.64% and 30% acid detergent fibre (ADF), making it an unattractive feed as sole diet. Goats were fed maize husk + 1%

urea, making a control ration with 1.1% N. They were also offered maize husk + *L. leucocephala* or *C. calothyrsus* in the ratio of 3:2, or a mixture of the two legumes in the ratio 3:1:1. *L. leucocephala* and *C. calothyrsus* had N content of 4.0 and 3.9% respectively, and ADF of 22.6 and 46.3%. Goats fed legume rejected some of the maize husk (Table 9.11), but increased their total DM intake. The mixtures containing legumes also exhibited higher digestibility, and large differences in LWG occurred. Table 9.2 was based on Nigerian data which assumed an additional 50 or 60 g LWG per kg legume DM feed; in this study 1 kg *L. leucocephala* DM produced an additional 130 g LWG. Feed efficiency in terms of kg DM intake per kg LWG was increased by a factor of three or four when legumes were fed.

This example with goats could be multiplied by numerous examples throughout the literature of tropical animal production. *G. sepium* has been especially successful as a protein supplement, for example in Sri Lanka with Bannur ewes fed *Brachiaria miliiformis* (Chadhokar and Kantharaju 1980) and with milk production of MRY cows fed grass mixtures or rice straw (Chadhokar 1983). In many instances expensive purchased concentrates may be replaced by shrub legumes grown on the farm, and the immediate cash return from increased milk yield makes this practice very attractive to farmers.

Anti-nutritive factors

The preceding discussion has mentioned condensed tannins, which may protect protein so that tannin–protein complexes reach the abomasum, where in low pH conditions dissociation may release the protein for absorption by the animal (Norton 1993). Alternatively, the protein may be so firmly bound with tannins that it is excreted unused by the animal, with perhaps an additional depressive effect on appetite. The role of tannins in nutrition is reviewed by Mangan (1988) and Kumar and Vaithiyanathan (1990); these polyphenolic compounds have high molecular weight in the range 500 to 3000, and are conveniently divided into the hydrolysable tannins, which may cause toxic manifestations in the rumen, and the condensed tannins (proanthocyanidins or procyanidins), which have no carbohydrate core and which are usually derived from the condensation of flavenoid precursors (Paterson 1993). In forage legumes they usually exist as leucoanthocyanins, which form complexes with proteins, as mentioned earlier. Tannins are probably implicated in the prevention of bloat by hindering the formation of stable protein foams in the rumen, a hazard of many temperate (but not tropical) legumes. There is considerable variation in the content of condensed tannins according to legume species and legume provenance, and great opportunity to select for a desirable (and moderate) level of tannin in cultivated shrub legumes.

Table 9.11 Effects of the addition of shrub legume to a diet of maize husks on the performance of goats in Zambia (from Phiri et al. 1992).

Component	Treatment additive			
	Urea	L. leucocephala	C. calothyrsus	Both legumes
Intake g day^{-1}				
Maize husk	250	149	147	149
Legume	–	182	168	168
Total	250	332	315	317
DM digestibility (%)	47	63	59	60
LWG (g day^{-1})	4.8	28.5	19.0	22.6
Feed efficiency (kg DM kg^{-1} LWG)	52	12	17	14

The content of condensed tannins is radically altered by drying the forage which then has detrimental effects on animal performance; sheep exhibited voluntary intake of *C. calothyrsus* of 59 or 37 g DM kg^{-1} W$^{0.75}$ respectively if fed material green or dried (Palmer and Schlink 1992).

Mimosine is the best known toxic amino acid, because of its occurrence and study in *L. leucocephala*; this limits the proportion it may contribute to the diet of monogastric animals. In Australia and Papua New Guinea ruminants lacked bacteria which avoided the degradation of mimosine in the rumen to 3-hydroxy-4(IH)-pyridone (DHP), which is a potent inhibitor of thyroid peroxidase, needed in the synthesis of thyroid hormones; these bacteria have since been introduced (Jones and Megarrity 1986) with remarkably positive effects on cattle performance (Quirk *et al*. 1988).

Cyanogenetic glycosides have been located in several *Acacia* spp., and saponins, which occur in *Sesbania sesban*, are glycosides that may be deleterious to monogastrics but have been less implicated in ruminant dysfunction. Various alkaloids occur in *Sesbania* spp. and *Acacia* spp. *G. sepium* when mixed with cooked maize is used as a rodenticide, presumably due to the conversion of coumarin to dicoumerol, a haemorrhagic compound (Simons and Stewart 1993).

Much research on the nutritive value of shrub legumes is relatively recent. The role of particular condensed tannins in affecting intake and N utilization is of special moment, and plant breeding and selection activity has been little directed to providing material élite in its nutritive value. Nevertheless the excellent responses to the incorporation of many shrub legumes in ruminant diets, which have been expressed in superior animal growth, reproduction and milk yield, are providing a strong rationale for the focus of agricultural development on the readier availability of these plants.

9.3 Alley farming practice

Hedgerow establishment and density

Inputs the farmer makes to effect successful shrub legume establishment have a high return, since the shrubs are long-lived, and if the farmer has long-term occupancy of the land the cost of establishment may be amortized over a long period. Many shrub seedlings are initially slow growing and therefore vulnerable to competition from volunteer weeds or companion grasses; in *L. leucocephala* this is associated with a lower root length density than C$_4$ grasses (Brandon 1993).

Direct planting of seed (at *c.* 2 cm depth for *L. leucocephala* and *G. sepium*) is feasible in humic environments (Reynolds *et al.* 1988), but in drier environments (less than 1200 mm annual rainfall) seedlings are often raised in perforated bags and transplanted to the field. Hot water treatment of hard seed (80° C for 4 min for *L. leucocephala*) may be needed to promote germination. Inoculation with *Rhizobium* is desirable, and if commercial cultures are not readily available soil from an established stand of the legume species in question should be sprinkled in the planting row. The growth of the *L. leucocephala* seedling is highly dependent on good P nutrition, and there is a positive effect of the density of VAM on the P uptake of the seedling. The slow growth of *L. leucocephala* seedlings on soils deficient in VAM may be overcome by high levels of P application close to the seedling (Brandon 1993). There is also a positive interaction between nodulation and the occurrence of VAM. Fertilizer inputs are usually needed to promote high levels of N fixation.

For some shrub species it is customary to plant stakes that readily root from stem tissue, and *G. sepium* is established either from seed or from stakes, which may range in length from 0.1 to 2 m, depending on the purpose of establishment (Atta-Krah and Sumberg 1988); a living fence may be rapidly achieved if long stakes are planted. Shrubs should be established on contour lines, unless the purpose of establishment is linked to defining a property boundary. Clean weeding should be carried out during the establishment year.

In section 9.2 (Table 9.6) the influence of row spacing on N output from prunings was discussed. Higher shrub densities lead to greater N fixation and to inter-row weed suppression, but also greater competition with annual intercrops. In alley farming situations hedgerow spacings of 4–5 m are common. Farmers may favour more informal arrangements of shrubs in cropping fields, plant high density shrub stands to maximize legume output in one section of the farm, plant shrubs in contour lines on sloping lands to control erosion, or restrict shrub plantings to boundaries or house compounds. In humid environments there is a positive association between shrub density and leaf output, as discussed earlier (Budelman 1989), and this relationship is especially strong during the main wet season, as for example in South Sulawesi, Indonesia for *C. calothyrsus*, *G. sepium* and *L. leucocephala* (Ella *et al.* 1989).

In alley cropping situations the key decision is the spacing between hedgerows. However if there is a choice of spatial arrangement a reduction in rectangularity will favour yield and N fixation. For example, in the Moyamba District of southern Sierra Leone (Karim and Savill 1991) the N yield from leaves of *G. sepium* plants grown at 0.5 m^{-2} for the establishment year was 96, 128 and 271 kg ha^{-1} as the ratio of between-row spacing to within-row spacing was reduced respectively from 16:1 to 8:1 to 2:1 (i.e. 8 × 0.25 m, 4 × 0.5 m or 2 × 1 m spacing). These differences were also reflected in height, collar diameter, branch number and biomass. Such flexibility is not available in alley cropping. Competition between hedgerows and crops has been discussed in section 9.2, and the choice of a suitable width between hedgerows is a powerful factor in manipulating the outcome of these relationships.

To this point in this chapter biological relationships have been emphasized; the labour inputs involved are often an overriding consideration in determining both the feasibility of the system and its attractiveness to farmers. The labour of pruning is directly related to closeness of hedgerow spacing. For *G. sepium* planted alternately with *L. leucocephala* in Nigeria at 4–5 m spacing between hedgerows producing 3 t DM ha^{-1} year^{-1}, pruning to 60 cm height four to six times annually requires *c.* 18 days ha^{-1} year^{-1} (Reynolds *et al.* 1988).

Cutting and grazing management

The system of cutting or grazing adopted by the farmer depends upon the emphasis placed upon different objectives:

- maximization of leaf and N output,
- production of farm timber or fuel,
- reduction of intercrop shading,
- economy of labour inputs.

The principles which have been enunciated for the defoliation of herbage legumes (Humphreys 1991, chapter 4) apply to shrubs, with the reservations that:

- stem production may be valued,
- re-establishment of the leaf surface after defoliation is more insecure in many shrubs than in herbage legumes,
- non-structural carbohydrate probably plays a more significant role in determining the initial rate of regrowth after defoliation.

The simplest generalization is that defoliation management determines DM production and N fixation through its effects on the size of the leaf canopy available to intercept radiation, integrated over time. Production then needs to be partitioned into the plant material actually harvested on the farm, the material which accumulates as standing biomass, and the senescence and decomposition of material which occurs most rapidly during periods of rapid growth. Simple experiments which report the harvested yield when cutting height and frequency are varied are difficult to conceptualize, since deleterious effects on growth may be compensated by greater efficiency of use.

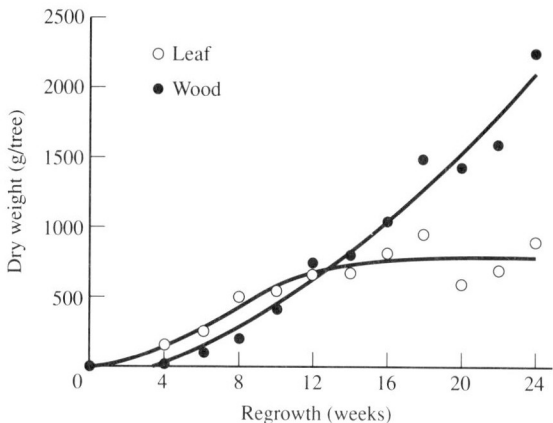

Figure 9.8 Production of leaf and wood after cutting *C. calothyrsus* at Mt. Cotton, southeast Queensland (from Stür *et al.* 1993).

One important fact is that the ratio of leaf/wood production decreases with age since cutting. This is illustrated for a study with *C. calothrysus* (Colour Plate 10) at Mt. Cotton, southeast Queensland (Stür *et al.* 1993). Yield of wood (Figure 9.8) showed a lag in production relative to leaf growth, but wood continued to be laid down from 12 to 24 weeks (Figure 9.9) and reached 50% after 12 weeks. In this particular study the shoot growth rate was maximal 7 weeks after cutting but accumulated growth increments were greatest at 10–12 weeks, which might in these terms have merit as an ideal cutting frequency. Studies of other species give varying results; Gutteridge and MacArthur (1988) found higher leaf yields of *G. sepium* at more infrequent cutting. Table 9.7 suggests a small advantage in output of leaf N from 6-monthly cuts relative to 3-monthly cuts with *L. leucocephala*, and Ferraris (1979) and Hutton and Beattie (1976) obtained maximum output from this species with 4-monthly cuts.

The ability of forage shrubs to coppice is an important selection characteristic, and is well developed in *L. leucocephala* (Plate 9.2). Plant recovery from defoliation may depend upon the presence of buds below cutting height, and the relative significance of lateral buds at the base of the shoot and of axillary buds at the base of the petiole. In some species it is suspected that uninterrupted growth leads to an elevation of the basal buds, so that delayed cutting may leave few buds for recovery growth. Thus, branching of *Sesbania sesban* may benefit from cutting within 3 months of seedling establishment

(Evans and Rotar 1987) and the survival of *S. grandiflora* may be prejudiced by delaying the first cut at 1 m height until the shrub is 5.5 m high (Ella *et al.* 1991). The latter study was carried out at Gowa, South Sulawesi (2700 mm annual rainfall, monsoon climate) on a Typic Ustifluvent; shrubs were established at successive times and the first cut applied at ages from 13 to 21 months. The leaf yield at first harvest (Figure 9.10) increased with increasing age since planting, and *S. grandiflora* and *C. calothyrsus* outyielded *G. sepium* and *L. leucocephala*. In subsequent cuttings many *S. grandiflora* shrubs died, and the rate of leaf production of *C. calothyrsus* has only slightly affected by age at first cut. On the other hand was delaying age at first cut to 19 or 21 months had a profound positive effect on subsequent leaf growth of *G. sepium* and *L. leucocephala* which was associated with greater branch number and the capacity for rapid canopy development. A more moderate defoliation regime is one which leaves some lateral branches intact so that the carbon shortage induced by defoliation is less extreme, but this is inimical to formal hedgerow development directed to minimizing shading of companion crops. Cutting *G. sepium* at the end of the wet season produces a mid-dry season flush of growth. The beneficial effects of severe defoliation regimes on inter-row crop yield was discussed in relation to Figure 9.7 where avoidance of shading was a more important objective than maximizing N output from prunings. The timing of crop demand for N will also influence decisions about the timing of pruning, as well as the imperative to develop a dry season fodder bank.

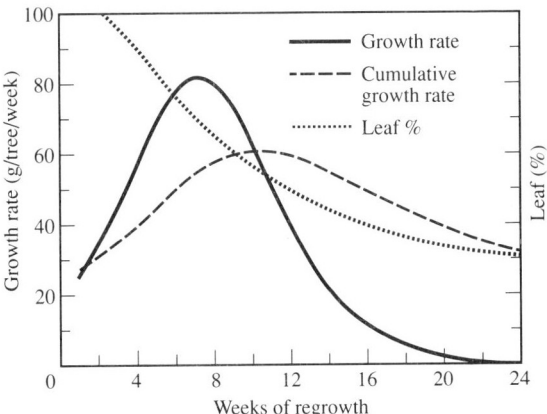

Figure 9.9 Regrowth characteristics of *C. calothyrsus* after cutting at Mt. Cotton, northeast Queensland (from Stür *et al.* 1993).

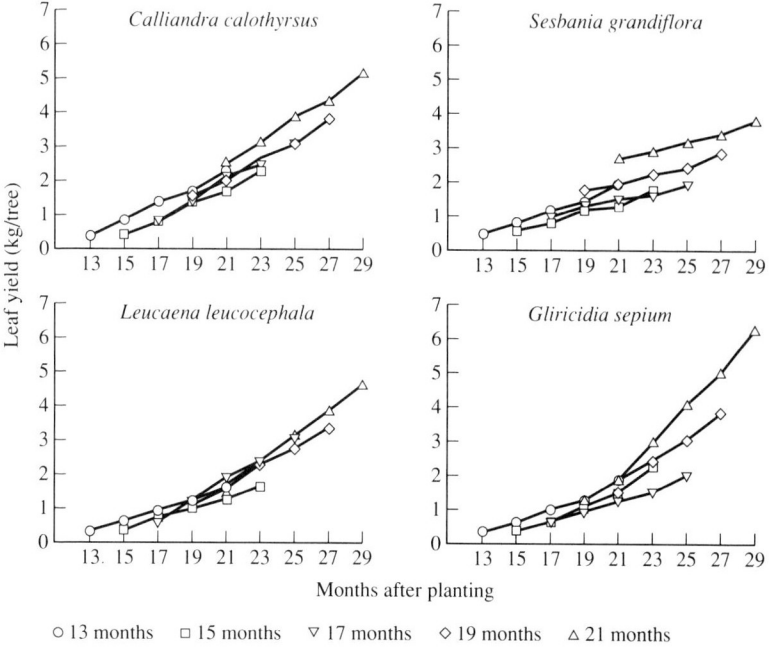

Figure 9.10 Cumulative leaf yield of four shrub legumes in South Sulawesi, Indonesia as influenced by age at first cutting (from Ella *et al.* 1991).

Hedgerow pruning of *L. leucocephala* and *L. pulverulenta* has been successfully mechanized (Felker *et al.* 1991), but in most tropical countries hand pruning is the norm and constitutes a significant farming input. Darnhofer (1992) studied at Machakos, Kenya the factors influencing the labour requirement for pruning *L. leucocephala* and *Cassia siamea* hedges; wood of the latter species was more difficult to cut. The time taken to prune a row to 30–50 cm height was greater using hand secateurs than using a sickle, and this in turn was greater than using a machete or panga, which caused more splitting and damage to trees. A model was developed relating pruning time to shrub density within the row and biomass present. When hedges were spaced 4 m apart (2500 m hedgerow ha^{-1}), (1) biomass varied from 0.5 to 1.2 kg fresh weight per m hedgerow and (2) spacing between plants in the hedgerow varied from 0.3 to 1 m, and the time taken to prune 1 ha ranged from 62 to 189 person h ha^{-1} according to the levels of the two variables identified.

An economic analysis of current prices in Kenya suggested that the return to labour from pruning *L. leucocephala* was greater by a factor of about four if prunings were fed to dairy cows rather than placed on the soil to promote increased maize yield at Machakos.

Choice of shrub legume species

Shrub legumes are needed which are ecologically successful in the particular farming systems that occur in the tropics and subtropics. Conditions may combine extremes of heat and cold in the continental subtropics, and whilst moisture stress is a rare feature of humid equatorial tropics, long dry seasons occur in monsoon climates which may combine 5–6 months of heavy rainfall with little rain at all in the other half of the year. Shrub legumes are successfully grown in low rainfall zones which are marginal for cropping and in which deep rooted shrubs are a stabilizing factor in the agricultural ecosystem. Poorly drained areas unsuited to annual cropping may be reserved by farmers for forage production using shrubs. Whilst much shrub planting has been most successful on soils of neutral reaction or of medium density, plants are especially needed for low pH soils which have toxic levels of monomeric aluminium or of manganese, and for saline situations. Biotic challenges, such as that of the psyllid *Heteropsylla cubana*, need to be met.

Plants achieve dominance in the different situations through their superior resistance to environmental stress and/or their capacity to interfere with the supply of environmental growth factors to their neighbours (Humphreys 1981). These may be conveniently considered in terms of climatic, edaphic, and biotic factors. Shelton (1993) has collated a good deal of literature to produce a tabulation of the resistance to stress of many of the shrub legumes of principal interest (Table 9.12). These comparative data are expressed in relative terms, often based on information from folk-lore; they need to be viewed in this context and also in the expectation that useful variation may be found of ecotypes within each species which will extend the current range of adaptation (Simons and Dunsdon 1992, Cobbina and Atta-Krah 1992 for *G. sepium;* Hughes 1993 for *Leucaena* spp.; Macqueen 1993 for *Calliandra* spp.).

Climatic adaptation

Chamaecytisus palmensis, which originates in the Canary Islands, is the pre-eminent shrub legume for cold highland situations, and one of the few cultivated subtropical shrubs which exhibits true frost resistance, although many species such as *L. leucocephala* will survive severe frosting despite loss of foliage. *L. diversifolia*×*L. leucocephala* hybrid showed superior tolerance of 27 frosts at Blenheim, southeast

Table 9.12 Climatic and edaphic resistances of shrub legumes (VG very good, G good, M medium, P poor, VP very poor) (from Shelton 1993).

Species	Climatic resistance					Edaphic resistance			
	Rainfall (mm)	Drought	Cool temp.	Frost	Low fertility	Water-logging	Acidity	Alkalinity	Salinity
Acacia aneura	200–500	VG	G	G	G		G		
Acacia villosa	600–3000	G	M	P	G	G	G		
Albizia chinensis	600–3000	G	M	M	G	P	G	G	
Albizia lebbek	600–2500	G	M	M			G	G	G
Calliandra calothyrsus	700–4000	P	M	P	G	P	G		
Chamaecytisus palmensis	350–1600		G	G	M	P	M	G	
Codariocalyx gyroides		P	P		P	G			
Desmanthus virgatus	> 700	VG	M	M	G	G	M	G	
Faidherbia albida	300–3000	G	G	M	G	G	VG		
Flemingia macrophylla	1100–3000	P	M		G	M	G	G	
Gliricidia sepium	900–3500	P	P	VP	G	P	M	G	M
Leucaena leucocephala	650–3000	G	P	P	P				
Leucaena diversifolia			M						
Sesbania grandiflora	> 800	G	VP	VP		VG	M	G	G
Sesbania sesban	>1500	P	G	P	M	VG	M	G	G

Queensland (lat. 28° S) where *S. sesban* died and *C. callothyrsus* had 50% mortality (Gutteridge and Sorensson 1992). Shelton (1993) recalculated some data of Swasdiphanich (1993) which compares 16 weeks' growth of nine species in a range of temperature regimes (Table 9.13). In this group *S. sesban* is notable for its growth in cool conditions, which was superior to that of *L. leucocephala*. The poor performance of *S. scabra* and *Albizia chinensis* at 25/21 °C suggests truly tropical adaptation, whilst at Mt. Cotton, southeast Queensland, *G. sepium* drops its leaves when night temperatures reach 12 °C.

The field drought resistance of *S. scabra* is well known, and a study by Swasdiphanich (1993) quoted by Shelton (1993) in which seedlings of five shrub legumes were grown under conditions of good moisture supply and then subjected to moisture stress confirms its superiority; water was extracted to a lower soil content, young leaves reached a lower percentage relative water content and were retained on the plant (Table 9.14). In this comparison seedlings of *C. calothyrsus* and *G. sepium* performed more poorly than *L. leucocephala*. *Desmanthus virgatus* is showing excellent drought resistance in western Queensland, and *S. grandiflora*, *Faidherbia albida*, and many *Prosopis* spp. are adapted to semi-arid regions.

Shade tolerance is of value when shrubs are planted in a 'three-strata system', in plantations or in forest margins. Benjamin *et al.* (1991) have data indicating *G. sepium* shows superior shade tolerance. The percentage reduction in growth in low radiation regimes relative to growth at higher radiation levels was 23, 38, 39, 40, 56 and 70% respectively for *G. sepium*, *S. grandiflora*, *Albizia chinensis*, *C. calothyrsus*, *L. leucocephala* and *A. villosa*.

Table 9.13 Shoot growth of shrub legumes (% of species maximum) in relation to temperature (from Shelton 1993).

Species	Temperature regime (day/night °C)			
	20/17	25/21	30/27	34/32
Albizia chinensis	1	3	63	100
Acacia villosa	1	4	61	100
Calliandra calothyrsus	10	19	100	93
Gliricidia sepium	5	20	69	100
Leucaena leucocephala	6	11	46	100
L. diversifolia	2	22	76	100
Sesbania sesban	36	67	85	100
S. grandiflora	4	13	75	100
Stylosanthes scabra cv. Seca	1	9	100	53

Table 9.14 Indices of drought resistance of shrub legume seedlings (water use efficiency WUE, relative water content RWC) (from Shelton 1993).

Species	Average weekly WUE prior to stress (g kg^{-1})	Drought resistance indices after severe stress		
		Soil moisture (%)	Lowest RWC of young leaf (%)	Fallen leaf (%)
Calliandra calothyrsus	1.6	13.0	49.5	90
Gliricidia sepium	4.2	12.1	43.3	55
Sesbania sesban	1.6	12.5	38.2	55
Leucaena leucocephala cv. Cunningham	2.6	12.0	33.0	25
Stylosanthes scabra cv. Seca	0.6	11.9	22.8	2

Edaphic adaptation

Tolerance of waterlogging and capacity to grow on soils of poor internal drainage is a feature of *S. sesban* (Colour Plate 12), *Acacia villosa* and *S. grandiflora*. On the other hand *L. leucocephala* and *C. calothyrsus* are especially sensitive to poor drainage, and this limits considerably their zone of adaptation. Tolerance of soil acidity has many components, according to local conditions. It is a severe limitation to the use of *L. leucocephala*, whose mineral characterization is available (Ruaysoongnern et al. 1989), although breeding activity is showing promise in improving the breadth of its distribution. *Flemingia macrophylla* has been successful on acid soils of high Al status (Thomas and Schulze-Kraft 1990 for Colombia; Matthews et al. 1992a for Zambia), whilst there are many reports of *G. sepium* growing in soils of medium acidity (Szott et al. 1991 for Peru). *Desmodium strigillosum* is adapted in Carimagua, Colombia and is better accepted by cattle in the dry season than other shrub legumes (Thomas and Schulze-Kraft 1990). The low-growing shrub *S. scabra* has been highly successful on low fertility acid soils in north Queensland.

Biotic factors

The capacity of shrubs to branch freely in response to defoliation by grazing or cutting was emphasized earlier as a primary selection characteristic. Farmers also look for resistance to diseases and pests, since it is rarely practicable to use chemical control measures on shrubs used for fodder, fuel or green manure. Two notable problems have dominated recent agricultural research. The anthracnose disease *Colletotrichum gloeosporioides* (Chakraborty *et al.* 1993) threatens the eventual success of the shrub *S. scabra*, and genetic manipulations at the molecular level as well as conventional breeding programmes are directed to maintaining plant resistance in the face of rapid disease evolution.

The psyllid *Heteropsylla cubana* has been especially devastating to *L. leucocephala* production in humid areas of many tropical countries (Bray and Woodroffe 1991), although an equilibration with the pest has occurred in subhumid regions of 900 mm annual rainfall. Successful breeding programmes, especially with tetraploid *L. pallida*×*L. leucocephala* hybrids are overcoming this problem and providing greater tolerance than possessed by cultivated *L. leucocephala*. The narrow genetic base of much *L. leucocephala* which has become naturalized throughout the tropical world has exacerbated this problem. One reservation to be expressed is that psyllid-resistant lines usually exhibit higher levels of condensed tannins which may depreciate the high animal performance previously reported from *L. leucocephala*.

9.4 Conclusions

Farmers seek solutions to widespread decreases in the yield of annual crops, the inconvenience of soil erosion and the need for food and income security in the tropics and subtropics; shrub legumes constitute part of the solution. However the forms in which shrub legumes are used will not be restricted to alley farming, which to date is poorly adopted in farm practice.

This book is mainly concerned with the biological processes which determine the sustainability of farm systems, but this emphasis is not intended to deny the overriding significance of the decision making of farmers and of institutions, and the cultural, sociological and economic circumstances from which these decisions emerge.

Land tenure is a primary consideration in the adoption of alley farming (Francis 1987). Communal ownership of land, or the right of extraneous access with grazing animals, operates against the planting of shrubs and the appropriate management of hedgerows subsequently. Shrub planting is readily accepted where farmers have long-term rights of land use. A further consideration is the separation of ownership of trees and use of land. If different people have the right of use of the tree and of the land for cropping the concept of spreading prunings on the adjacent land requires an accommodation to be reached. This problem is exacerbated if there exist traditions of gender access; males may own the trees where women grow the crops. Particular trees may vary in this gender access; in parts of western Kenya *Sesbania sesban* is regarded as 'a woman's tree' and therefore has especial utility in improved fallow systems. The choice of ruminant species may also be linked to the farm tasks of men and women; goats, which tend to browse near the home, may be more readily managed by women in southwest Nigeria than sheep, which are more prone to graze fields and roadsides (Okali and Sumberg 1985).

The material presented in this chapter indicates that the adoption of alley farming is favoured by:

- long-term rights of land and shrub use,
- shortage of high fertility cropping land, also associated with high population density,
- good annual rainfall, perhaps in excess of 1200 mm,
- sloping land under annual crops,
- availability of shrub legumes adapted to local soil conditions,
- farm requirement for poles and fuel,
- incentives for increased ruminant production,
- seasonal availability of labour for pruning.

Shrub legumes will continue to play increasing roles in contour plantings on sloping land, in defining field boundaries and in augmenting the diet of livestock. On flat land farmers will often choose to plant blocks of forage shrubs in otherwise unwanted land, or close to the home as part of an intensive feed or crop garden. When hedgerows are established and pruned vigorously, the spreading of prunings on crop land (especially at the time of planting maize) will be favoured by the factors listed above, but many farmers will give priority to feeding legume foliage to their livestock. A more careful conservation of livestock excreta and its return to cropping fields may prove to be the primary pathway by which the N fixation of shrub legumes is used to sustain crop production.

CHAPTER 10

Pastures with annual crops

10.1 Introduction
The philosophy of the ley

The ley, or the planting of short-term pastures in rotation with crops, has long been an essential part of the European tradition of farming. Mills (1759) wrote about agriculture in England: 'The general custom in this country is, to sow the clover seed with barley, in the spring: and when the barley is taken off the ground, the clover spreads and covers it, and remains two years: after which the land is ploughed again for corn, and is thought to be greatly enriched by the clover'.

This statement describes three important aspects of the ley:
- the undersowing of a crop to establish a pasture inexpensively,
- the use of a legume,
- the capture of the benefits of the legume in the yields of succeeding crops.

Lawes et al. (1861) referred to the last concept in their experiments: 'it should be particularly observed that, after taking 206.8 lbs of nitrogen from an acre in the clover crop of the first year, the wheat crop of the succeeding year was about double that obtained in the same season on a plot which had grown the crop for a series of years without manure'.

The early chapters of this book set out the objectives of integrating forages in cropping systems; the benefits revolve about:
- the maintenance of the chemical, physical and biological fertility of the soil,
- the control of soil erosion,
- the more efficient use of environmental resources and the control of pollution,

- the development of crop protection,
- increased levels of animal and crop production,
- better and more stable levels of farm income.

The failure of past attempts at ley farming in the tropics

Ley farming continues to be practised in temperate agriculture, although the increasing specialization of crop production, the availability of cheap nitrogenous fertilizer and changes in economic incentives have reduced the incidence of the mixed farming of crops and animals. Leys based on clovers and on lucerne are still planted in Europe and North America, and in southern Australia wheat yields in that Mediterranean environment have been sustained by the growth of the annual legumes *Trifolium subterraneum* and *Medicago* spp.

Ley farming has never been widely adopted in the tropics and subtropics, despite the efforts of workers in agricultural research and extension. Land development to planted grasses may pass through a pioneering phase; large areas of scrub in Brazil have been cleared and planted to rice by share farmers prior to the establishment of permanent pastures for cattle which are based on *Panicum maximum* or *Brachiaria decumbens*. In West Africa land is allowed to revert to a natural 'bush fallow' as crop yields decrease, but pastures are seldom planted in this situation.

The planting of pastures in rotation with crops was earnestly advocated in the decades between 1920 and 1960 in eastern and southern Africa (Martin 1944). We have now come to understand the misplaced emphases which in part led to the failure of these efforts. They hinge upon

- the focus on grass planting,
- fertilizer policies,
- cut-and-remove systems of use.

Grasses are known to be more effective in fostering crumb structure in soils than are legumes, as discussed in Chapter 3. For instance Martin (1944) noted that at Serere, Uganda the proportion of water-stable aggregates greater than 0.5 mm diameter was marginally higher after a *Cynodon plectostachyus* ley, than after the legume *Stizolobium deeringianum*. Water-stable aggregates increased as the duration of the grass ley increased and cultivation frequency decreased. In South Australia after a 3-year ley aggregation was 51% in a grass ley and 28% in a clover ley (Clarke *et al.* 1967). However, much of the benefit of improved soil structure is lost in the first season after cultivation, and on some soils constraints to crop growth other than physical structure may determine yield. At Mazowe, Zimbabwe (lat. 17° S, 910 mm annual rainfall, 1200 m altitude) the grain yields of maize after 4 years of grass ley were no greater than those of maize grown for 12 successive years on the same land, fertilized adequately and having maize stover ploughed in (Rodel et al. 1981). On these silty clay soils the justification for using a grass ley would rely on the animal productivity of the ley phase and the protection from soil erosion the grass ley would provide and not on any benefits to following crops.

A second question is the adequacy of the fertilizer policy which has been adopted. Intensively managed grass leys which have received high levels of N fertilizer may require liming if soil acidity is to be controlled. The grass used in the ley may tolerate the low pH conditions induced, but the crops following the ley may suffer. At Marondera, Zimbabwe (lat. 18° S, 950 mm annual rainfall, 1650 m altitude) the acid tolerant *Cynodon nlemfuensis* var. *robustus* was grown for 4 years and received different levels of N fertilizer. If no lime was applied subsequent yields of maize were considerably depressed from 2.2 to 0.4 t ha^{-1} as the previous level of N application increased from 0 to 212 kg ha^{-1} year^{-1}. Maize yields were 3.7–4.5 t ha^{-1} where lime was applied (Barnes 1981a). There was no positive residual response to previous N level in maize yield on these sandy loam and sandy clay loam soils (orthoferralitic and paraferralitic) derived from granite.

A parallel case occurred at Kawanda, Uganda where *Pennisetum purpureum* was grown for 3 years and received a high rate of N and P fertilizer. A significant decrease in the subsequent yield of cotton, sweet potatoes and maize was attributed to depletion of soil potassium (Foster 1971), and this also occurred at Marondera (Barnes 1981a). Depressed cotton yields in another study (Brockington et al. 1965) were attributed to S deficiency induced by the ley.

A third area of concern is the utilization of the ley. Nutrient depletion by the ley is associated with cut-and-remove systems. The grass harvested in a well-managed and productive ley might well contain 1.5% N, 0.2% P and 1% K; a harvested yield of 10 t DM ha^{-1} removes 150 kg N, 20 kg P and 100 kg K ha^{-1} year^{-1}. This rate of depletion has seriously detrimental effects on subsequent crop yields, especially on light soils, unless the nutrients are replaced by other means.

Tropical grasses are highly efficient in their conversion of CO_2 and light energy to carbohydrate and in their use of water and nutrients for this purpose. The rapid growth of tropical grasses, associated with the C_4 dicarboxylic acid pathway, leads to the rapid accumulation of carbonaceous material. Under conditions of low or moderate N supply surplus carbohydrate is allocated to the crown and root system, as discussed in Chapter 2. A lenient system of management in which grass accumulates or is only removed at infrequent intervals leads to the production of greater plant biomass than occurs under grazing or frequent cutting (Humphreys 1991). The physical handling of this material when preparing a seed bed for crops poses difficulties, but greater problems arise from the immobilization of soil N associated with the decomposition of residues. At Marondera the low fertility soils with which Barnes (1981a) worked contained only 0.05% total N and the second test-crop of maize after the grass ley was acutely N-deficient. In a companion experiment (Barnes 1961) the biomass of the surface roots and crown of *P. maximum* fertilized with N increased from 10.5 to 15.5 t ha^{-1} in a single season; this material had a content of *c*. 1% N; material less than 1.7% N would be expected to immobilize soil N during its transformation to humus. Grazed leys are less likely to cause this problem (Barnes 1981b), as discussed subsequently.

The failure of the thrust to ley farming is associated with these factors; more cogent reasons have also arisen as increasing population pressure has accentuated the need for land to be in crop rather than pasture, and as a greater basic demand for crop rather than animal products has arisen in some impoverished tropical societies.

New prospects for ley farming

Technology is now available for many regions of the tropics and subtropics which makes ley farming more attractive. This has arisen from (1) the recognition that the availability of soil N is usually the crucial factor in maintaining crop yields, (2) the identification of well-adapted forage legumes capable of good levels of N fixation and (3) new approaches to soil surface management which preserve the soil *in situ* in a condition of biological health.

Three case studies illustrative of these successes are mentioned.

Pasture–rice in the llanos of Colombia

The high plains of the eastern llanos are extensively farmed savannas with beef and rice as major products. The soils are mainly of good physical structure, but have a highly acid reaction (pH 4.1–5.1) and aluminium saturation ranging from 29 to 86% (Guerrero 1975). The research organization CIAT has successfully developed a rice variety well adapted to upland acid soils. The economics of rice growing in Colombia in 1990 indicate a break-even point between the cost of inputs and the farmer's return from rice at a level of rice grain yield of 1.6 t ha^{-1}.

At Carimagua, Colombia (lat. 5° N, 2160 mm annual rainfall, 175 m altitude) a comparison was made of rice yields planted after ploughing (1) the native savanna, (2) 10-year-old pastures of *Brachiaria decumbens* and (3) 10-year-old pastures of *Brachiaria decumbens–Pueraria phaseoloides*. Three levels of N fertilizer and two levels of P fertilizer were applied. Following native savanna it was necessary to use high levels of fertilizer to attain rice yields in excess of 1.6 t ha^{-1} (Figure 10.1; CIAT 1990). Following *B. decumbens* rice yields exceeded 3 t ha^{-1} if N fertilizer was used, but after a legume-based pasture there was no response to N fertilizer, and the productivity of rice makes this option economically attractive.

Pasture–maize in the highlands of northern Thailand

Slash-and-burn agriculture in the highlands of northern Thailand can lead to a grassland disclimax as *Imperata cylindrica* colonizes farmer fields devoted to maize/opium. It seems that the intensive weeding

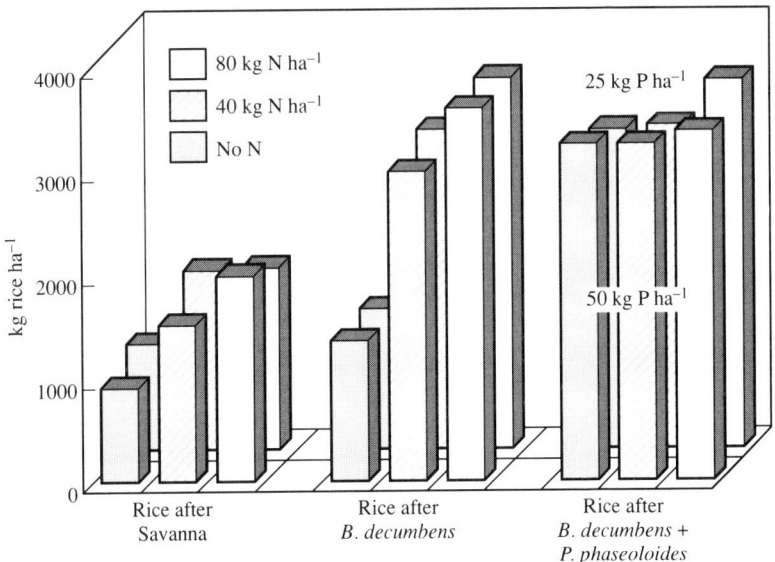

Figure 10.1 Rice yields (kg ha^{-1}) at Carimagua, Colombia after 10 years of improved pastures and native savanna in response to different levels of N and P fertilization (from CIAT 1990).

practised in this system and the removal of shade in the seedling phase favours the invasion of *I. cylindrica* (Gibson 1983). This tall grass is difficult to control by hand-cultivation, but is a poor competitor if fertilized, vigorous legumes are grown and grazed.

I. cylindrica grassland may be burnt late in the dry season, grazed heavily, and broadcast with 2 kg ha^{-1} of *Desmodium intortum* seed; 20 kg S and 34 kg P ha^{-1} are also needed to maximize N fixation by the legume on these granitic soils. Controlled grazing with cattle and slashing of weeds gives a highly productive pasture, from which a cash seed crop can also be taken. The land after 2–3 years of pasture may be hand-cultivated to maize for 2 years before re-establishing the legume. Maize cultivation is simplified after the reduction of *I. cylindrica*, soil fertility is restored and soil erosion is minimized. At Doi Sam Mun, Thailand (lat. 19° N, *c.* 1500 mm rainfall, 1400 m altitude) grain yield of local maize varieties was only 0.7 t ha^{-1} when grown after unfertilized *I. cylindrica* grassland but reached 1.3 t ha^{-1} after *D. intortum* pastures, despite parrot damage to the cobs.

Socio-economic conditions may not currently favour cattle production in these relatively inaccessible highlands, but an alternative agricultural system is now available which would improve food supply and protect the environment.

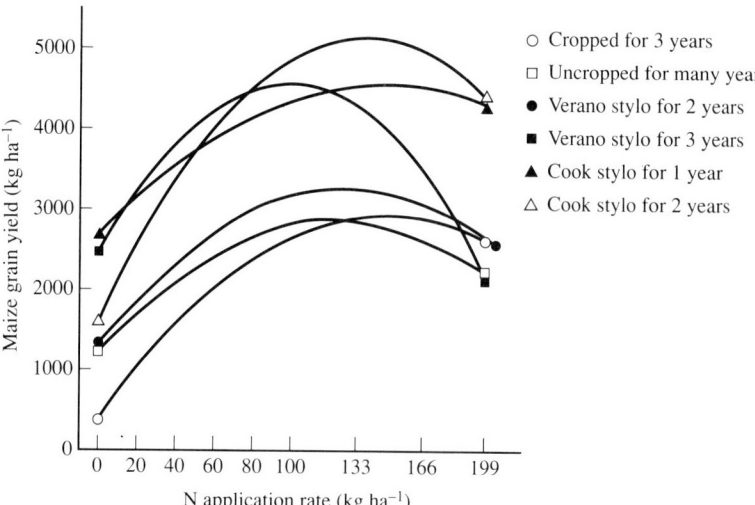

Figure 10.2 Effects of N application on maize grain yields following different land use histories at Kachia Grazing Reserve, Nigeria (from Mohamed Saleem and Otsyina 1986).

Stylo fodder banks–maize in the subhumid zone of Nigeria

Pastoralists in West Africa increasingly are planting fodder banks based on the legumes *Stylosanthes hamata* cv. Verano and *S. guianensis* cv. Cook. As sedentary cropping systems are introduced the use of the traditional bush fallow as a means of maintaining crop productivity becomes less appropriate in view of the rising population pressure on land resources.

One option is to use the enhanced soil fertility in the fodder banks to provide good yields of food and cash crops. A study by Mohamed Saleem and Otsyina (1986) indicates the level of benefits which may accrue.

Their experiment was done at the Kachia Grazing Reserve, Nigeria (lat. 10° N, 1150 mm annual rainfall) on a ferruginous soil containing 0.075% total N. Maize was grown with basal P and K fertilizer, and nine levels of N fertilizer (as urea) were tested on six different areas having different histories of use.

The N response curves (Figure 10.2), fitted as quadratic equations, indicate that the agricultural system which was practised modified both the level of rice production and the response to N. Continuous cropping for 3 years led to a low maize yield of 400 kg ha^{-1} in the fourth year if no N was applied; a bush fallow gave 1275 kg ha^{-1} of grain. Maize yields on the stylo fodder bank areas were substantially higher than on the bush fallow land, depending on the length of time the legume ley was grown. Three years of Verano stylo at the zero N fertilizer level or 2 years of Cook stylo gave yields equivalent to 90 or 110 kg N ha^{-1} fertilizer application on the continuously cropped area. The higher level of production that was feasible on the legume areas at all levels of N application may have been partly associated with the better water-holding capacity of soils with higher OM content.

These three case studies provide grounds for optimism concerning the practicality of ley farming systems based on pasture legumes, and we now turn to the options for managing the ley.

10.2 Duration of the pasture and crop phases

Soil nitrogen

It is evident from these examples that decisions farmers make about the desirable length of the period the land is under a legume-based pasture and the period of cropping will hinge first of all upon the soil N status.

Estimates of gains and losses of N

We need to know the level of soil N, the rates at which N is being added to the soil in the differing phases of the legume ley, and the depletion of N under cropping, as discussed in Chapters 1 and 2.

The simplest approach (Greenland 1971) depends upon the estimation of the values in the following equation:

$$\frac{t_c}{t_p} = \frac{(A_p - kpN_e)}{(k_c N_e - A_c)}$$

where t_c and t_p are the periods of time under crop and pasture respectively required to maintain equilibrium soil N (N_e) estimated after long periods of cereal cropping, A_c and A_p are the rates of N addition during the crop and pasture phases respectively, and k_c and k_p are the rates of N loss during the crop and pasture phases respectively.

This calculation is illustrated by an example (Dalal et al. 1991) from a brigalow clay soil at Warra, Queensland (lat. 27° S, c. 650 mm annual rainfall). The equilibrium soil N value (N_e) is 800 mg N kg^{-1} soil (0–0.1 m depth), A_c and A_p are 20 and 80 kg N ha^{-1} year^{-1}, whilst the N loss under crop (k_c) is 0.1 year^{-1}, and k_p is 0.025 year^{-1} (k_c − net k_p or 0.1 − 0.075).
Thus

$$\frac{t_c}{t_p} = \frac{(80 - 0.025 \times 800)}{(0.1 \times 800 - 20)}$$

$$= \frac{60}{60} = \frac{1}{1}$$

A duration of cereal cropping:legume ley of 1:1 would maintain soil fertility. If the rate of N accretion under the legume ley is only 50 kg N ha^{-1} year^{-1} the ratio becomes cereal cropping:legume ley of 1:2.

The rate of N fixation

A key question is the rate of N fixation by the legume. As discussed in Chapter 2 this rate will be sensitive to (1) the level of soil N, (2) the adaptation of the legume to the farm environment and its rate of growth, which controls the flow of carbohydrate through the roots to the site of N fixation in the nodule, (3) the effectiveness of the

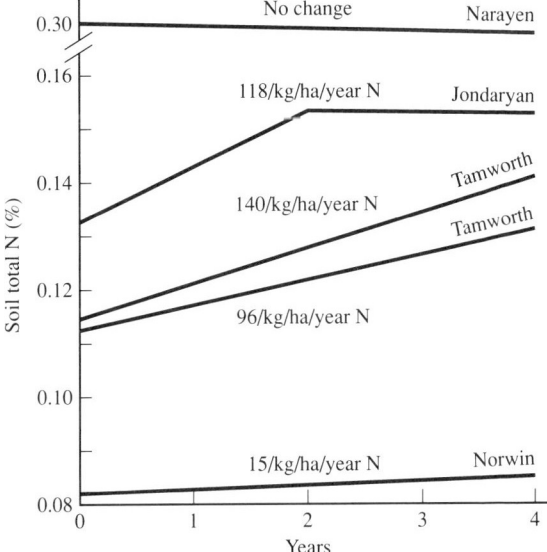

Figure 10.3 Changes in the soil total N status during *M. sativa* leys for five sites in the Australian subtropics (from Lloyd *et al.* 1991).

Rhizobium symbiosis, (4) the availability of environmental growth factors – water, nutrients, light and biotic factors such as mycorrhiza, and (5) the management of the ley.

We can expect higher levels of N fixation on low fertility soils than on high fertility soils, other things being equal, and provided nutrients other than N are available on the former. For example, at Samford, Queensland (lat. 27° S, 1100 mm rainfall) a red latosolic soil containing 0.16% total N showed no change in total N after 4 years of *Macroptilium atropurpureum*-based pasture. On an adjacent soil of 0.089% total N the increment under *M. atropurpureum* was 129 kg N ha^{-1} year^{-1} (Jones 1967). We know that N fixation is lessened if the soil N level is sufficiently high for the legume to use soil N rather than symbiotic N fixation, and nodulation is in any event depressed by soil nitrate.

Changes in total soil N during *Medicago sativa* leys for five sites in southern Queensland and northern New South Wales (Figure 10.3; Lloyd *et al.* 1991) indicate a range of values for N accretion from zero on high fertility brigalow clay soil at Narayen to 140 kg N ha^{-1} year^{-1} on a black earth at Tamworth. In this study wheat yields were increased by an additional 0.4–0.5 t ha^{-1} year^{-1} in the first four crops after the ley, and at Jondaryan the benefit extended to 8 years.

It is of interest that the gains from *M. sativa* on a black earth at Jondaryan were made in the first 2 years of the ley (Whitehouse and Littler 1984), suggesting that short duration leys are most efficient in terms of N fixation. Much would depend on the success of legume establishment and on legume vigour in its first years. *S. humilis* oversown into North Queensland savanna contributed more N to the soil after 2 years of legume build-up than in the first 2 years (Crack 1972). In dry years the pasture ley will contribute less N, and the net N balance may even be negative (Vallis 1972).

As indicated in Chapter 2, the level of N fixation is directly dependent upon the rate of legume growth and successful legumes are those which use environmental resources for growth most efficiently. Pasture legumes with promiscuous affinities for indigenous *Rhizobium* have a less exigent technology and are more usually successful in tropical countries. The effects of fertilizer practice and grazing management of the ley on N accretion are discussed later in the chapter.

The persistence of N effects

The duration of the effect of the legume ley on subsequent crop yields varies considerably, and the reasons for this variation are not well understood. Obviously the degree of nitrate leaching, which is greater in high rainfall areas and on particular soil types, is influential. The benefit of the ley may be related to factors other than N supply. The following examples indicate the range in experience.

In Zimbabwe Clatworthy (1986) considered the effects of different pasture legumes on the subsequent yield of maize. These legumes were grazed by cattle and grown for 4 years. On a clay soil (0.092% total N) at Mazowe there was a linear relationship between the total yield of legume over 4 years (t ha^{-1}, x) and the yield of maize grain (t ha^{-1}, y) from the first crop after pasture as follows:

$$y = 3.348 + 0.0628\ x$$

Thus *Desmodium intortum*, the best legume, which yielded 24 t ha^{-1} over 4 years, apparently increased maize yield by 1.5 t ha^{-1}.

At Marondera on a sandy loam (0.088% total N) there was a similar linear relationship between legume and maize yields, as follows:

$$y = 5.953 + 0.0373\ x$$

Table 10.1 Grain yield and N% for wheat grown after *M. sativa* ley at Jondaryan, Queensland (mean of 4 years) (from Littler 1984).

Years following pasture	Grain yield of wheat (t ha^{-1})	N% in wheat grain
1	2.09	2.49
2	1.88	2.28
3	2.05	2.20
4	1.92	2.20
5	1.90	2.16
Continuous wheat	1.65	1.82

In this case *D. intortum* yielded 32.3 t ha^{-1} over 4 years, and this led to an apparent increase of 1.2 t ha^{-1} of maize grain. However there were no detectable effects of previous legume yield on maize yield in the second crop after pasture.

By contrast Littler (1984) found long-term effects of an *M. sativa* ley on subsequent wheat yields which were evident in the seventh wheat crop after the pasture was ploughed in. On this black earth soil at Jondaryan, Queensland (lat. 28° S, *c*. 680 mm annual rainfall) *M. sativa* was sown with *Bromus unioloides* in each of 4 years and each field was grazed by sheep for 4 years after planting pasture. Thus the average effect of pasture was estimated over four successive cropping sequences. The fifth wheat crop after pasture was still 0.25 t ha^{-1} higher yielding than crops from a continuous cropping sequence (Table 10.1) and pasture substantially increased grain N concentration, a strong indicator of quality and price.

Legume N vs fertilizer N

Many of the comparisons made so far in this chapter relate to cropping systems including legume leys and those based on continuous cropping without N fertilizer inputs. In many farming situations it is better to contrast the economics of the legume system relative to a cropping system relying on the purchase of N fertilizer, and preferably incorporating surface retention of crop residue.

Responses to fertilizer N were estimated in the closing phases of the Jondaryan study above (Littler and Whitehouse 1987). Urea was applied at levels of 0, 33, 67 and 101 kg N ha^{-1}. The continuous wheat treatments produced grain yield responses of 0.36 t ha^{-1} up to the level of 67 kg N ha^{-1}, but in this season the level of yield did not match grain yield in crops grown after pasture. This was attributed to the more even distribution of nitrate down the soil profile following a legume ley. Fertilizer N was concentrated in a band near the surface; its uptake was no doubt limited by the capacity of roots to absorb N, especially under the conditions of surface drying which prevail in this particular environment and in which crop growth depends heavily on stored soil moisture. In this study the critical soil nitrate level (0–60 cm soil depth) at planting was 7.5–10 p.p.m. for grain yield, and 25–30 p.p.m. for grain N concentration.

Soil organic matter

OM and N accretion are closely linked in the effects of legume leys, but there are instances where there are separate benefits of the legume ley due to effects of OM. The most obvious benefits relate to (1) increasing CEC capacity of light soils, and complexing micronutrient elements (for example, Cu and Mn), thus favouring the retention of nutrients for crop use, and (2) improving soil structure, increasing moisture infiltration and perhaps moisture holding capacity.

Table 10.2 Soil nutrient status and maize yield following *C. nlemfuensis* var. *robustus* leys of differing duration at Marondera, Zimbabwe (from Barnes 1981b).

	Duration of ley (years)		
	2	3	4
Carbon (%)	0.54	0.57	0.60
Total nitrogen (%)	0.051	0.054	0.055
Mineral nitrogen after incubation (p.p.m.)	42	45	47
Available phosphorus (resin extract) (p.p.m. P_2O_5)	6	7	8
Exchangeable bases (mEq%)			
Calcium	0.58	0.73	0.80
Magnesium	0.28	0.28	0.32
Potassium	0.18	0.21	0.22
Total bases	1.04	1.21	1.36
Base saturation (%)	28	33	32
Maize yield (t ha^{-1})	3.08	3.27	4.27

Duration of the ley

Effects of the length of the pasture phase on soil chemical characteristics are illustrated from a study at Marondera, Zimbabwe (Barnes 1981b). *Cynodon nlemfuensis* var. *robustus* No.2 was planted on a well-drained sandy loam derived from granite. OM content of the soil was low (Table 10.2) but the degree of base saturation of the soil apparently increased as the duration of the ley increased from 2 to 4 years. The levels of Ca and of K were highly significantly increased in sympathy with the length of the previous ley phase. These small increases on this light-textured soil had a profound effect on subsequent maize yield, and the 4-year ley led to an increased yield of 1–2 t ha^{-1} relative to that following a 2-year ley.

In northern Nigeria Jones (1971) noted at Samaru (lat. 11° N, 1100 mm annual rainfall, 690 m altitude) that a bush fallow based on *Andropogon gayanus* augmented OM, which increased from 0.30% C after 2 years to 0.53% after 6 years. However, growth was reduced in the later years of a long bush fallow by the accumulation of surface residues, which also hamper the preparation of a seed bed for crops. For these and other reasons a 3-year grass fallow followed by 3 years cropping was recommended. A similar policy was advocated in Uganda (Jones 1972).

The effects of increased OM on the water-holding capacity of soil are relatively small in relation to crop demand for water. In reviewing this topic Lal and Kang (1982) report that the largest effect was a linear increase in percentage moisture at field capacity of 5.6 × organic C%. Thus an increase in soil OM of 0.2% would only increase percentage moisture by *c.* 1%.

On the other hand the management of the surface residues and the effect of OM on the infiltration rate affect surface runoff profoundly, as discussed in Chapter 3.

Table 10.3 Effects of length of *A. gayanus* ley and subsequent cropping on the equilibrium infiltration rate (cm h^{-1}) at Samaru, Nigeria (from Wilkinson 1975b).

Length of grass ley (years)	Length of cropping sequence (years)	End of ley period	End of cropping sequence
2	3	2.90	0.58
3	3	2.39	0.94
6	3	5.84	0.84

Table 10.4 Some characteristics (0–0.1 m depth) of virgin southern Queensland soils (from Dalal and Mayer 1986b).

Soil	Classification	Organic C (%)	Clay %)
Waco clay	Typic pellursterts; black earth	1.63	72
Langlands-Logie clay	Typic chromusterts; grey, brown and red clays	2.23	49
Cecilvale clay	Typic chromusterts; grey, brown and red clays	1.73	40
Billa Billa loamy clay	Typic chromusterts; grey, brown and red clays	1.48	34
Thallon clay	Typic chromusterts; grey, brown and red clays	0.77	59
Riverview sandy loam	Rhodic paleustalfs; red earth	1.28	18

An infiltrometer study at Samaru, Nigeria (Wilkinson 1975b) showed how the length of an *A. gayanus* ley positively influenced infiltration rate at the end of the ley when crops were planted, and there was a much smaller but an apparently persistent effect after 3 years cropping (Table 10.3).

Duration of cropping

Factors affecting OM depletion during the cropping phase were enumerated in Chapter 2; the most significant are temperature, rainfall, clay content, cultivation, and the nature and management of the crop residues.

The rate of depletion reflects the balance between the rate of addition of organic materials from litter, the residues of fauna and microorganisms, and dead and living roots on the one hand, and the rates of OM loss through decomposition, and in some circumstances, through leaching and erosion, on the other.

The net balance of OM under cropping is illustrated for a series of soils growing cereals for varying durations in southern Queensland. This example from the subtropics shows C values characteristic of semiarid zones (Table 10.4; Dalal and Mayer 1986b); annual rainfall varied from 480 mm at Thallon to 670 mm at Waco and Cecilvale and the C content of virgin soils reflects increasing values with increasing rainfall. For each soil series the duration of cropping varied from 20 to 70 years.

The relationship between organic C and time (t, years) is best described by the equation:

$$C_t = C_e + (C_o - C_e) \exp(-kt)$$

where C_t, C_e and C_o are the carbon concentrations at the time of observation, in equilibrium with long-term cropping and in the initial virgin condition respectively, and k is the rate of loss.

The equilibrium C values (Table 10.5; Dalal and Mayer 1986b) are also positively related to annual rainfall and are considerably lower than values for soils in the humid tropics, as discussed in Chapter 2. The rate of C loss on a percentage organic C basis in the top layer of the soil, which measures how rapidly the organic C concentration moves towards a new equilibrium level, varied from 0.04 at Thallon to 1.211 at Riverview. This may be expressed as a half-life of loss (i.e. the time to 50% loss of C): at Riverview this was only 0.6 years, whilst at the dry Thallon site it was 16.9 years.

The addition of OM required to maintain organic C at equilibrium is calculated by multiplying C_e by the rate of loss of C corrected for changes in soil bulk density. Assuming organic materials contain 40% C, the calculated annual OM requirement (Table 10.5) varies from 0.8 to 5.4 t ha^{-1} year^{-1}, and is considerably higher in the lighter textured Riverview soil. The rate of organic C loss was negatively related to clay content.

This study indicates that within a region the duration of cropping causing a significant run-down in OM varies greatly with rainfall and soil type. Long durations of cropping might be tolerable at Thallon, Waco and Langlands-Logie, but would be less acceptable at Cecilvale, Billa Billa and Riverview.

Table 10.5 Equilibrium values (C_e), overall rate of loss (k_c), half-life of loss of organic C, and addition (A) requirement of organic matter (40% C) of southern Queensland soils (0–0.1 m depth) (from Dalal and Mayer 1986b).

Soil	C_e (%C)	k_c (year^{-1})	$t^{1/2}$ (years)	A (t ha^{-1} year^{-1})
Waco	0.97	0.063	11.2	1.4
Longlands-Logie	0.64	0.060	11.6	1.6
Cecilvale	1.00	0.178	3.9	4.6
Billa Billa	0.82	0.161	4.3	5.4
Thallon	0.34	0.041	16.9	0.8
Riverview	0.77	1.211	0.6	–

This conclusion is then modified by the crop husbandry practices employed. Removal of crop residues from the field or the burning of crop residues will accelerate OM run down; retention of crop residues or their grazing *in situ* will slow down the loss of OM. Incorporation of cereal residues in the soil accelerates initial decomposition relative to the rate of loss of residues left on the soil surface (Cogle *et al.* 1987). Management of the pasture–crop system is discussed later in this chapter.

Soil structure and soil loss

We know that soil aggregation, macropore space and infiltration increase following a planted pasture phase (Bridge *et al.* 1983b); these benefits may be lost following cultivation for the first crop after pasture (Pereira *et al.* 1954; Littler 1984) or soil physical factors may not represent the primary constraints to crop production, as suggested earlier in this chapter. Soil structure is not therefore regarded as a first consideration in determining the duration of the pasture and crop phases.

A more cogent question is the need to control soil erosion on sloping lands (Lal 1990). Special measures may be taken to minimize soil loss from a cropping system: surface stubble retention, minimum tillage, contour operations, terracing or contour grass strips. A further alternative is to reduce the frequency of cropping and to alternate cropping with a protective pasture phase.

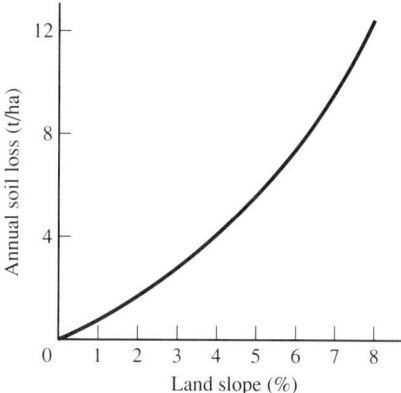

Figure 10.4 Expected annual soil loss during tobacco production on differing land slopes in Zimbabwe (from Hudson 1957).

One concept is to base land use on a tolerable level of soil erosion, and to design the crop–pasture system so that this value is not exceeded. In subhumid zones, where the rate of soil formation is slower than in humid zones, a loss of 1–5 t ha^{-1} year^{-1} may be tolerated. In the moist tropics some scientists accept target values up to 13 t ha^{-1} year^{-1}.

This approach is illustrated from an early study of soil erosion in tobacco areas in Zimbabwe. Measurements on the sandveld at Kutsaga near Harare indicated steeply increasing soil loss as land slope increased (Fig. 10.4; Hudson 1957). Land slopes of up to 3% would be regarded as satisfactory for tobacco cultivation if a notional annual soil loss of 3 t ha^{-1} was regarded as tolerable. For slopes in excess of 3% the introduction of a pasture in the rotation would be desirable to reduce soil loss, if not for other reasons.

A farmer growing tobacco on sandveld sloping at 5% might expect from Figure 10.4 to lose 5.5 t soil ha^{-1} year^{-1}. What is the ratio of tobacco production to pasture production which would reduce average soil loss to 3 t ha^{-1} year^{-1}? If well managed grass on 5% sloping sandveld gives a notional soil loss of 0.4 t ha^{-1} year^{-1}, a rotation of 3 years tobacco/3 years grass might provide an acceptable system.

Crop protection

Diseases and pests

Pastures may be alternated with a cropping phase in order to break the life cycle of organisms which damage crop production, as discussed in Chapter 5.

This approach may be directed to the control of pathogens, in which a pasture is chosen which denies the presence of alternative hosts, so that further build-up of inoculum is precluded. The same approach may be directed to insect pests, or to root knot nematodes (e.g. *Meloidogyne* spp.).

Weeds

Monoculture of particular crops inevitably leads to invasion by weeds whose life cycle matches that of the crop. Another problem occurs when nitrophilous grass weeds invade a legume-based ley. At Katherine, northern Australia (lat. 14° S, 930 mm annual rainfall) leys using *S. hamata* cv. Verano, *Centrosema pascuorum* or *Alysicarpus vaginalis* show promise in rotation with maize and sorghum. However, although the first year of pasture is well dominated by the legume, grass weeds invade in the second year (Table 10.6; McCown *et al.* 1986). Well-adapted legumes also volunteer. Grasses may be partially

Table 10.6 Botanical composition and yield of 1st and 2nd year leys at Katherine, Australia (from McCown *et al*. 1986).

	Planted legume		
	S. hamata	*A. vaginalis*	*C. pascuorum*
1st year ley			
% Composition			
Planted legume	85	85	72
Volunteer legume	10	11	6
Grass	2	0	0
Yield (t ha^{-1})	6.6	4.8	7.1
2nd year ley			
% Composition			
Planted legume	44	32	22
Volunteer legume	13	39	47
Grass	39	29	29
Yield (t ha^{-1})	5.8	6.2	6.5

controlled by heavy mid-summer grazing of the ley, or the planting of a grass–legume mixture may provide a more stable pasture; the alternative is to use a short-duration legume ley.

Similarly in Nigeria the invasion of grasses in *Stylosanthes* fodder banks reduced subsequent maize yields. Maize grain yield was 1390, 780 and 220 kg ha^{-1} respectively as grass density in the previous fodder bank increased from 0 to 50 to 100% (ILCA 1989).

Economic considerations

Farmers are interested in maximizing net income; small farmers may be more concerned with stability of income and with maximizing returns from scarce cash investment. The balance of crop and animal production can be readily changed in response to changing market prices and costs, provided a mixed farming system is in place. The duration of the ley and of the cropping phase can be simply adjusted by bringing more or less of existing forage areas into crop.

Three case studies are mentioned to illustrate farmers' options.

Kanpur, India

At Kanpur, India (lat. 26° N) a 2–year ley of *M. sativa-Chloris gayana* was incorporated (A) with 2 years or (B) with 4 years annual cropping

(Dixit and Jain 1979). A fallow–wheat system was contrasted with maize–wheat at three levels of fertilizer input. These ley systems were also compared with continual cropping (C). The ley pasture significantly increased crop yields by 6–19%, and this was reflected in greater gross and net returns in the ley systems; overall net returns averaged 3730, 3500 and 3050 rupees ha^{-1} respectively in systems A, B and C. The highest net return of 4075 rupees ha^{-1} occurred in the maize–wheat ley system A where substantial fertilizer inputs were applied. However, the less intensive fallow–wheat ley system gave higher marginal returns per rupee invested in farm costs. Credit or cash availability will often be a decisive factor in farmers' decision making.

Darling Downs, Queensland

Modelling of the economics of large scale (2500 ha) farms in the western Darling Downs of Queensland contrasted a conventional mixed enterprise of wheat–sorghum production, breeder cattle and wether sheep (A) with a somewhat similar system modified by the incorporation of a *M. sativa* 2-year ley (B) or of a ley using the annual medic *M. scutellata* and *Lablab purpureus* (C) (Lloyd *et al.* 1991). Gross income and variable costs were greatest in system C, but the operating profit was 10–11% greater in systems B and C than in system A. The sustainability and hedge against risk were least in system A. However, currently only a minority of farmers have adopted a ley system in this region.

Northeast Thailand

Considerable success has attended the planting of the legumes *S. hamata* cv. Verano and *M. atropurpureum* cv. Siratro in the Ubolratana district (lat. 16° N, 1150 mm annual rainfall) of northeast Thailand (Gibson 1987). The increasing population pressure in this region has led to the cropping of light upland soils and yields of cassava and kenaf (roselle) decline rapidly under monoculture. These may be restored by a fertilized legume ley, as discussed in section 10.4.

The burgeoning dairy industry in Thailand provides attractive opportunities for smallholders. Self-sufficient systems of feeding dairy animals that avoid the costly purchase of concentrates and combine the grazing of pastures with feeding locally grown crop products and crop residues has been devised. Village health is improved, off-farm labour movement is reduced, and the runoff water from the uplands contains less soil.

An economic analysis by Gibson (1987) compared dairy production with the option of using ley pastures for beef production and with continuous cassava production. The analysis was based on 1.3 ha land, half of which was planted to pasture in the ley systems. The area was grazed by two dairy cows and their calves, or by three beef cows and their calves. Milk was priced at \$US0.25 l^{-1}, beef output was given a cash value of \$US0.87 kg^{-1} LW and cassava was priced at \$US0.02 kg^{-1} at the farm gate. Some increase in labour associated with dairying was absorbed on the basis that family labour is underutilized and may not have an alternative opportunity use.

This analysis estimated a net annual income of \$US777 from dairying plus cassava, \$US319 from beef plus cassava, and \$US260 from cassava alone. The dairying option has become so attractive to many farmers that the trend is to extend the duration of the ley phase and to reduce the dependence upon crop income.

Legume N vs fertilizer N

An important overall economic consideration affecting the adoption of the ley is the value of legume N relative to fertilizer N. Pricing policy of fertilizer N is subject to heavy government intervention, and much depends on whether a country is self-sufficient in oil, or whether fertilizer N requires the consumption of scarce forcign exchange.

This section has mentioned a complex of factors which enter the farmers' decision making processes when the duration of the pasture ley phase and of the cropping phase are determined. Often the soil N status is the overriding factor.

10.3 Pasture and crop establishment

Farmers have a multitude of options available from which they may choose the techniques of pasture and crop husbandry most appropriate to their farm system. Differences in farmer goals, managerial skills and access to credit will influence the level of intensity of technology which is suitable, and this must also be judged in terms of the returns from the inputs applied. In subhumid areas a system with lower inputs than applies in the humid tropics is usually required.

Management of the soil surface

The objectives are to provide soil conditions favourable for (1) seedling emergence, and/or (2) plant survival, (3) plant growth unrestricted by the competition of unwanted species, (4) conservation of the soil, and (5) protection of the environment.

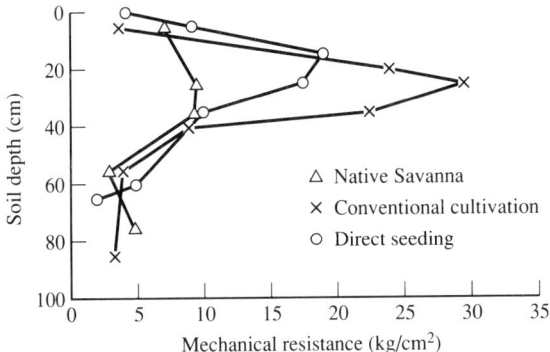

Figure 10.5 Soil compaction as influenced by system of cultivation in Brazilian cerrado (from Luchiari *et al.* 1985).

Tillage

Cultivation represents an energy input designed to modify soil structure and hence nutrient availability, and to kill weed species. Tillage is necessary to break soil seals which impede germination and seedling emergence and reduce moisture infiltration and aeration; it promotes mineralization of nutrients and may create soil tilth favourable for the seedling.

In recent years recognition has grown that many of these objectives can be met by the use of organic mulches and the consequential activities of mesofauna such as earthworms (for example *Millsonia anomala*; Lamotte 1982).

It has also become evident that frequent cultivation, which is expensive of energy, may degrade soil structure, reducing macroporosity and infiltration and rendering soils vulnerable to erosion. Repeated cultivation at the same depth tends to produce a 'plough layer' or a narrow compacted horizon which restricts root exploration, thereby rendering the plant more vulnerable to drought and to nutrient deficiencies. This is illustrated for a red latosol in the cerrado of Brazil (Figure 10.5, Luchiari *et al.* 1985), where conventional use of an offset disc plough has broken down the rather weak macrostructure, causing rapid soil disaggregation and the formation of a plough pan; the surface soil also seals under the impact of raindrops.

In developing countries least tillage is represented by the use of a planting stick in which seed is dibbled into a hole and covered and firmed by the heel of the operator. Hand hoes are used to break the surface and invert the top soil layer. Animal traction increases the area which can be cultivated in a timely way, but many indigenous ploughs with a single tine and metal shoe are energetically less efficient than modern designs. The Ethiopian wooden plough or *maresha* may be modified by the addition of a metal mouldboard wing for terracing (Jutzi *et al.* 1987), or two *mareshas* may be linked together with added wings and a chain harrow to produce a broadbed maker; the latter is especially appropriate to reduce waterlogging of vertisols. A ridge–furrow pattern developed on the contour is a common method of reducing runoff on sloping land.

The mouldboard plough which inverts the upper soil layer, perhaps followed by harrows which further break down the clods, and the disc cultivators are generally considered to be more damaging to soil structure than tined implements. The latter operate with a shattering action, and broad sweeps may be attached to the base of the tines to cut through the roots of weed species present.

When operating in soils with surface trash a narrow type planter, preceded by a rolling coulter to cut surface mulch and followed by a narrow in-furrow press wheel, has given good results (McCown *et al.* 1985).

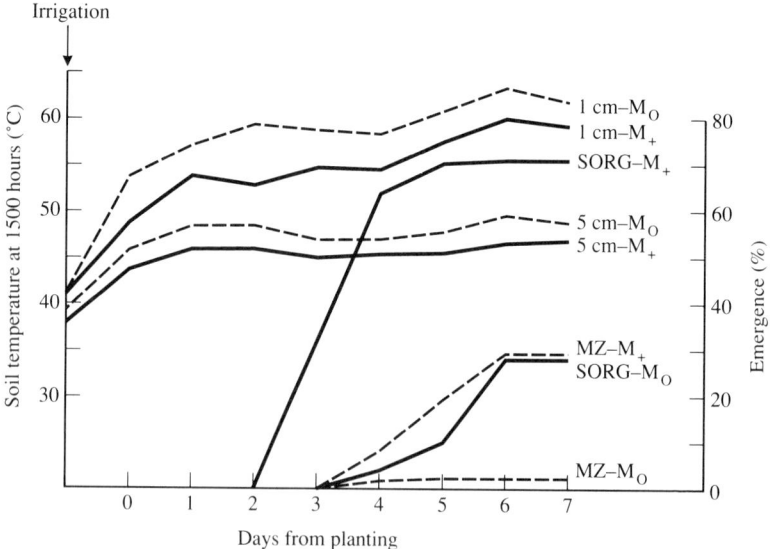

Figure 10.6 Effects of mulch (M_O ——— no mulch; M_+ --- one layer of hessian) on soil temperature and on percentage emergence of maize (MZ) and sorghum (SORG) at Katherine, Australia (from McCown *et al.* 1985).

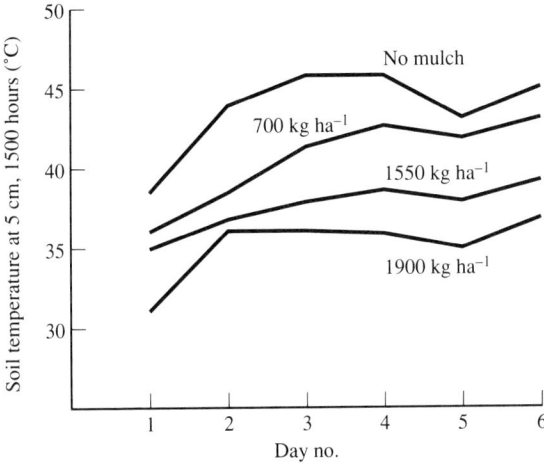

Figure 10.7 Effects of varying amounts of mulch (dead *S. hamata*) on soil temperatures at Katherine, Australia (from McCown *et al.* 1985).

Surface mulch and minimum tillage

A major trend of modern farming is towards the retention of crop residues or pasture litter on the surface of the soil, and the avoidance of exposure of a bare soil surface to rain and sun. The primary objectives of this thrust are:

- reduced runoff,
- reduced soil loss,
- lowered soil temperatures,
- increased biological activity.

These then have consequent positive effects on soil moisture supply, arising both from greater surface detention of water, better infiltration and less evaporation, which have been discussed earlier. Soil structure may also be maintained by precluding raindrop impact and by maintaining soil moisture content above the point where rain causes an explosive action of entrapped air (Lal 1985).

In hot environments the temperature of bare soil limits crop establishment. This is especially common if methods of seedbed preparation expose the soil surface to high insolation through the construction of hillocks, mounds, raised beds and ridges. At Katherine, Australia the imposition of a single layer of hessian as a simulated organic mulch reduced temperatures of a light-textured soil (Figure 10.6; McCown *et al.* 1985). In the absence of mulch soil temperature at 1 cm depth exceeded 60° C; this had deleterious effects on sorghum seedling emergence and was even more disastrous for maize seedlings. In other studies at Katherine a combination of no-till and mulch retention led to an increase in maize yield of 35%.

A level of surface mulch of crop residues in humid tropical situations of 4–6 t ha^{-1} has been considered desirable (Lal 1985). Lower levels of mulch have beneficial effects. At Katherine soil temperatures are significantly reduced by a mulch of dead *S. hamata* cv. Verano of 700 kg ha^{-1} (Fig. 10.7; McCown *et al.* 1985), and higher levels of mulch provide a cooler environment and a longer lasting mulch.

The presence of a surface mulch changes the ecology of the system. The added organic substrate nourishes different populations of organisms from those present in a bare surface soil system. This is illustrated from a study near Warwick, Queensland (lat. 28° S) on a black earth used for cereal production. After 12 years of stubble retention or stubble burning, and of conventional tillage or a zero-tillage system employing the herbicides paraquat, diquat and glyphosate, the microbial biomass of C and N in the topsoil was greater in the stubble retention system (Table 10.7; Thompson

Table 10.7 Earthworm population and microbial biomass in relation to cultural treatment at Warwick, Queensland (from Thompson 1991).

	Stubble retained		Stubble burnt	
	Cultivated	No-tillage	Cultivated	No-tillage
Earthworms				
Density	23	54	9	9
(no. m^{-1} 0–12.5 cm depth)				
Biomass	156	530	25	129
(kg ha^{-1} 0–12.5 cm depth)				
Microbial biomass				
Carbon (mg kg^{-1} soil)				
0 fertilizer N	697	703	644	644
69 kg ha^{-1} fertilizer N	656	718	589	552
Nitrogen (mg kg^{-1} soil)				
0 fertilizer N	137	112	114	102
69 kg ha^{-1} fertilizer N	133	125	122	95

1991). This occurred at differing levels of N fertilizer input. The method of weed control – herbicides or cultivation – had little effect, indicating that herbicide use was compatible with microbial activity.

The mesofauna present, as represented by earthworms, reflected both the stubble retention and tillage decisions. Mechanical cultivation, even with stubble retention, was inimical to earthworm activity and burning of cereal residues also had a negative effect (Table 10.7). Earthworms circulate nutrients to the surface layers where they are accessible to plant roots, and have significant positive effects on soil aggregation and on infiltration, as discussed previously.

Regrettably the presence of earthworms undisturbed by cultivation leads to the formation of continuous channels down the soil profile, and this and other factors add to the leaching of nitrate to the lower layers of the soil profile or to nitrate exit to groundwater and streams. Additionally the breakdown of carbonaceous surface residues in a no-till system immobilizes soil N in microbial organisms, and both factors add to crop demand for N. The no-till, surface mulch retention system of cereal production works best where fertilizer N augments soil N supply (Thompson *et al.* 1987), or where legume N fixation is especially successful.

A further concern is the effect of stubble retention on insect pests and disease, as discussed in Chapter 5.

Live mulch

An alternative concept extends the use of leguminous cover crops in plantation crops (Chapter 8) to their planting in conjunction with annual crops. 'Live mulch' has been defined as a crop production technique in which a food crop is planted directly in the living cover of an established cover crop without tillage or destruction of the fallow vegetation (Akobundu 1982). Some workers emphasize that the cover crop is neither grazed nor harvested (Lal *et al.* 1979), and some disadvantages of this stricture are outlined later in this chapter. The concept of live mulch is extended to self-regenerating, short lived legumes such as *S. hamata* cv. Verano (Jones *et al.* 1991) and to systems involving minimum tillage which escape complete dependence on herbicide use.

The main thrusts of this practice are directed to:

- N accretion from a long-term leguminous cover,
- the provision of cover which conserves soil and increases infiltration,

302 *Pastures with annual crops*

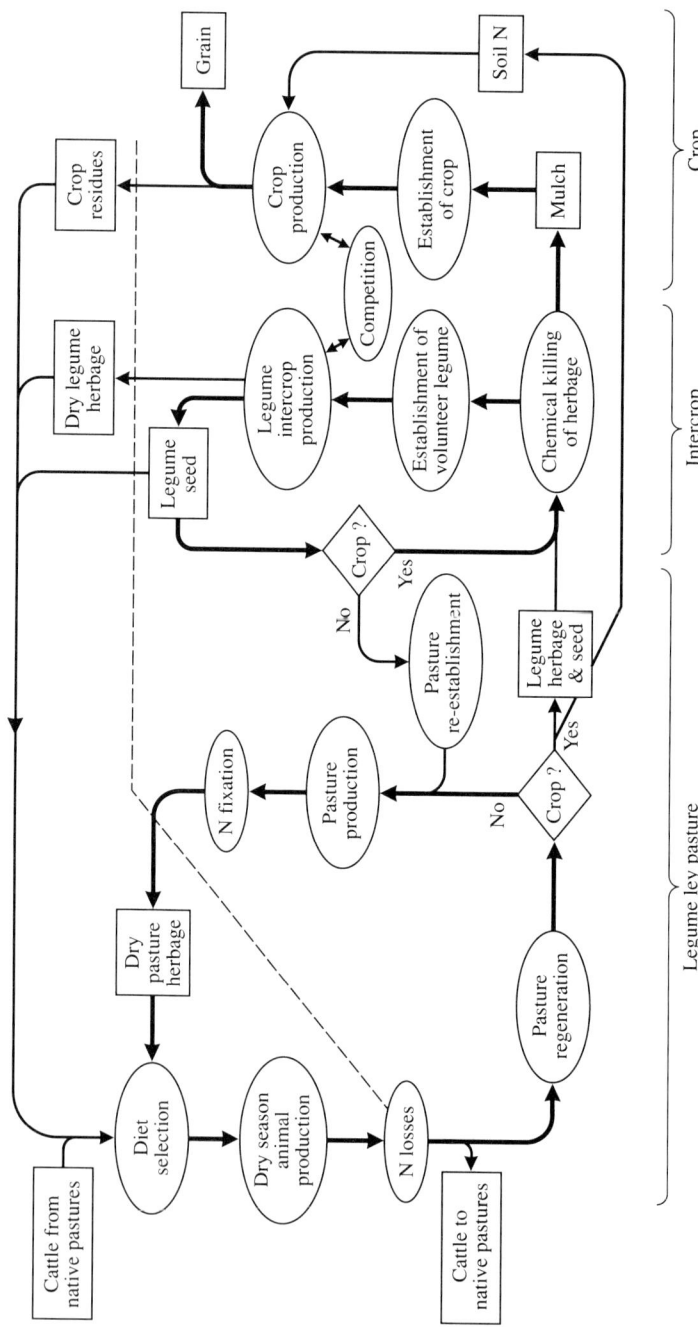

Figure 10.8 A hypothetical farming system involving live mulch at Katherine, Australia, showing the flow of inputs and outputs among subsystems (boxes) and some processes researched (ovals) (from McCown *et al.* 1985).

- the control of weeds through the competitive effects of well-adapted covers.

Subsidiary benefits may be:

- increased animal production from consumption of the pasture legume, and the consequent improved utilization of crop residues, together with
- improved nutrient cycling.

The main disadvantages are:

- competition between the herbage and the food crop, which requires careful management if crop yields are to be increased,
- invasion of leguminous covers by nitrophilous weeds,
- the cost of herbicides.

Kannegieter (1967) was an early exponent of this system. At Kumasi (lat. 7° N) in the forest zone of Ghana culture of the vine legume *Pueraria phaseoloides* was perceived as an alternative to the conventional bush fallow system. Hand planting of maize by punching holes with a stick through the trash cover was carried out 20 days after spraying a 3-year-old stand of *P. phaseoloides* with 2, 4–D. This led to a maize yield of 1.92 t ha^{-1}, relative to yields of 1.79 and 1.52 t ha^{-1} where *P. phaseoloides* was respectively burnt *in situ* and the ash incorporated in the soil or slashed and ploughed in; maize yield was reduced to 0.82 t ha^{-1} if the legume were removed and conventional tillage applied.

Table 10.8 Effect of various cover crops on soil characteristics at Ibadan, Nigeria (from Lal *et al.* 1979).

Species	Organic C (%)	CEC (mEq /100 g)	Total N (%)	Bulk density (g cm^{-3})	Infiltration rate (cm h^{-1})
B. decumbens	1.57	8.5	0.19	1.34	19
P. notatum	1.45	8.2	0.17	1.35	14
C. nlemfuensis	1.70	8.9	0.19	1.30	18
P. phaseoloides	1.50	7.7	0.17	1.32	16
S. guianensis	1.63	8.8	0.21	1.33	16
S. deeringianum	1.57	10.5	0.21	1.33	21
P. palustris	1.57	10.9	0.20	1.14	42
C. pubescens	1.53	10.0	0.18	1.33	18
Control	1.37	8.4	0.17	1.42	13
LSD ($P < 0.05$)	0.23	3.5	0.03	0.04	17

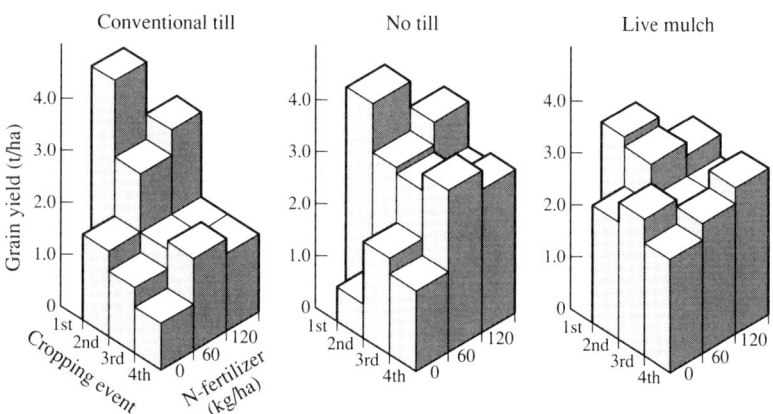

Figure 10.9 Effect of land management and N fertilizer level on maize yield at Ibadan, Nigeria (from Akobundu 1982).

A hypothetical farming system involving the use of live mulch at Katherine, Australia is illustrated (Figure 10.8) to demonstrate the components of the system and the principal processes involved in its maintenance. In this particular system maize or sorghum is cropped for a varying number of years, and crops are planted into a mulch of *S. hamata* cv. Verano, *A. vaginalis* or *C. pascuorum*, which is maintained as an intercrop or alternatively may be suppressed until a ley phase is resumed. Cattle grazing on separate areas of native pasture are introduced to graze the crop residues and the ley during the dry season.

The effects of different covers on soil characteristics are now illustrated by a study at IITA, Ibadan, Nigeria (lat. 7° N; Table 10.8; Lal *et al.* 1979). The sloping site bore an alfisol (Oxic Paleustalf) which had been degraded by continuous cultivation for 5 years. It was planted to different covers: three grasses (*Brachiaria decumbens*, *Paspalum notatum*, or *Cynodon nlemfuensis* cv. IB8) or five legumes (*Pueraria phaseoloides, Stylosanthes guianensis, Stizolobium deeringianum, Psophocarpus palustris* or *Centrosema pubescens*). Soil was sampled 2 years after planting the covers, before cropping the area.

Organic C increased in the 0–10 cm layer (Table 10.8), *C. nlemfuensis* and *S. guianensis* showing high values. N accretion was evident in *S. guianensis, S. deeringianum* and *P. palustris*. Cation exchange capacity (CEC) was variable, but showed high values after *P. palustris* and *S. deeringianum*; this was associated with increased Ca and Mg concentrations.

Plate 10.1 Cattle on *Setaria sphacelata* var. *sericea*, south Thailand.

These chemical changes were complemented by physical changes. Soil bulk density was decreased by all covers, and this decrease was especially evident following *P. palustris*, which also exhibited strongly positive effects on final infiltration rate.

What beneficial effects on crop yields can be expected from the use of these covers? Two other IITA studies are quoted in illustration. The first employed *P. palustris* as the soil cover in continuous maize cropping, in which two crops per year are grown in the bimodal rainfall environment of Ibadan. The initial crop received an application of 400 kg ha^{-1} compound fertilizer (15–15–15 equivalent to 60 kg N ha^{-1}), whilst subsequent crops were grown at three levels of N fertilizer input (Figure 10.9; Akobundu 1982). Conventional tillage was compared with a no-till system of stubble retention and with a live mulch in which narrow planting strips (15 cm wide) of *P. palustris* were burnt with paraquat at 0.4–0.6 kg ai paraquat ha^{-1} (2–3 l ha^{-1} product).

Under conventional tillage the yield of the second maize crop was maintained by N fertilizer input, but the latter was quite ineffective in promoting good yields in the third and fourth maize crop (Figure 10.9). In the no-till system fertilizer inputs of 60 and 120 kg N fertilizer ha^{-1} were able to sustain yields, presumably since the OM status was maintained and the surface mulch had beneficial

Table 10.9 Crop yields (t ha^{-1}) as affected by cover crops at Ibadan, Nigeria (from Lal et al. 1978).

Cover crops	Maize	Cowpea	Pigeonpea	Cassava
(a) First arable crop				
P. maximum	3.13	0.37	1.19	–
S. sphacelata	5.77	0.49	0.86	–
B. ruziziensis	5.17	0.76	1.27	–
M. minutiflora	5.18	0.63	1.19	–
C. pubescens	5.79	0.76	1.23	–
P. phaseoloides	4.77	0.84	1.12	–
N. wightii	5.05	0.65	1.04	–
S. guianensis	5.17	0.61	1.17	–
Control	4.87	0.50	1.06	–
LSD ($P < 0.05$)	2.21	0.05	0.20	–
	Cowpea	Maize	Soybean	Cassava
(b) Second arable crop				
P. maximum	0.62	1.69	0.50	3.50
S. sphacelata	0.71	2.97	0.91	7.90
B. decumbens	1.04	3.80	1.14	17.39
M. minutiflora	0.87	3.43	0.77	18.85
C. pubescens	0.76	3.73	0.75	15.01
P. phaseoloides	0.79	3.44	0.80	19.49
N. wightii	0.71	3.02	0.93	14.12
S. guianensis	0.67	3.11	0.91	19.83
Control	0.43	2.06	0.51	8.05
LSD ($P < 0.05$)	0.06	0.53	0.23	2.53

effects on the moisture and temperature regimens. However, the use of herbicides at the pre-plant and pre-crop stages was necessary. The pre-eminence of the live mulch system was evident in the zero N fertilizer treatment, where good maize yields were maintained for the second, third and fourth crops.

The second IITA study (Lal et al. 1978) illustrates different cropping sequences at Ibadan following cover crops of four grasses (*Panicum maximum, Setaria sphacelata* (Plate 10.1), *Brachiaria ruziziensis* or *Melinis minutiflora*) or four legumes (*Centrosema pubescens, Pueraria phaseoloides, Neonotonia wightii* or *Stylosanthes guianensis*). The site was a medium-textured alfisol with 10% slope. Two years after their establishment the covers were sprayed with paraquat twice one week apart to give a total application of 4 kg ha^{-1}. Narrow strips

(10 cm wide) were then cut and crops sown with a hand planter in rows 75 cm apart. Crops in this bimodal rainfall environment were grown in 1-year sequences: maize–cowpea, cowpea–maize, pigeonpea–soybean, or a single crop of cassava. Maize received 120 kg N ha^{-1} plus PK fertilizer, leguminous crops received 30 kg N ha^{-1} plus P, and cassava received no fertilizer. The control plants had been in weed fallow for 2 years and were tilled conventionally.

Maize yields (Table 10.9; Lal *et al.* 1978) in this relatively high N fertilizer system were depressed in the *P. maximum* and control treatments. Cowpea crops were reduced after *P. maximum*, *S. sphacelata* and the control weed fallow. Soybean yields were low in the control and following *P. maximum*. The unfertilized cassava crop benefited especially from *P. phaseoloides* and *S. guianensis* covers, and was depressed by *P. maximum* and *S. sphacelata*. *B. ruziziensis* and *S. sphacelata* sods were especially difficult to suppress and to manage for cropping.

The use of fertilizer on crops in this study limited the correlations with the performance of covers which might have been made. However it was notable that both grain and tuber yields were positively associated with rate of infiltration and negatively correlated with soil bulk density.

Plate 10.2 *Stylosanthes scabra* cv. Seca.

Plate 10.3 *Stylosanthes guianensis* var. *intermedia* naturalized in a well grazed situation in Zimbabwe.

The concept of live mulch offers much to the sustainability of agriculture, and technologies which are feasible in local farming systems need to be developed.

Choice of pasture species

The suite of pasture legumes available for planting in leys has expanded considerably in the past three decades and well-adapted cultivars have been domesticated for most cropping situations in the humid and subhumid tropics and subtropics. There are various general works describing the characteristics of these species, including Skerman (1977), Crowder and Chheda (1982), Skerman and Riveros (1990) and CIAT (1992), and more specialist texts about particular genera: Stace and Edye (1984) for *Stylosanthes* and Schultze-Kraft and Clements (1990) for *Centrosema*.

Farmers hope to plant pasture species well adapted to their local farm environment. These are judged primarily by their ecological success, as reflected in their ease of establishment, growth and capacity for plant replacement, which determines the persistence of yield. The farm environment is conveniently categorized in terms of climatic, edaphic and biotic factors. The most suitable plants show superior resistance to the stresses of the environment and/or a superior capacity to modify the availability of environmental growth factors to their neighbours (Humphreys 1981).

Resistance to moisture stress is the primary aspect of climatic adaptation required in the subhumid zone; *S. hamata* and *S. scabra* (Plate 10.2) are adapted to drier zones than *S. guianensis*, *C. pubescens* or *P. phaseoloides*. In the cool subtropics the growing season of *Lotononis bainesii* and *Lotus major* will be longer than that of *S. hamata*. Shade tolerance, as discussed in Chapter 8, is advantageous for pastures established by undersowing in a crop or used as an intercrop; *Arachis pintoi* and *Desmodium intortum* are more shade tolerant than *S. guianensis* or *D. uncinatum*.

Emphasis has been given in plant improvement problems to adaptation to low pH soils and resistance to Al or Mn toxicity; *S. capitata* and *C. acutifolium* cv. Vichada are better adapted than *N. wightii*. Resistance to salinity is better developed in *Chloris gayana* than *Cenchrus ciliaris*. Responsiveness to improved nutrient availability is desired for high input systems. Response to improved P supply is greater for *D. intortum* than for *L. bainesii* (Blunt and Humphreys 1970); the latter, and *S. hamata* and *S. scabra*, are suited to low input, low fertility situations.

The changing pattern of the biotic factors operating in the sward with respect to diseases and pests causes many farmers to plant a mixture of legumes, as an insurance measure. Anthracnose (*Colletotrichum gloeosporioides*, Plate 5.1) may devastate *S. guianensis*, whilst *M. atropurpureum* will fall victim to other diseases. Larval feeding of the weevil *Amnemus quadrituberculatus* may destroy *D. intortum* and *D. uncinatum* but not *M. atropurpureum* or *N. wightii*. The value of vine-type climbing legumes such as *M. atropurpureum* has in recent years been discounted by the recognition of their vulnerability to overgrazing, since growing points are accessible; their tendency to climb on and smother companion crops is also a disadvantage in an intercrop situation. Legumes with growing points close to the ground surface, such as *Arachis pintoi*, *S. guianensis* var. *intermedia* (Plate 10.3) and *S. hamata*, are resistant to grazing, whilst erect plants such as *Crotalaria juncea* and *S. guianensis* are conveniently used in cut-and-remove systems.

Legumes are judged primarily by the amount of N fixation they contribute to the farm system and, as discussed in Chapter 2, N fixation in well-nodulated legumes is primarily a function of their growth rate. Some scientists argue that legumes with specific *Rhizobium* requirements (such as *L. bainesii*), which benefit from inoculation with high performing strains, are to be preferred; in many developing countries a need for *Rhizobium* inoculation and its associate technology will be avoided and legumes with promiscuous rhizobial affinities such as *S. scabra* are chosen.

The ease of N cycling is a further consideration. *M. atropurpureum* residues will release N to the system faster than *D. intortum* residues (Vallis and Jones 1973).

The persistence of the planted pastures for the length of the ley and the capacity of the pastures to resist weed invasion are significant considerations. Short-lived species such as *S. hamata* need to exhibit good seedling regeneration, especially if grown as an intercrop. This is accommodated by a combination of high levels of seed production and a suitable rate of hardseededness which provides long-lived seed reserves in the soil and germinable seed which emerges as opportunities arise. At Katherine, Australia seed production of each of the species *S. hamata*, *A. vaginalis* and *C. pascuorum* exceeds 100 kg ha^{-1} in the intercrop, so that re-establishment of the legume species in a subsequent year is satisfactory (Jones *et al.* 1991); however the large seed (15–20 mg) of *C. pascuorum* is less reliable in establishment from a position on the soil surface than the smaller seeded (1.5–2 mg) legumes; the thicker radicle is less successful in forcing itself down cracks in the soil surface or through a surface seal.

Ease of eradication is desired in species used for short-term pastures. Grasses with a high rate of carbohydrate allocation to the roots and crown provide a mass of carbonaceous residues whose breakdown immobilizes N and limits crop growth. *C. ciliaris* accumulates greater root and crown material than *P. maximum* var. *trichoglume* (Humphreys and Robinson 1966), and yields of crops are higher following the latter. Sod forming grasses with a high density of underground buds are also more difficult to plough out than bunch grasses.

Adaptation to the farm system also implies successful animal outputs which are influenced by the nutritive value of the pastures. The final criteria for pasture selection are the levels of animal production and of crop production supported by the ley system.

Techniques of establishment

This topic is also covered in Part 4 of *Tropical Pastures and Fodder Crops* (Humphreys 1987). Quality of pasture seed needs to be reviewed in terms of its:

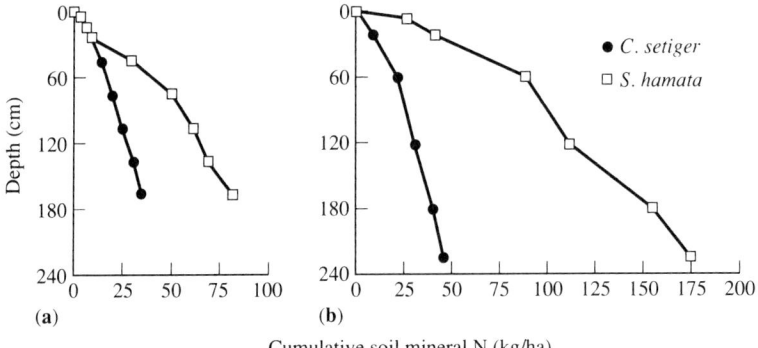

Figure 10.10 Cumulative amounts of mineral N (nitrate + ammonium) with depth just prior to planting sorghum after (a) a short ley or (b) a long ley (from Jones et al. 1991).

- viability,
- freedom from contamination,
- longevity,
- rate of germination,
- associated vigour,
- dormancy,
- seed weight,
- provenance.

The quality of vegetative material used for planting also requires control.

Failure of establishment arises from the physical loss of seed or of seed viability, the inability of germinated seed to emerge from the soil, and from the mortality of emerged seedlings. The success of establishment depends upon:

- the water relations of the seed and seedling,
- gas exchange,
- temperature,
- light (for light-demanding seeds),
- soil physical characteristics,
- the incidence of pathogens, pests and predators,
- the competitive relations of the seedlings with existing plants or emerging volunteers (Cook 1985).

These factors may be manipulated by the attention the farmer gives to:

- land preparation for sowing,

- seed treatment (such as rhizobial inoculation of legumes),
- sowing practice,
- rate of sowing,
- fertilizer placement and level,
- direct weed control.

Time of sowing

It is usual to plant both crop and pasture species under the most assured moisture conditions which are feasible at the beginning of the main growing season.

A particular problem relates to the synchrony of crop establishment with the release of N from decaying residues of pasture legumes. Appreciable rains may fall before crops can be successfully established, N in legume crop residues may be rapidly mineralized in warm, humid conditions, and mineral N may be leached to depths below the reach of the root system of crop seedlings, especially on lighter soils. This is illustrated for leys at Katherine, Australia (Figure 10.10; Jones *et al.* 1991). Appreciable amounts of N were released after *S. hamata* leys, but much of this occurred below 100 cm depth at the time of planting, where it would be unavailable for crop uptake.

Deep rooted pasture legumes such as *M. sativa* make efficient use of soil water throughout the whole soil profile. This may cause moisture deficiency in the first crop following *M. sativa*, unless the length of fallow period before crop planting is sufficient to replenish the profile. Littler (1984) believed this problem might be overcome in southern Queensland by commencing the fallow period in the spring and planting a winter cereal crop in the subsequent autumn; this exposes the bare soil to summer rains and may generate erosion. The available water-holding capacity per metre of soil depth is estimated as 220, 150 and 80 mm respectively for black earth, medium brigalow clay and loamy duplex soil (Lloyd *et al.* 1991); the probability of bringing these soils to field capacity at planting in this subhumid environment is sufficiently low to cause a grain yield penalty following *M. sativa* in most cases. An alternative is to succeed *M. sativa* with a fodder crop which does not need to be brought to full maturity.

One danger of the long cultivated fallow is its detrimental effect on the population of VAM, upon which many crops depend for improved mineral nutrition. VAM favour living roots to complete their life cycle and to increase their inoculum. Crops such as linseed and sunflower which are VAM dependent may exhibit P or Zn deficiency in southern Queensland if the fallow period is extended (Thompson 1987, 1991).

Sowing practice

Depth of sowing is crucial for the establishment of most tropical grasses; the perennial species (except *Sorghum* spp.) are small-seeded and a maximum sowing depth of 1.3 cm is indicated. Larger seeded legumes may be planted more deeply and placed within a band of stored moisture, from which seedling emergence is more assured.

The early growth of the sward is linearly and positively related to the density of seedlings, and as growth proceeds the yield–density relationship forms a plateau in which further increases in density do not increase herbage yield. A high seeding rate of good quality seed increases the first year rate of N fixation of legume sowings, and the latter is often a primary objective of leys. Weed suppression and control of soil erosion is favoured and the earliness with which the sward can be grazed (and hence give returns for the cost invested by the farmer) is enhanced. In humid areas the total mixture sowing rate of good quality seed should not be less than 7 kg ha^{-1} and is preferably 10–12 kg ha^{-1}. In subhumid areas, where returns from the pasture are less, a total seeding rate of *c.* 3 kg ha^{-1} is common. Much depends on the success of seed bed preparation, the successful choice of sowing time, and the degree of weed control.

Vegetative planting is sometimes carried out with 0.7 m between rows, and 0.2–0.3 m between plants in the row, depending upon the capacity of plants to spread and rapidly occupy ground space.

Control of weeds

The weed problem encountered in the cropping phase depends upon its duration, as illustrated previously, and upon the management of the ley phase in minimizing weed invasion. This question was discussed in Chapter 5.

Companion sowing of pastures and crops

The sowing of a pasture in the last crop of a cropping sequence is a widespread (Poultney 1963) and efficient practice, since the costs of land preparation may be attributed to the crop, and the pasture is ready for grazing by the time the crop is harvested; there is no unused land area.

This practice may also be applied to land development for permanent pasture; great areas of Brazil have been planted to *Panicum maximum* or to *Brachiaria decumbens* following the clearing of forest or of cerrado. This has been often achieved for cattle production by smallholder sharecroppers who grew rice crops in exchange for their land development activities and subsequently returned the land to the original owners as pasture land. The practice of undersowing leguminous pasture in crops in slash-and-burn agriculture has been advocated as an alternative to the forest fallow phase (Shelton and Humphreys, 1972).

A further alternative is to sow forage species as a continual intercrop with a food or cash crop, as discussed earlier in this chapter with respect to the concept of live mulch.

This topic is explored in more depth in Chapter 11, and at this stage brief mention is made of some basic principles. The farmer needs to manipulate the competitive relations of the main crop and the forage crop so that main crop yields are not reduced or are reduced minimally. Some loss of main crop yield may be tolerated in the interests of the value of grazing from the forage, of the legume N accretion to the productivity of the total system or the reduced incidence of weeds; these perceptions are often not cogent for smallholders or farmers oriented to subsistence production. Main crop yield may be protected as follows.

Table 10.10 Effect of fertilizer and duration of leys on crop yields in northeast Thailand (from Gibson 1987).

Ley treatment			Applications of P,S,K fertilizers,	Roselle yield, 1983 DW bast fibre	Cassava yield, 1984 FW tubers
1980	1981	1982	1980–82	(t ha^{-1})	(t ha^{-1})
Cassava	Cassava	Cassava	+	0.83	7.1
			0	0.76	7.0
Weeds	Weeds	Weeds	+	1.00	8.9
			0	0.70	6.4
Cassava	Cassava	Siratro	+	1.44	8.3
			0	0.87	5.3
Cassava	Siratro	Siratro	+	1.55	10.5
			0	0.86	7.7
Siratro	Siratro	Siratro	+	1.51	12.9
			0	0.91	9.3

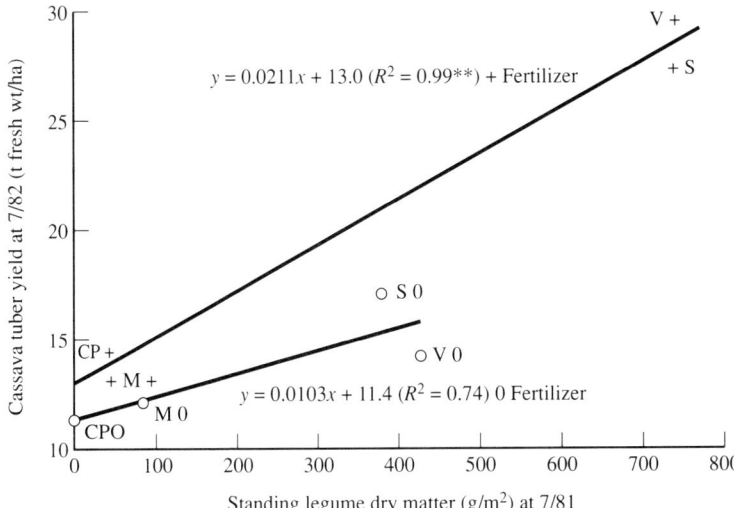

Figure 10.11 Cassava yield as related to legume dry matter present in previous year in northeast Thailand (CP = continuous cassava, V = *S. hamata* cv. Verano, S = *M. atropurpureum* cv. Siratro, M = *Mimosa invisa*, + plus fertilizer, 0 zero fertilizer) (from Gibson 1988).

- Time of sowing of forages may be delayed until the main crop is well established. This delay may extend from 14 to 60 days after planting the main crop.
- Density of forages may be controlled to a low level by reducing the seeding rate or planting rate.
- Grass competition will be more acute than competition from legumes, since the latter will exhibit some self-sufficiency for N requirement.
- Fertilizer policy is directed to supplying the needs of main crop and forage species, so that competition for nutrients is obviated; competition for moisture or light may still require modification.

These themes are developed subsequently.

10.4 Fertilizer practice

Agricultural development in the tropics has emphasized the use of low input systems; these have usually been driven by low prices for agricultural commodities and a shortage of farm capital. Options such as the incorporation of new élite plant material in the farm system are inexpensive and are readily adopted where farmers perceive the performance of these plants as in tune with their goals.

Nevertheless the real constraints to more efficient production require analysis, and this analysis may challenge previous assumptions cast within a framework where grassland is viewed as a waste product, or as the outcome of natural phenomena unrelated to farming intervention. This historical basis of grassland use in many developing countries is such that changed cultural perceptions of grassland are basic to its improvement.

Fertilizer use on the legume ley

Farmers prepared to recognize soil N as a major constraint to production may need also to recognize that fertilizer application to pasture legumes is necessary for adequate levels of N accretion and that fertilizer is best applied to the legume ley and not to the crop.

Mention was made earlier in this chapter of the favourable economics of fertilized legume leys for dairy production in northeast Thailand. Light upland soils of the Korat series (Oxic Paleustults) bear open dipterocarp forest. When this is cleared initial yields of cassava are of the order of 25 t ha^{-1}; after a few years of continuous cultivation cassava yields decrease to 6–10 t ha^{-1}. The soils have a surface texture of sandy loam overlying loam to sandy clay loam with mottled layers below 60 cm. Organic C of a long-cultivated study site (Gibson 1987) was only 0.2–0.4%, total N 0.01–0.02%; and CEC 1.6–2.0 meq%.

The effects of continuous cassava production were contrasted with those of weed fallow or grazed legume leys of varying duration. These cultural treatments were either not fertilized or received 30 kg P ha^{-1} and 20 kg S ha^{-1} in the first year, and 10 kg P ha^{-1} and 10kg S ha^{-1} in each of the next 2 years; K was also applied but was later considered inessential. The area was then test-cropped without fertilizer application to roselle (kenaf), followed by cassava. The results involving *M. atropurpureum* cv. Siratro (Table 10.10; Gibson 1987) showed that although Siratro grew and persisted in the absence of fertilizer, fertilizer application to the legume ley was necessary if crop production were to benefit appreciably. Application of P, S and K to cassava had little residual effect on subsequent crop production. However, a 1-year ley of fertilized Siratro lifted roselle yield from 0.8 to 1.4 t ha^{-1}, and only small benefits accrued in the first test crop from a longer ley. The second test crop (cassava) again benefited from the fertilized legume ley, and a longer ley gave better

Table 10.11 Effects of grazing intensity of a grass ley and of level of N fertilizer application to crop on the yield of maize (t ha^{-1}) at Marondera, Zimbabwe (from Barnes 1981b).

Fertilizer N application to maize (kg ha^{-1})	Previous grazing intensity on grass ley		
	Nil	Light	Heavy
0	1.61	2.08	2.62
72	3.27	4.15	4.05
144	4.37	4.66	5.00
Mean	3.08	3.63	3.89

results than a short ley; the 3-year fertilized ley lifted cassava yield from 7.1 to 12.9 t ha^{-1}.

The benefit of the fertilized legume to a succeeding crop is illustrated from another component of this study, which contrasted the legumes *M. atropurpureum* cv. Siratro, *S. hamata* cv. Verano and *Mimosa invisa* (which was unsuccessful). The linear functions relating standing legume DM in the previous season to the subsequent yield of a first test-crop of cassava (Figure 10.11; Gibson 1988) had differing slopes; Verano and Siratro produced similar effects on cassava yield, but the addition of P and S fertilizer increased both the yield of legume and the influence of unit legume yield, since N fixation was enhanced by fertilizer, as discussed in Chapter 2.

This concept is further illustrated from a study of a ley based on *Centrosema pubescens* at Ngetta, northern Uganda. The ley received annual applications of single superphosphate or not, and this led to annual steer LWGs of 419 and 345 kg ha^{-1} respectively from the fertilized and non-fertilized pastures (Stobbs 1969a). Total soil N increased from 0.075 to 0.086% in the fertilized treatment. Previous fertilized application to the ley increased the yield of subsequent test crops by 190 kg ha^{-1} of seed cotton and by 960 kg ha^{-1} of finger millet. The legume ley may require exigent management, but the gains to the farm system are substantial.

Crop fertilizer needs following the legume ley

Legume N fixation will meet the N demand of some subsequent crops, as previously illustrated in Figure 10.1 for rice production in the eastern llanos of Colombia. Crops with a high N demand such as maize may require supplementary fertilizer N, and in this situation the ley operates to reduce the level of fertilizer N, as illustrated for the

Plate 10.4 *Setaria sphacelata* var. *sericea* and *Centrosema pubescens*, south Thailand.

subhumid zone of Nigeria in Figure 10.2. Further studies in Nigeria (Tarawali 1991) showed that following *Stylosanthes* fodder banks maximum yield of maize was obtained with application of 60 kg N ha^{-1}, whilst a rate of 120 kg N ha^{-1} is recommended for maize following bush fallow or under continuous cultivation.

Policy for N fertilizer use on grass leys

By contrast the predominant theme which has emerged from studies of fertilized grass leys is that the N fertilizer applied should be justified in terms of the animal production from the pasture and not in terms of expected benefits to subsequent crop yields. Experience in Zimbabwe indicates that residual effects of N fertilizer on crop yields are small, as discussed earlier in connection with Barnes' (1981a) experiment, and supported by other studies (Barnes 1981b; Rodel *et al.* 1981).

Many of the cultivated tropical pasture grasses are tolerant of low pH and of high levels of soil Al and Mn (Humphreys 1981). Plants such as *Cynodon nlemfuensis*, *C. dactylon* and *Melinis minutiflora* grow well at pH 4–4.5, but crops of maize or sorghum are less tolerant; a policy of high N fertilizer use on the grass ley is inimical to subsequent crop production unless heavy liming which will restore soil pH to reasonable levels is accepted (Barnes 1981a).

N fertilizer use on grass leys is only appropriate in intensive (usually subsidized) animal production systems.

10.5 Utilization of the ley

The high level of nutrient transfer from the pasture which occurs in cut-and-remove systems of use was mentioned early in this chapter, together with the problems of incorporating carbonaceous material from a lightly used ley into a cropping phase.

Many writers have emphasized the risks associated with grazing, due to the spatial transfer of nutrients to centres of stock concentration, and to the losses of nutrients by volatilization and leaching, as discussed in Chapter 2.

There are few studies which determine the field situation. In Zimbabwe Fenner and Rattray (1966) found on a heavy clay loam near Harare that the management of a grass ley of *P. coloratum* var. *makarikariense* did not significantly influence subsequent maize yield.

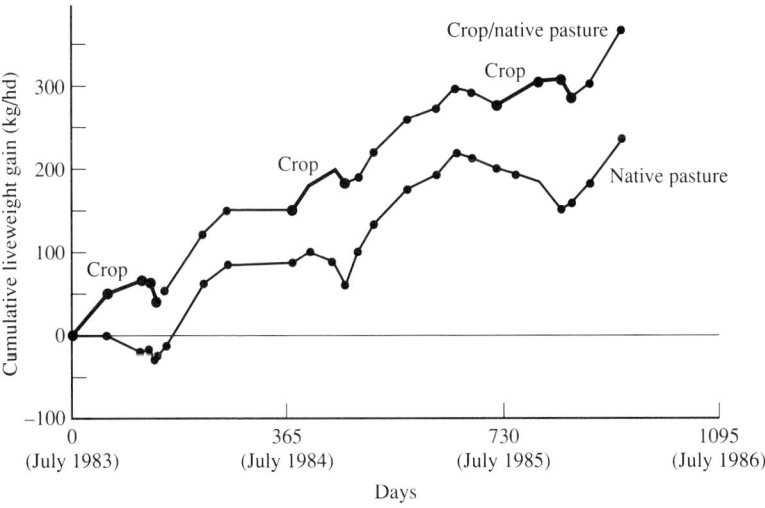

Figure 10.12 Cumulative liveweight gain for cattle grazed on native grass pastures or on a combination of crop lands with leguminous pastures for dry season grazing and of native grass pastures at Katherine, Australia (from Jones *et al.* 1991).

Pastures were grazed, cut for hay, or cut and the grass left *in situ* over a 5-year period. Similarly at Mazowe, Zimbabwe variation in SR did not affect subsequent maize yield (Rodel *et al.* 1981), although in a companion study pastures of *P. coloratum* cv. Bushman Mine which provided additional grazing relative to *P. repens*, *C. nlemfuensis*, and *C. gayana* leys gave marginally higher maize yield subsequently. By contrast on the sandveld at Marondera, Barnes (1981b) found real effects of grass utilization on the maize yields which followed. This study was referred to earlier, and soil characteristics are indicated in Table 10.2. The ley of *C. nlemfuensis* var. *robustus* cv. No.2 was mown and removed, grazed with cattle intermittently at a light intensity, or grazed with an additional 50% stock days. The test-crop yields of maize (Table 10.11; Barnes 1981b) received differing levels of N fertilizer and were on average 26% greater following heavy grazing. There was a significant interaction between treatments; the response to heavy grazing was greatest at zero N fertilizer input, and the response to N fertilizer was greatest following the cut-and-remove system. A companion experiment also showed a positive response to grazing.

These results may be interpreted to indicate that nutrient cycling was more effective under grazing, and this favourably influenced crop yields where nutrients were in short supply. The deleterious effects of cut-and-remove systems are easy to comprehend (Foster 1971; Scaife 1971). An additional consideration is that coarse carbonaceous residues whose breakdown immobilizes N are reduced under heavy grazing, and that bases are recirculated from deeper layers to be deposited in the upper soil layers where they are accessible to crop roots.

A further illustration of the benefits of grazing the ley occurred in another study from Ngetta, Uganda. Leys based on *S. guianensis*, *C. gayana* and *Hyparrhenia rufa* were used over 3 years for day grazing, for night grazing, or were not grazed (Stobbs 1969b). The area was then test-cropped for 3 years to a rotation of cotton, millet, beans, groundnut and sorghum. The overall crop yields were increased by 19 and 10% respectively for night and day grazing, relative to crop yields from the ungrazed areas. The higher yield after night grazing was associated with greater stock density, and it is feasible that the night grazed area benefited from transfer of nutrients from the day grazed area. The utilization of farmyard manure, which is a significant aspect of crop production in developing countries, was discussed in Chapter 2.

The SR of a legume ley needs to be in synchrony with forage availability (Plate 10.4), and adjusted in the recognition that the level of N fixation depends upon the rate of legume growth (Humphreys 1991).

The attractiveness of the ley to farmers is enhanced if systems of use are developed which are seen to enhance the production of the agricultural system and to contribute to environmental protection. The concept of the pasture ley as an unused resting phase does not survive the population pressures of the developing world or the enterprising spirit of farmers seeking returns from their land.

This chapter is concluded by an example of cattle production from the integration of native pasture areas with areas of legume ley and crop residues at Katherine, Australia. In this wet/dry monsoonal climate cattle gain liveweight during the main wet season at similar rates on native grass pastures and on planted leguminous pastures, and the latter show to advantage during the dry season during which dry legume forage has reasonable nutritive value, since the dry season is reliably dry. The animal density in this region of low human population density is low, and the native pastures are therefore an underutilized resource which can absorb seasonal fluctuations in animal density. In this study cattle were grazed continuously at 0.07 b ha^{-1} on native pastures, or were removed in the dry season (July–October) to use ley pastures based on *S. hamata*, *C. pascuorum* and *A. vaginalis* and crop residues of sorghum and maize at an SR of 3.3 b ha^{-1} for this period.

During the main dry season animals grazing native grasses lost an average of 250 g head^{-1} day^{-1} whilst animals grazing the leys gained c. 460 g head^{-1} day^{-1} (Figure 10.12; Jones *et al.* 1991). At the beginning of the wet season animals lost more weight grazing the leys than those grazing native pastures, since the legumes deteriorated sharply in nutritive value and acceptability, and there were subsequent compensatory weight gains in the native pasture only cattle group which reduced the dry season advantage of grazing the leys. The net advantage (Figure 10.12) of the incorporation of the ley in this harsh environment was c. 135 kg LW head^{-1} after three seasons of ley/crop grazing, which represents a substantial advantage both in terms of overall rate of LWG and reduced age at turn-off.

10.6 Conclusions

This chapter has indicated the many scientific gains which have been won in recent years in developing the technology for ley pastures with annual crops in the tropics and subtropics. Many of these gains have yet to be translated to farmer acceptance; population pressure on the land exacerbates this problem in many regions.

The greater incorporation of leys in farming hinges upon a number of factors, of which the relative prices of crop and animal products, and of N fertilizer predominate. As the demand for protective foods such as dairy products increases, and as a sufficient proportion of the population of tropical countries can translate this into an economic demand, so will the pressure towards the planting of ley pastures increase. The production of N fertilizer depends upon resources of fossil fuel; the energy conversion of these to N fertilizer exacerbates the 'greenhouse effect' and countries lacking fossil fuel require scarce resources of foreign exchange to make N fertilizer available to farmers. These two factors point to legume N fixation in the ley system as a desirable pathway towards sustaining crop production. The severe environmental degradation associated with cropping of marginal lands can be moderated by the introduction of pasture leys; more leys will be planted as awareness of environmental issues is translated into government policies for amelioration and as farmers perceive that ley farming leads to the sustainability of agriculture.

This thrust to the planting of leys will founder in regions where technology is insufficiently robust and free of risk to lead to farmer acceptance. The availability of élite germplasm which meets farm requirements is patchy across environments, and the knowledge to guide the management of legume N fixation is incomplete for many farm situations. Some scientists are sceptical about the adequacy of N fixation to meet crop needs; in farm practice N fixation is more likely to be in the range 15–80 kg N ha^{-1} year^{-1} than in the range 100–200 kg N ha^{-1}/y which is required for intensive farm production.

Finally, it should be recognized that mixed farming for crop and animal production which incorporates legume leys demands a high level of managerial skills relative to those needed for monocultural systems. The transition to the ley will be driven by economic opportunity, policies of environmental protection and the evolution of farmer skills to utilize technology in which the efficiency of legume use is apparent.

CHAPTER 11

Annual Crops with Forage Crops

11.1 Introduction
The rationale for intercropping annuals with forage

The fourth type of agricultural system in which forages contribute to sustaining the yield of crops is one in which annual crops grown for food or cash are grown mixed with forage crops. A high proportion of the diet of ruminants in the tropics and subtropics is drawn from the residues and otherwise unused by-products of annual crops (Plate 11.1), so that 'forage' may be regarded as including these residues as well as the forage species, such as *Stylosanthes hamata* cv. Verano, which are deliberately planted as an intercrop. Relay cropping, in which forage crops are planted in sequence or overlapping the growth of other crops, is also considered in this chapter, together with the practice of undersowing food or cash crops with forage species in order to establish a pasture, as mentioned in section 10.3. The main objectives of farmers in planting intercrops are:

- increased resource use, giving higher and diversified yields,
- N fixation of leguminous crops,
- control of weeds,
- increased animal production.

There are many general reviews of this topic, including Bradfield (1974), Papendick (1977), Gliessman *et al.* (1981), Beets (1982), Gomez and Gomez (1983), Francis (1986), Humphreys (1986), Vandermeer (1989) and Loomis and Connor (1992). Regional illustrations of intercropping theory and practice are given in this chapter, and the following are mainly additional examples inserted for the local interest of readers.

- Africa: Burkina Faso (Stoop 1981); Ethiopia (Nnadi and Haque 1988; Kahurananga 1991); Ghana (Haizel 1974); Ivory Coast (Talineau et al. 1976); Malawi (Kanyama and Edje 1976); Nigeria (Heide et al. 1985; Yayock 1981; Powell and Waters-Bayer 1985); South Africa (Kruger and Van Den Berg 1993); Tanzania (Sarwatt 1993); Zimbabwe (Avila 1988).
- America: Brazil (Faris et al. 1983a; Gianluppi et al. 1983); Colombia (Mason et al. 1986a); Honduras (Sinclair et al. 1991); Paraguay (Glatzle and Ramirez 1993); St Lucia (Rao and Edmunds 1984).
- Asia: India (Natarajan and Willey 1980; Agarwal et al. 1986; Hosmani et al. 1986; Willey et al. 1989); Indonesia (Nitis 1977; Siregar and Semali 1982); Laos (Shelton and Humphreys 1972, 1975a); Malaysia (Beets 1979); Philippines (Zandstra 1979; IRRI 1983; Carangal and Calub 1987); Thailand (Shelton and Humphreys 1975b; Wilaipon et al. 1981; Pongskul et al. 1982).
- Oceania: Australia (Chamberlin et al. 1986); Papua New Guinea (Bourke 1974).

Increased resource use

A convenient assessment of the effectiveness of intercropping is the land equivalent ratio (LER, Mead and Willey 1980) which compares the area of land which would be needed to produce the yields of crops grown as sole crops with the yields of crops grown as intercrops. For the yields (Y) of two crops a, b

$$\text{LER} = \frac{Y_a \text{ as intercrop}}{Y_a \text{ as sole crop}} + \frac{Y_b \text{ as intercrop}}{Y_b \text{ as sole crop}}$$

This may be further adjusted to the area time equivalent ratio (ATER) where in humid environments it is feasible to grow more than one crop per season as sole crops, and the LER is amended by the relative durations in the ground of a crop in a sole cropping system and in an intercropping system. LER can be calculated in terms of total production, yield of economic products, and of cash return; the advantage is calculated in terms of the amount by which LER exceeds unity.

LER measures the efficiency with which intercropping captures the environmental resources which determine crop growth, light, moisture and nutrients. In India Patil and Pal (1985) compared the energy relations of pearl millet grown as pure stands or when intercropped with *Vigna mungo, V. radiata, V. unguiculata, Cyamopsis tetragonoloba* and other legumes. The additional energy cost of growing an intercrop varied from c. 10 to 13 GJ ha^{-1} compared with the cost of growing sole

Introduction 325

Plate 11.1 Stored teff straw, Ethiopia.

stands. However, intercropping systems increased the energy capture by 21–59 GJ ha^{-1}, and also enhanced the N economy of the system. The greater energy capture by intercropping results from increased interception of radiation through the maintenance of a greater leaf canopy integrated over time and through intercepted light being used more efficiently through being spread over a larger leaf surface with a differing architecture; these in turn reflect a greater exploitation of the moisture and nutrients available in the soil, and the N accretion of forage legumes. These themes have been considered in different circumstances in Chapters 8 and 9.

The advantage of intercropping is often derived from greater energy capture early in the cropping cycle if a fast growing crop is included. This is illustrated by a study (Mason *et al.* 1986a,b) at Santander de Quilichao, Colombia where cassava (*Manihot esculenta*), cowpea (*V. unguiculata*) or peanut (*Arachis hypogaea*) were grown alone or as cassava intercropped with either legume on a Typic Dystropept clay soil of low fertility. The LAI of the sole cassava and cowpea crops and of the combined intercrops (Figure 11.1a) in 1981 did not differ from 80 days after planting onwards, but the early canopy development of cowpea substantially enhanced the total canopy of the intercrops in the early period of crop development when cowpea occupied the space between cassava plants which would otherwise have been bare ground.

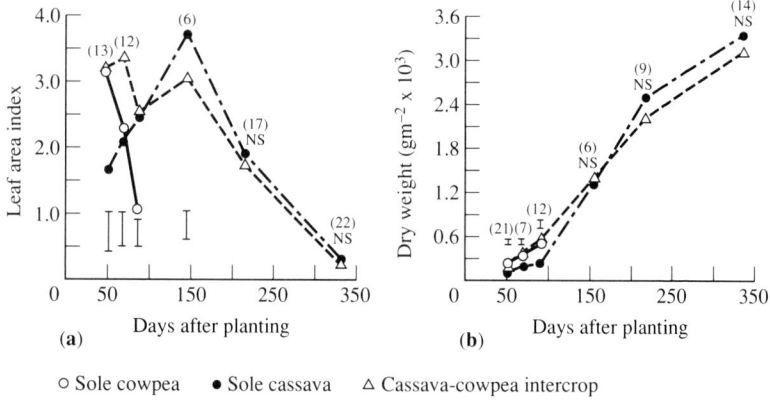

Figure 11.1 Leaf area index (a) and crop dry weight (b) of sole cowpea, sole cassava and cassava–cowpea intercrops in Colombia (bars indicate LSD at $P = 0.05$) (from Mason *et al.* 1986b).

The DM production of the combined intercrop was initially greater than that of the sole cassava (Figure 11.1b) but the long-term growth of cassava in the intercrop was not significantly affected. Intercropping with cowpea slightly reduced cassava storage root yield in 1981, but the LER was 1.48, indicating a substantial advantage of 48% from intercropping. In assessing the latter it should be recognized that an additional sole crop of cowpea might have been grown in this rainfall environment where two pronounced dry periods occur, and a more realistic assessment of the gain is *c.* 30%.

Intercropping results in greater access of crops to soil moisture and to nutrients. In the last example, the nutrient uptake (Mason *et al.* 1986c; Table 11.1) from intercropped systems was initially much greater than from sole crops, and at cassava harvest the numerical differences were suggestive but did not reach significance. Much of the increased N uptake may be attributed to biological N fixation by cowpea, but the increased drain of P and K suggest that either fertilizer inputs would need to be adjusted or nutrient cycling (through livestock?) would need to be carefully managed to avoid soil depletion. Intercropping with cowpea reduced P% by 0.08, 0.06 and 0.03 respectively in cassava stems, leaves and storage roots at 50 days after planting, and this may have been associated with the reduced cassava growth occurring in the intercrop in this early phase. No effect of intercropping on concentration of N, P or K in cassava occurred at later harvest dates.

Superior use of natural resources also reduced the instability of

Table 11.1 Uptake of N, P and K (kg ha^{-1}) by cassava and cowpea sole crops and cassava–cowpea intercrop in Colombia (from Mason et al. 1986c).

Cropping system	N	P	K
	80 days after planting		
Cassava sole crop	66	51	46
Cassava–cowpea intercrop	120	110	98
Cowpea sole crop	89	92	84
LSD (0.05)	14	17	17
	335 days after planting		
Cassava sole crop	204	272	179
Cassava–cowpea intercrop	234	302	230
LSD (0.05)	NS	NS	NS

yield associated with single cropping. Faris et al. (1983b) compared the intercrop patterns of maize or sorghum with either cowpea or *Phaseolus vulgaris* against their component sole crop production over a 5 year period. They concluded that intercropping or cereals alone had better performance stability than pulses alone in northeast Brazil.

A further option is the use of waste land for legume planting, to be grazed in conjunction with cereal residues. This has been well adopted in many tropical areas, and the planting of the paddy bund between the rice bays with *Stylosanthes hamata* has been especially successful in northeast Thailand (Gutteridge 1983).

N fixation of leguminous crops

This topic has been dealt with in a general way in section 2.5. Leguminous crops are particularly adapted to intercropping because of their relative independence of soil N supply, their capacity to add to the N economy of the system or at least reduce the rate of N run-down, and their positive effects on the feeding value of crop residues for livestock. This question is elaborated further in section 11.2.

Control of weeds

Edible volunteer weeds are often valued by farmers though their contribution to the seasonal feed supply, especially at times when the land available for grazing is restricted since more of the area is growing crops (for example Moog 1980 at Batangas, Philippines; see

Humphreys 1991, figure 2.3). However, much agronomic practice is directed to the suppression of weeds, which is costly of labour and herbicidal inputs. The damaging effects of weeds on crop yield were discussed in section 5.2 and illustrated in Table 5.1 for a situation at Morogoro, Tanzania where green gram or soybean grown as an intercrop with millet substantially ameliorated weed damage. Similarly in Kerala State, India, cassava intercrops with cowpea, groundnut, soybean, *V. radiata* and *V. mungo* were studied (Ashokan et al. 1985). Tuber yields of cassava were similar as sole crop or as intercrop, but the cost of weeding was reduced 50% by growing cowpeas with cassava. There were also ancillary benefits to OM and N status in intercropped plots.

Increased animal production

It is commonly believed that the path to wealth for smallholder farmers is the greater off-take of animal products (Colour Plate 8). A more fundamental view is that where animals are used for draught the success of crop production requires the maintenance of animals in good working condition, especially at the beginning of the cropping season when the area which may be ploughed in a timely fashion depends upon the nutritive value and availability of the fodder for ruminants before the rains begin. This topic is well reviewed by Little and Said (1987), Dixon (1987), Reed et al. (1988) and Said and Dzowela (1989).

Most cereal crop residues are high in fibre and low in protein. For example Arias et al. (1980) found that sorghum crop residues could hardly act as a maintenance ration for grazing cattle. Supplementation of low protein diets with leguminous forage (Colour Plate 16) was suggested to give responses in small ruminants of 50–60 g LWG kg^{-1} legume DM in Table 9.2, whilst in section 9.2 the example quoted in Table 9.11 gave a response of 130 g LWG kg^{-1} legume DM. Legume crop residues often provide good levels of cattle performance. At Morogoro, Tanzania the growth of black head Persian sheep increased from 34 to 69 g LWG hd^{-1} day^{-1} when *Chloris gayana* hay was supplemented with *Crotalaria ochroleuca* (Sarwatt 1993). In Venezuela straw of *Arachis hypogea* contained 2.3–2.4% N, and gave 52–56% OM digestibility and daily intake of 58–61 g kg^{-1} $LW^{0.75}$ (Velásquez and González 1972). These observations suggest considerable benefits to animal performance by incorporating legumes in the cropping system.

At Lynne East in the Transvaal Lowveld of South Africa (lat. 25° S, 970 mm annual rainfall, 720 m altitude) Kruger and Van Den Berg (1993) grew maize with a high input of fertilizer of 150 kg N or without N fertilizer but with five leguminous intercrops. *Canavalia ensiformis* was the highest yielding of these. Total seed production of the intercrop was

Table 11.2 DM yield and N concentration in stubble and chaff of maize or maize and *Canavalia ensiformis* intercrop in Lynne East, South Africa (from Kruger and Van Den Burg 1993).

Attribute	Maize + 150 N	Maize + C. *ensiformis*
Stubble		
DM (t ha^{-1})	5.9	9.7
N (%)	0.89	1.40
Chaff		
DM (t ha^{-1})	3.2	5.2
N (%)	0.55	1.42

c. 10% lower than the sole maize crop, but yield of stubble and of chaff (maize head leaves, maize husks, legume pods after grain removal) (Table 11.2) was substantially increased. The N concentration of maize residues was suboptimal as a feed, but the stubble and the chaff from the maize–C. *ensiformis* intercrop each provided a satisfactory level of 1.4% N.

The benefits to animal production from intercropping cereals with forage legumes is illustrated by a study (Sinclair *et al.* 1991) at Juticalpa, Olancho, Honduras (15° N, 1400 mm annual rainfall, 500 m altitude) which compared cattle performance utilizing maize stubble or stubble of maize intercropped with *Lablab purpureus* (Table 11.3). *L. purpureus* is a late-flowering legume known for its capacity to extend the grazing season (Hendricksen and Minson 1985). High individual animal weight gains, approaching 1 kg hd^{-1} day^{-1} were recorded from cattle grazing the stubble, since animals had opportunity to improve the

Table 11.3 Cattle performance either grazing or fed cut stubbles of maize or maize plus *Lablab purpureus* at Juticalpa, Honduras and economic returns (from Sinclair *et al.* 1991).

	Maize alone		Maize + *L. purpureus*	
Attribute	Grazed	Cut	Grazed	Cut
LWG (g hd^{-1} day^{-1})	929	548	1019	810
Carcass weight (kg ha^{-1})	68	78	157	176
Variable costs ($US ha^{-1})	27	39	61	101
Gross margin ($US ha^{-1})	142	157	329	336

quality of their diet by selective grazing, and costs were lower than those of a cut-and-remove feeding system. Carrying capacity was 3.2–3.8 AU ha^{-1}, and a cut-and-remove system enabled the feed supply to be rationed over a longer period than was available under grazing. The gross margin per hectare was similar for grazing and cutting, but was more than doubled by incorporating *L. purpureus* as an intercrop. This example can be duplicated in farmer experience in many tropical regions. The value to animal production of incorporating legumes as intercrops provides a cogent imperative for farmer adoption of this practice.

11.2 Relations between crops

Light

The concept of analysis of growth based on the level of radiation, the proportion intercepted by a crop, and the efficiency with which light energy is used in the production of carbohydrate was introduced in section 8.2. The light relations of forages in plantations differ from other situations where forage plants are not always overtopped by the companion crop. In sections 9.2 and 9.3 the light relations of legume hedgerows with intercrops were discussed, and Figures 9.5 and 9.6 and Table 9.8 illustrate the complex interaction which occurs between competition for light, moisture and nutrients; the capacity of *L. leucocephala* to reduce the availability of moisture and nutrients to a millet crop (as evidenced by the effect of inserting a below-ground barrier between the two species) was expressed through a shorter millet crop being denied radiation by its taller companion shrub, which resulted in reduced C fixation by the millet crop. This in turn might be expected to reduce the root system of the millet crop and its capacity to access moisture.

In this section the light relations of intercropped annuals are discussed, and the many cropping practices available to farmers to manipulate the outcome of competition for light are elaborated further in section 11.3. The main factors are:

- the growth rate and canopy structure of the crop varieties chosen,
- the crop time to maturity,
- the density and spatial arrangement of plants,
- the time of sowing each crop component,
- the fertilizer practice adopted,
- the cutting management of the crops for forage production.

The effect of choice of crop variety and of crop species on the light regime for crop growth is illustrated from two experiments (Cenpukdee and Fukai 1992a) conducted at Redland Bay, southeast Queensland (lat. 28° S) which received supplementary irrigation. Solar radiation increases from October to December, when weekly averages are $c.$ 28 MJ m^{-2} day^{-1}, and declines to $c.$ 14 MJ m^{-2} day^{-1} in June–July. On this deep, fertile red loam cassava cultivars of differing growth and canopy characteristics were planted in the first experiment with 1.5 m inter-row spacing, and simultaneously pigeonpea (*Cajanus cajan*) cv. QPL 65 was planted (or not) in four rows 0.2 m apart between each pair of cassava rows. The performance of three cassava cultivars of differing vigour and habit illustrates the responses.

Pigeonpea seedlings grew faster than the vegetatively planted cassava and overtopped the cassava plants throughout the period until pigeonpea harvest 103 days after planting. This reduced the PAR received by the cassava, as indicated in Figure 11.2; the short, compact MAUS 19 variety was especially disadvantaged, but all cassava varieties experienced reduced illuminance when this was measured 40 and 60 days after planting. Canopy height (Table 11.4) of pigeonpea 78 days after planting was substantially greater than that of cassava and the intercropped cassava grew taller than sole crop cassava. MAUS 19 was significantly shorter than SM1-150, and canopy width in the former was also less. In all varieties intercropping with pigeonpea greatly reduced the canopy width of cassava.

Total DM production of sole cropped cassava was highly correlated with cumulative intercepted radiation, and the shading of intercropped

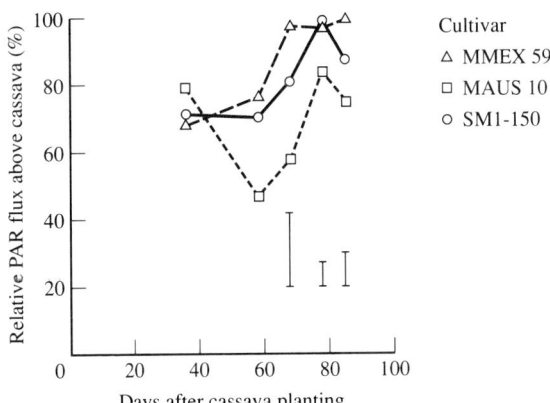

Figure 11.2 Relative flux density of PAR at the top of three intercropped cassava cultivars using different cassava cultivars in cassava/*Cajanus cajan* intercropping systems in southeast Queensland (bars indicate LSD at $P = 0.05$) (from Cenpukdee and Fukai 1992a).

Table 11.4 Plant height and canopy width of sole cassava, intercrop cassava and intercrop *Cajanus cajan* for three varieties of cassava 78 days after planting in southeast Queensland (from Cenpukdee and Fukai 1992a).

Attribute	Cassava cultivar		
	SM1–150	MMEX 59	MAUS 19
Plant height (cm)			
Sole cassava	99	93	82
Intercrop cassava	105	98	94
Intercrop C. *cajan*	134	133	131
Canopy width (cm)			
Sole cassava	155	52	118
Intercrop cassava	152	67	120
Intercrop C. *cajan*	102	41	122

cassava was reflected in total DM production only c. 25% of that of sole cropped cassava, MAUS 19 being especially disadvantaged. Tuber initiation was delayed in the intercrop and relative tuber yields at 230 days after planting were reduced even more. Root sampling indicated that the top 25 cm of the soil in the intercropped area was mostly occupied by pigeonpea roots. LER calculated on the basis of cassava tubers and pigeonpea seed for the combined intercrop was 1.02, 0.74 and 0.89 respectively for SM1–150, MMEX 59 and MAUS 19 cultivars, indicating that the system of simultaneously planting these cassava varieties with pigeonpea at the chosen densities led to competition for light which failed to give any advantage for the intercropping system and led to pigeonpea dominance which greatly reduced cassava yield. This conclusion is modified slightly if the forage value of the residues is taken into account.

The second experiment (Cenpukdee and Fukai 1992a) illustrates the differing effects of the legume intercrop chosen. The sole legume crops were grown at twice the legume density of the intercrop. The pigeonpea cultivar QPL3 and the short-statured, early-maturing soybean cultivar Fiskeby V were planted 35 days after cassava planting, and two (rather than four) legume rows were planted between each pair of cassava rows. This led to a radically different situation from the first experiment, since the more vigorous cassava varieties overtopped pigeonpea and soybean throughout the course of crop development. Intercropping with soybean had only a minor effect on the canopy width of cassava, and the effect of pigeonpea in reducing cassava canopy width was much less than in the first experiment shown in Table 11.4.

The relative radiation received by the legume intercrop (Figure 11.3) was affected by choice of cassava cultivar; the less vigorous and shorter MAUS 19 had no measurable effect on the light received by either legume intercrop whilst pigeonpea had higher relative PAR than soybean. However, the seed yield of earlier maturing soybean was less sensitive to intercropping than was pigeonpea. With the lower yielding MAUS 19 the yield of the legume intercrop relative to sole legume was 36% and 39% for soybean and cassava; with the vigorous SM1-150 the relative yields were 23% and 1% for soybean and cassava.

The overall cumulative interception of radiation was similar in the three cropping systems, but was up to 20% less in MAUS 19 than in other cassava cultivars. Radiation use efficiency averaged 0.76, 0.71 and 0.63 g MJ^{-1} respectively for the sole cassava, soybean intercrop and pigeonpea intercrop combinations. Cassava cultivars also differed in radiation use efficiency, which was 0.79, 0.55 and 0.61 g MJ^{-1} respectively for SM1-150, MMEX 59 and MAUS 19. When the cassava tuber yield and legume seed yields were taken into account, the LER of the cassava/pigeonpea system was less than unity, but the three selected cassava cultivars with soybean gave LER values varying from 1.15 to 1.53, indicating successful intercropping practice and

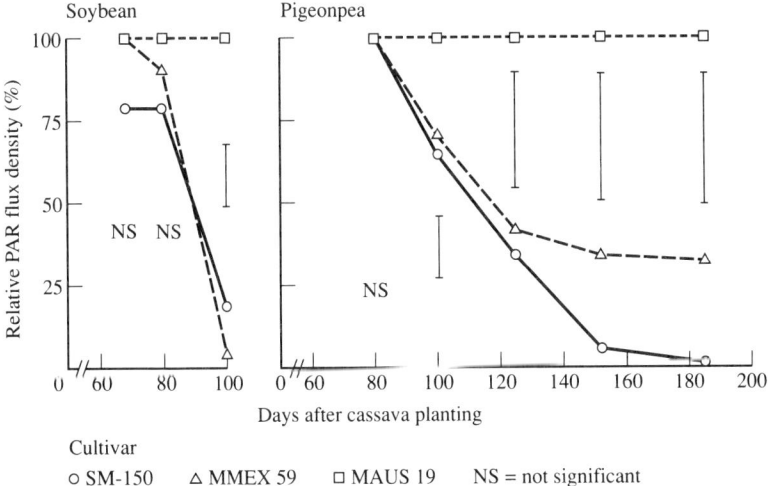

Figure 11.3 Relative flux density of PAR at the top of intercropped (a) soybean and (b) pigeonpea in cassava/pigeonpea intercropping systems using different cassava cultivars in southeast Queensland (bars indicate LSD at $P = 0.05$) (from Cenpukdee and Fukai 1992a).

improved efficiency of resource use.

These two experiments indicate how variation in genotype and crop husbandry alters the light relations of component species to produce negative or beneficial yield outcomes. Although emphasis is often placed on height development, it should be appreciated that when the shortest cultivar of a tall species is taller than the associated species, canopy width is more important than height in determining the yield of the taller species (Cenpukdee and Fukai 1992b). If leaf area is the same for two cultivars, the taller species may permit more radiation to reach the shorter statured crop (Wooley and Rodriguez 1987). If the intercrops are of similar height it is the canopy width of each which mainly determines the amount of light intercepted. The timing of the interference which one species exerts on another may be crucial for the progress of developmental events; in these studies the rapid growth of an early maturing legume crop had less effect on cassava yield than interference occurring during later phases.

Moisture

Crop competition for moisture is diminished if the root systems of crops predominate in different soil strata, or are active at different times. These conditions did not appear to apply in the cassava/pigeonpea intercrop of Experiment 1 discussed above (Cenpukdee and Fukai 1992a) where the dominance of the pigeonpea canopy in reducing the light environment of cassava was also expressed in active root competition for moisture and nutrients. This was evident from measurements of combined root length density, which is the best measure of the degree to which the soil volume is accessed. These measurements were taken at 25 cm laterally away from the centre of cassava rows, which was the middle position between the cassava and pigeonpea rows in intercropping, and at 50 cm from the cassava rows, which was the pigeonpea position in intercropping. In sole cropped cassava (Table 11.5) the root length densities at 25 and 50 cm from the cassava row centre were similar at each depth, indicating that cassava roots had spread evenly across rows by 61 days after planting. However in intercropping root length density at 50 cm (pigeonpea row) was substantially higher than that of sole cropped cassava, and at 25 cm between the cassava and pigeonpea rows root length density was also higher in intercropping than in sole cassava. These values suggest that pigeonpea roots would mostly occupy the soil volume at the 25 cm position, and help to explain pigeonpea dominance.

A second study illustrates a different situation where an undersown forage (*Stylosanthes hamata* cv. Verano) developed a root system at deeper soil layers than were occupied by the roots of the main crop

Table 11.5 Root length density (cm cm^{-3}) in sole cropped cassava and cassava/pigeonpea intercropping at 25 and 50 cm away from the centre of cassava rows 61 days after planting in southeast Queensland (from Cenpukdee and Fukai 1992a).

System and position	Soil depth (cm)		
	0–20	40–60	80–100
Sole cassava			
25 cm	8.5	0.75	0.25
50 cm	8.1	0.70	0.32
Intercrop cassava			
25 cm	9.8	1.76	0.87
50 cm	10.9	2.20	0.67

(maize). This work (Hulugalle 1989) was undertaken at Kamboinse Research Station near Ouagadougou, Burkina Faso (lat. 12 ° N, 300 m altitude). Annual rainfall in the successive years of the study was 574, 814 and 648 mm and mainly fell in the 4 months June-September. The soil was a kaolinitic, hyperthermic, Oxic Paleustalf with a sandy loam topsoil and a sandy clay subsoil. Maize was planted in tied or open ridge cultivation systems with high fertilizer inputs of 144 kg N ha^{-1} and of PK. The system was not designed to use the forage legume as an N source, and Verano was cut and removed four or five times each season, yielding an average of 4.5 and 2.6 t DM ha^{-1} year^{-1} in the tied and open ridge systems respectively. The first-year Verano yield was small, but Verano did not require resowing in the second and third years. Maize grain yield was greater in the tied ridge system; it was independent of undersowing in two years but reduced in the tied ridge system from 2.5 to 0.9 t ha^{-1} by the presence of Verano in 1987.

This result was associated with changes in moisture availability to the maize crop. The soil water content of the profile during crop development in the three years (Figure 11.4) shows that much less water was available in each year in the open ridge system, where considerable runoff occurred. In the tied ridge system the presence of Verano caused a drier profile by a seasonal average of 26 and 28 mm respectively in 1986 and 1987; the intensity of competition was especially high during periods of low rainfall (20–35 days after planting in 1986, and 0–20 and 50–75 days after planting in 1987). The latter period, associated with tasselling and grain formation was crucial for the development of maize yield.

Root growth of maize was limited to the 0–50 cm soil depth (Table 11.6); the subsoil had a high bulk density (> 1.7 Mg m^{-3}) which would be restrictive, but this layer was penetrated by the tap-rooted *S. hamata* cv. Verano. The tied ridge system, which trapped more rainfall, gave superior root growth relative to the open ridge system. The presence of Verano gave more effective use of the water resource, but reduced maize grain yield when rainfall was crucially deficient in one of the three seasons studied, despite the repeated defoliation of Verano which must have reduced its competitive status. The maize yield reduction needs to be set against the value of 4.5 t DM ha^{-1} year^{-1} of Verano fodder.

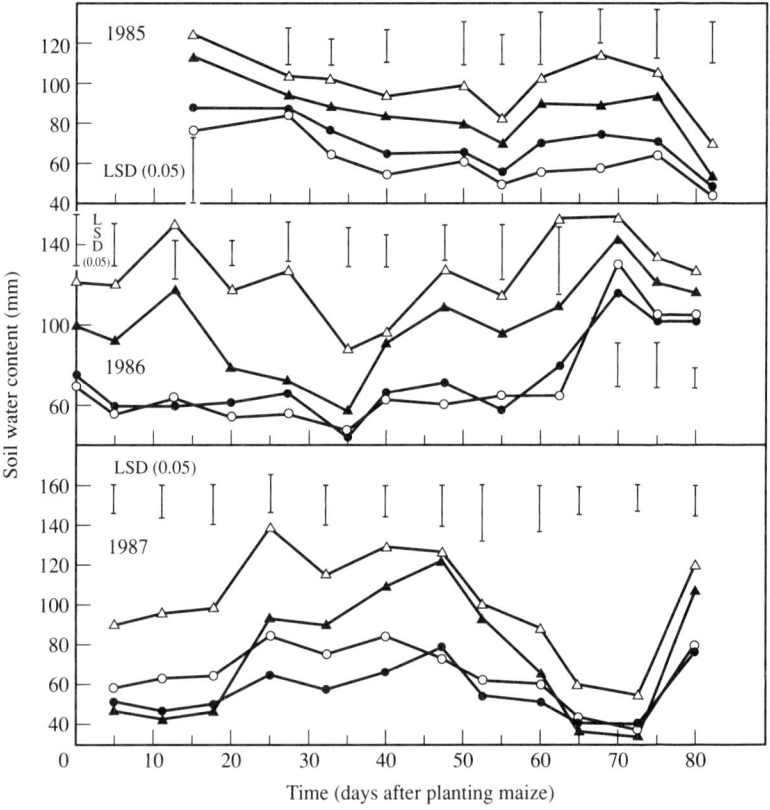

△ Tied ridges/sole maize
▲ Tied ridges/undersown
○ Open ridges/sole maize
● Open ridges/undersown

Figure 11.4 Profile water content as influenced by ridging and undersowing *S. hamata* in Burkina Faso (from Hulugalle 1989).

Another study (Shelton and Humphreys 1975c) of the effects of an undersown forage legume on water relations was carried out at Khon Kaen, Thailand (lat. 16 ° N, 1260 mm annual rainfall), and this illustrates the effects of water stress on upland rice. *Stylosanthes guianensis* cv. Endeavour was sown simultaneously with upland rice and where no N fertilizer was applied it produced 3.4 t DM ha^{-1} 132 days after planting (Plate 11.2). The presence of *S. guianensis* reduced rice grain yield by 12% (2.4 to 2.1 t ha^{-1}).

This reduction was associated with a higher number of spikelets aborting per panicle (22 relative to 10) and a consequent decrease in number of grains per panicle from 74 to 68. Planting *S. guianensis* led to more effective use of water, and possible water stress periods (as evidenced by differences in pF measured at 15 and 45 cm depths, but not at 90 cm depth) were observed during early flowering from 105 days after planting and during the final grain filling stage. The lack of an effect of *S. guianensis* on the use of moisture until an advanced panicle differentiation stage agreed with the independence of panicle density and branching per panicle according to undersowing treatment, whilst increased spikelet abortion occurred during quite high pF levels. The independence of grain size to mild moisture stress is a common crop phenomenon. There were concomitant effects of *S. guianensis* in reducing the P uptake of the rice crop and in shading the flag leaves of rice in the final stages of crop development, indicating the complexity of interpreting plant interrelations. It is clear that intercropping with forages leads to more effective exploitation of soil moisture, but these effects need to be manipulated to meet farmer objectives as discussed further in section 11.3.

Nutrients and the nitrogen economy

The greater use of soil nutrients by intercrops compared to sole crops was shown from a Colombian study in Table 11.1. The importance of biological N fixation to the farm economy and its sensitivity to non-N fertilizer inputs has been developed in discussing the other three types of farming systems (Chapters 8, 9 and 10) whilst section 2.5 addressed the management of N fixation in farming systems; the reader is again referred to this section. Particular note might be taken of Table 2.14 and Figure 2.12 which show how the N fixation of rice bean growing as an intercrop with maize in Thailand may be managed to equal that of a sole rice bean crop, and how the dependence of the legume on N fixation may be increased if a grass plant such as maize is present to deplete the soil N present.

There are many studies in the scientific literature which produce the invariable finding that intercropping with legumes increases the

Plate 11.2 Strips of upland rice planted to *Stylosanthes guianensis*, northeast Thailand.

Table 11.6 Root weight (g m^{-2}) in sole cropped maize or maize undersown with *S. hamata* cv. Verano grown in a tied or open ridge system 54 days after planting maize in Burkina Faso (from Hulugalle 1989).

System	Soil depth (cm)		
	0–30	30–50	50–100
Tied ridges			
Sole maize	84	13	0
Maize + Verano	90	15	7
Open ridges			
Sole maize	75	3	0
Maize + Verano	53	11	0.2

total plant N produced by the system. In Table 2.14 (Rerkasem and Rerkasem 1988) maize grown in northern Thailand fertilized with 20 kg N ha^{-1} applied produced 74 kg N ha^{-1} as a sole crop, whilst an intercrop system of maize/rice bean in the ratio 25:75 produced 169 kg N ha^{-1}, of which 98 kg N ha^{-1} arose from biological N fixation. In achieving this type of result the successful farmer will grow legume crops which nodulate well in local farm conditions and give high yields as an intercrop, as may be compatible with their effects on the growth of the main companion crop. An example from the section on light illustrated the greater compatibility in southeast Queensland of soybean intercrops with cassava relative to that of pigeonpea, although intercrops using the latter will enhance N yield (Dalal 1974). At Pantnagar, India (lat. 29 ° N, 240 m elevation) Nair *et al.* (1979) found soybean was more successful with maize crops than cowpea or pigeonpea, and this finding was confirmed by Singh *et al.* (1986) at the same location, where the N nutrition of soybean/maize and black gram (*V. mungo*)/maize intercrops was superior to that of groundnut/maize intercrops.

Reference was made earlier in section 2.5 to the depressing effect of the nitrate ion on nodulation, the increased dependence of the legume on soil N if fertilizer N is applied, and the negative effect of N fertilizer application on the competitive status of the legume. Low levels of N fertilizer (to *c.* 20 kg N ha^{-1}) may enhance initial nodulation or not depress it, but the practice of applying fertilizer N to intercrops is usually wasteful of resources. In the northeast Thailand study of Shelton and Humphreys (1975c) sole crop upland rice yielded 33 kg N ha^{-1} whilst the intercrop with *S. guianensis* gave 96 kg N ha^{-1}. However, stylo yield in the intercrop was 3.4, 2.9, 2.2 and 1.2 t DM ha^{-1} respectively as N fertilizer levels of 0, 20, 40 and 80 kg N ha^{-1} were applied, and N fertilizer application did not increase N output from the system. Eaglesham *et al.* (1981) using ^{15}N with cowpea and Herridge and Brockwell (1988) using xylem ureides with soybean have shown how application of fertilizer N decreases legume dependence on N fixation.

The final comment in this section concerns the fate of N fixed by the legume. Emphasis has been given in earlier discussions to the N content of legume shoots as the main pathway by which N is added to cropping systems. However there are reports of active N transfer occurring from the legume intercrop to the main crop under specified soil conditions (such as low soil N of 0.07% in Nigeria, Eaglesham *et al.* 1981) and of increased bacterial activity and of nitrate and ammonium ion concentration in the rhizosphere of maize plants intercropped with legumes (Singh *et al.* 1986). In cut-and-remove systems the main benefit of a forage legume will come if animal excreta are carefully collected and returned to the cropping land, as discussed in section 4.4. Grazing of the crop stubbles is a simple system. Legume shoots retained at maturity as a mulch or incorporated by cultivation in the soil may provide only slow return to succeeding crops; for example at Kununurra, northern Australia only 11% of legume residue N was recovered in the subsequent rice crop (Chapman and Myers 1987). The long-term N economy inevitably benefits from the incorporation of legumes in the cropping cycle, but the increased demand for other nutrients to sustain the system should not be neglected. The greater competition for nutrients between main crops and non-legume crops predisposes the farmer to legume intercrops.

11.3 Cropping practice
Intercropping

Intercropping is such a flexible system that crop husbandry may be readily adapted to the changing objectives of farmers. New crops for which a market has suddenly emerged may be added to the system and intelligence about changing prices for crop products and for livestock products can be used to alter the balance of food or cash crop and forage crop. If the price of animal products rises farmers may adjust their planting schedules to favour forage production, or to accept a risk that a higher competitive status of a legume forage crop may reduce cereal yield if the late season rains are deficient in that year. In section 11.2 the main husbandry practices which influence the canopy development of the component crops were listed, and some further comment is now provided.

Choice of species and cultivar

Specific examples of the differing behaviour of soybean and pigeonpea with cassava, as modified by plant density and planting time, were illustrated in Figures 11.2 and 11.3 and Tables 11.4 and 11.5 (Cenpukdee and Fukai 1992a). The primary need is the selection of crops well adapted to the local environment, and farmer education and flexible market development can modify the established influence of cultural tradition; for example maize has an entrenched position in many farming cultures where the local climatic environment produces more variable maize production than occurs with the more drought resistant grain sorghum. In northeast Brazil sorghum competed less than maize with companion legumes (Faris *et al.* 1983a).

The modifying effect of crop variety on the outcome of intercropping relations was illustrated for cassava. Cenpukdee and Fukai

(1992b) affirm that the choice of ideal cassava cultivars depends on the competitive ability of the chosen associate species. When strongly competitive species such as pigeonpea are grown it is necessary to select for high tuber yield and height in cassava; where soybean is the companion species high tuber yield and narrow canopy width measured c. 90 days after cassava planting provide appropriate selection criteria.

The maturity characteristics of the main crop influence the accumulated yield of the forage crop. A late flowering perennial pasture legume such as *S. guianensis* cv. Endeavour will produce more forage if planted with an early maturity crop which does not compete with the legume in the later phase of the growing season. This is illustrated by another study (Shelton and Humphreys 1975b) at Khon Kaen, Thailand where *S. guianensis* was sown with rain-grown upland *indica* rice varieties of successive maturity; Khao Dor (116 days), Khao Hom Pa (127 days) and Khao Narn Moun (165 days). *S. guianensis* yield 165 days after rice planting averaged 4.4, 2.0 and 2.8 t DM ha^{-1} respectively. Baker (1981) suggests that yield is increased in intercropping by an amount proportional to the difference in plant size and length of season; it is also suggested that the indeterminate yield components of mixtures should show a rapid recovery from competition in the vegetative phase by compensation during the reproductive phase of development. Similarly Rao and Willey (1983) suggested that a short, early, high-yielding cereal combined with a legume that matures as late as possible without incurring undue risk of moisture stress may provide an ideal combination. The planting of non-leguminous intercrops is not usually recommended, as commented further upon later in this section.

Plant density and spatial arrangement

Planting density is a powerful tool in determining the outcome of competition between intercrops. The radically differing result of cassava–pigeonpea competition in the two experiments of Cenpukdee and Fukai (1992a) described in section 11.2 arose from reduced pigeonpea density and delayed sowing time of pigeonpea in the second experiment. Much depends on the relative rates of early seedling growth and of the capacity of plants to branch in order to compensate for low initial plant density. The optimum density of the main crop will usually be known for the local conditions of climate and soil fertility, as modified by fertilizer and weeding practice.

The influence of companion forage legume density on cereal production is illustrated from a study (Shelton and Humphreys 1975a) at Na Pheng, central Laos (lat. 18° N) during a year when 1909 mm rainfall were recorded during the main growing season. The site carried secondary forest which was hand cut, burnt *in situ*, and cultivated. The low fertility grey sandy loam overlay laterite at 1.5 m and sandstone. A local upland rice variety (130 days) Khao Non Deng was planted at densities of 0, 20, 40, 80 and 120 plants m^{-2} and *S. guianensis* cv. Endeavour was sown simultaneously at 0, 3, 9, 27 and 81 plants m^{-2}. Rice tillered freely to reduce differences in yield which were initially evident, and during early flowering (102 days after sowing) shoot yield only varied from 2.1 to 2.5 t DM ha^{-1} for the rice densities 20 and 120 plants m^{-2}, and at final harvest yield of rice grain was independent of rice density.

On the other hand the density of stylo sown simultaneously with rice influenced rice grain yield (Table 11.7), and at the highest stylo density rice yield was reduced by 36%, apparently through effects on both panicle density and individual panicle yield. The level of light transmission at early flowering (Table 11.7) suggests that this mainly arose from competition for nutrients or moisture, and some moisture stress was evident during early flowering. The amount of legume forage available at the time of rice harvest was positively associated with stylo density, but the capacity of the local upland rice to compensate for variation in rice plant density was noteworthy.

Time of planting

In a swidden rice planting at the same site (Shelton and Humphreys 1972) yield of upland rice was unaffected by *S. guanensis* undersown 30 days after rice was sown. However legume yield was much reduced by delayed sowing, and after a 60 day delay *S. guianensis* competed unsuccessfully with weed growth after the rice harvest. At Khon

Table 11.7 Effect of *S. guianensis* cv. Endeavour density intercropped with upland rice in central Laos on various attributes (from Shelton and Humphreys 1975a).

Attribute	*S. guianensis* density (plants m^{-2})				
	0	3	9	27	81
Rice					
Panicle density (no. m^{-2})	87	70	74	75	71
Grain yield/panicle (g)	1.1	1.2	1.1	1.0	0.9
Grain yield (kg ha^{-1})	980	850	820	720	630
S. guianensis			155	52	118
DM yield (t ha^{-1})	–	0.7	1.3	2.1	2.6
Light transmission (% at 102 days)	68	59	59	51	47

Kaen, Thailand (Plate 11.2), a delay of only 10 days in sowing *S. guianensis* was sufficient to avoid reduction in rice grain yield, but this short delay of 10 days reduced legume yield by a factor of two (Shelton and Humphreys 1975b). *S. hamata* cv. Verano sown 2 weeks after seedling emergence of kenaf (at the first weeding) led to no reduction in yield of kenaf (roselle) at Khon Kaen (Pongskul *et al.* 1982). Rather similar results were obtained with *S. hamata* cv. Verano and *S. guianensis* cv. Cook undersown in sorghum near Kaduna, Nigeria (Mohammed Saleem 1985), who recommended a delay in legume sowing of 3–6 weeks.

In the highlands of northern Thailand (lat. 20° N) Andrews (1981) planted *Cajanus cajan* (pigeonpea) with hill rice at three sites from 900 to 1300 m altitude. At low pigeonpea densities of 2–8 plants m^{-2} simultaneous sowing led to *c.* 50% reduction in rice yield. Delayed sowing of pigeonpea for 2 or 7 weeks obviated any decrease in rice yields, but substantially decreased yield of pigeonpea in this cool highland environment.

The final illustration integrates the effects of main crop variety, legume density and legume sowing time (Cenpukdee and Fukai 1992c) and shows how these may be manipulated to maximize LER or to increase the output of either component crop. At Redland Bay, southeast Queensland the cassava varieties MCOL 1468 (tall, spreading) and MAUS 19 (short, narrow canopy) were grown in a fertile latosol with supplementary irrigation. Pigeonpea cv. Quantum was planted simultaneously or 35 days later at 7 (low) or 27 (high density) plants m^{-2}. The course of canopy development was reflected in the percentage PAR above the intercropped cassava (Figure 11.5). For both cassava varieties a delay of 35 days in pigeonpea planting gave cassava dominance. However, simultaneous planting of both crops enabled high density pigeonpea to shade cassava MAUS 19 severely by 45 days after planting, and to exert some reduction in radiation receipt of cassava MCOL 1468; this effect was much less influential at low pigeonpea density.

Pigeonpea was harvested 160 and 174 days after cassava planting in the simultaneous and delayed planting treatments, whilst cassava was harvested after 219 days. The yield outcomes (Table 11.8) varied greatly according to treatment. Cassava MCOL 1468 outyielded MAUS 19 in every combination, and both total DM and tuber yield of the former variety were not reduced by the pigeonpea intercrop if this was grown at low density from a delayed planting. This combination gave an LER based on tuber and seed yields of 1.34, indicating a successful intercrop system.

These data also indicate how the farmer may swing the balance of competition towards pigeonpea if this is the economic or family imperative; choosing the less vigorous MAUS 19 cassava and delaying

Table 11.8 Yield (t ha^{-1}) in relation to cassava variety and sowing time and density of intercropped pigeonpea in southeast Queensland (from Cenpukdee and Fukai 1992c).

Crop System	Cassava		Pigeonpea	
	Total DM	Tuber	Total DM	Seed
MCOL 1468				
Sole cassava	14.6	6.5	–	–
Intercropped:				
35 day delay				
Low density	15.6	7.3	1.8	0.34
High density	11.6	5.3	2.7	0.39
Simultaneous sowing:				
Low density	6.7	1.3	6.8	1.11
High density	1.4	0.02	8.7	1.21
MAUS 19				
Sole cassava	8.5	2.9	–	–
Intercropped:				
35 day delay				
Low density	4.5	1.5	3.5	0.93
High density	3.3	0.7	4.7	0.93
Simultaneous sowing				
Low density	2.3	0.3	8.0	1.32
High density	0.4	0.0	10.5	1.58

planting of pigeonpea gave a seed yield of 0.9 t ha^{-1} and a tuber yield in this short season cassava crop of 1.5 t ha^{-1}; alternatively complete pigeonpea dominance gave a legume grain yield of *c.* 1.6 t ha^{-1}. These calculations have omitted the significance of the value of the crop residues for fodder or fuel, and knowledge of changed economic value for cassava and for pigeonpea can be used to modify the cropping system to achieve farmer objectives.

Fertilizer practice

In most agricultural districts information is available about crop fertilizer needs on particular soils and the probable crop responses in relation to fertilizer inputs. For intercropping systems there are three particular considerations. The first is that nutrient demand is high in an intercrop system which makes more effective use of the environmental growth factors of radiation and soil moisture than occurs in sole crop systems.

The second observation reiterates earlier emphases that biological

N fixation will only be maximized in any environment as the non-N soil deficiencies are rectified. In the Ethiopian highlands Kahurananga (1991) has demonstrated how the application of P fertilizer increased the growth of indigenous *Trifolium* spp. intercropped with wheat. At Ludhiana, India, Gill *et al.* (1987) found that P application to groundnut or to green gram greatly increased the yield of subsequent wheat crops.

Finally, the negative effects of N fertilizer on the legume intercrop and on the N economy of legume-based systems were elaborated in section 11.2.

Grazing and cutting management

The previously enunciated principles apply. Defoliation practice can be used to modify the light relations of intercrops. Growth and legume N fixation are intimately linked and depend upon the extent of leaf canopy maintained under particular defoliation regimes. Cut-and-remove systems deplete soil nutrients unless care is exercised in the collection and return of the excreta of the animals which eat the fodder.

Relay cropping

Delayed planting time of intercrops may extend the period under crop beyond the period of main cropping. An extension of this concept is that of relay cropping where different crops are grown in succession within the one cropping season. In unimodal rainfall environments there are two particular circumstances where relay cropping contributes to the total output from a mixed farming system.

The first is where unused soil moisture is present in the soil profile when the main crop is harvested. This is the situation where rainfed flooded bays are used for rice production. In the uplands of northeast Thailand Shelton (1980) showed that *Crotalaria juncea* and *Lablab purpureus* were well adapted to planting after rice, but the latter required rhizobial inoculation. On these puddled rice soils yield of *C. juncea* was greatly improved if a single cultivation was applied at planting, and this reduced soil bulk density from 1.65 to 1.19 Mg m^{-3}. *C. juncea* is best sown at 20 kg ha^{-1} into 5–10 cm deep rows spread 30 cm apart with the row partially backfilled. A first cut for hay may be taken 6–8 weeks after planting, and a subsequent ratoon crop will provide a combined legume hay yield of 2–3 t ha^{-1}. A cut with about 10 residual stubble leaves retained ensures good ratoon growth (Kessler and Shelton 1980). This system has been effective in smallholder dairying, where *C. juncea* is fed mainly from January to April (Gibson 1987). At Peradeniya, Sri Lanka (lat. 7° N, 430 m altitude) Sangakkara (1989) found that *C. juncea* relay cropped after rice, cut for fodder at 5 cm height, and the stubble ploughed in, led to increased rice yield in successive crops; this was associated with augmented soil N.

The second situation is where the opening rains are of uncertain occurrence. In some systems rice is planted in rainfed flooded bays, and rice planting is delayed until the bays are filled. The upper terraces may not fill until an advanced date in some seasons of late or intermittent opening rains so that in these situations rice is planted too late in the season for maximum production or rice may not be planted at all. Over this period sufficient rain may have fallen for an early maturing fodder or pulse crop such as cowpea (*V. unguiculata*) to be successfully grown.

A further consideration is that the incorporation of both pulse and fodder legume crops in a relay system improves the quality of the crop residues for livestock consumption. Mung bean (*V. radiata*) varieties exhibit considerable variability in grain/stover ratio and some produce grain yields of 1.6–1.8 t ha^{-1} as a crop before rice (IRRI 1983). Cowpea provided green fodder at 8.9–19.8 t ha^{-1} at Solana, Cagayan, Philippines. The use of legume crops in rotation systems improves the N economy; illustrations may be found in the highlands of Papua New

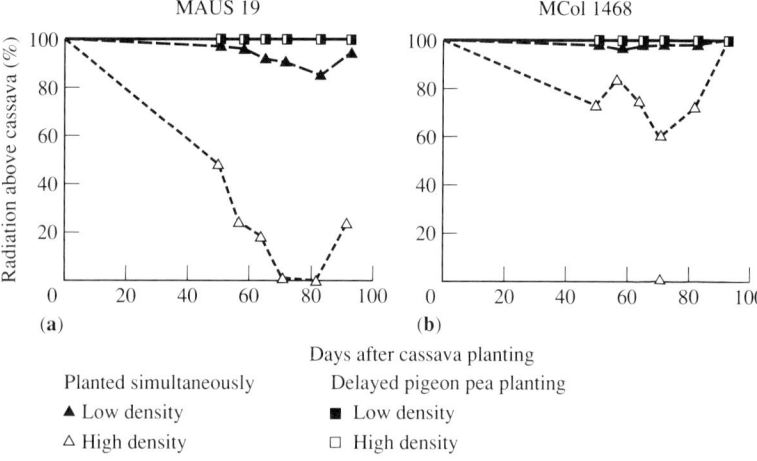

Figure 11.5 Flux density of PAR above two cassava cultivars intercropped with pigeonpea in southeast Queensland (planted simultaneously at low or high density; delayed pigeonpea planting at low or high density) (from Cenpukdee and Fukai 1992c)

Guinea (Kimber 1974) and in north India (Kharwara et al. 1987). At New Delhi Giri and De (1979, 1980) found that the effect of preceding pulse crops on subsequent grain yields of millet was equivalent to the fertilizer application of 60 kg N ha^{-1}.

Undersowing crops for pasture establishment

Section 10.3 discussed the techniques available for establishing pastures in a ley rotation. Undersowing the last cash or food crop in the rotation with pasture species is the most pragmatic method of pasture establishment. It has advantages over conventional pasture establishment procedures in many tropical situations. The cost of land development for pasture is absorbed by the cropping enterprise, which may also provide the motivation for land clearing. The N accretion and soil stabilization which arise from a legume-based pasture phase in a ley system are provided in a way which requires little additional input of capital or sophistication.

The principles of undersowing pasture species are the same as those which apply in the intercropping of cash or food crops with forages or pulses. The practice is long established, and requires the same attention to planting density, time of undersowing and choice of forage species. For example, at Kitale, Kenya Poultney (1963) undersowed maize with grasses and clover following the last weeding when maize was 80–90 cm high; grain yield of maize was unaffected by sowing with *Melinis minutiflora* or *Chloris gayana* but was reduced 8% by *Setaria sphacelata* var. *sericea* and yield of maize stover was reduced 21% by *M. minutiflora*. Goldson (1967), also in western Kenya, recommended undersowing pasture after the second weeding when maize was c. 50 cm high.

There is some controversy concerning the inclusion of grasses in the undersown mixture, since these may compete with the main crop for soil N. On a low fertility soil in central Laos Shelton and Humphreys (1972) found that the inclusion of *M. minutiflora* and *Panicum maximum with S. guianensis* severely reduced the yield of swidden rice. By contrast, Thomas and Bennett (1975a,b) in Malawi successfully established *C. gayana* and *D. uncinatum* by delayed undersowing (24 or 41 days after maize sowing) without detriment to maize yield. Their site was clearly on a fertile soil, since maize yielded 5.4 t ha^{-1} and did not respond to fertilizer N. The farmer has the option of undersowing a legume and introducing the grass in the next season, or on fertile soils using a delayed undersowing of grass at low density.

In northeast Thailand Shelton and Humphreys (1975c) suggested sowing *S. guianensis* at 2 kg ha^{-1} 10 days after sowing an early variety of upland rice without N fertilizer and with the expectation of obtaining an *S. guianensis* density of *c.* 25 plants m^{-2} and a legume DM of *c.* 3 t ha^{-1} at the end of the growing season. The grazing of the legume with rice stubble improves ruminant nutrition and provides an infection source by which a well-adapted legume might be spread by ruminants to other grazing areas where the legume was not formally sown.

The production by CIAT of rice varieties which are adapted to acid soils of low CEC and high Al saturation has expanded the possibilities for rice production and the use of rice in association with pastures in areas of North America which were previously regarded as unsuitable for rice production. Figure 10.1 illustrated the high yields of rice which are feasible in the Colombian llanos following legume-based pastures. Rice may also be used as a pioneer crop when converting native savanna to improved pastures. In an experiment at Matazul, Colombia (125 km east of Villavicienco) ploughed and fertilized savanna gave rice yield of 2.2 t ha^{-1} as monoculture. Planted simultaneously with *Brachiaria dictyoneura* and *Centrosema acutifolium*, or with *Andropogon gayanus* and *Stylosanthes capitata*, rice yield was 2.1 and 2.0 t ha^{-1} respectively. In a further experiment delaying sowing of the latter for 30 days after rice sowing still led to good pasture establishment (CIAT 1990). These results offer great promise for the more intensive agricultural progress of the llanos savannas, since a rice yield of 1.6 t ha^{-1} covers development costs at current prices.

11.4 Conclusions

Intercropping annual food or cash crops is well established in farm practice in most tropical countries. It may take the form of controlled planting of cereals with inter-row pigeonpea, as in India, or it may arise in informally arranged plantings of many crop species in a multipurpose garden used for both food and cash. The greater inclusion of forage legumes in these systems is following the global trend of increasing demand for ruminant products and the latter is also causing farmers to focus on the role of pulses in improving the nutritive value of crop residues.

The flexibility of intercropping systems is of great advantage to the farmer, since small shifts in crop husbandry alter the balance of outputs. This flexibility of response has been illustrated in this chapter with many outcomes of particular crop dominance. When the outcome is unexpected – perhaps drought late in the season aggravated by the presence of a forage intercrop and reduced grain production – a farm system containing ruminants reduces the potential wastage by converting the abortive crop to useful animal products, whose value relative to grain may well become positive. The reduced incidence of weeds and the added control of erosion give long-term benefits.

The forage legume provides cheap biological N fixation which does not deplete fossil fuel reserves or cause the pollution associated with the manufacture of fertilizer N. However if the farmer is not aware of the pathways of N transfer the products of biological N fixation may be lost to the cropping land. The concentration of animals in housed units, as occurs in European husbandry practice, may be quite unwarranted in tropical and subtropical conditions, and cut-and-remove feeding systems may result not only in spatial transfer of nutrients from crop land but may cause stream pollution and community nuisance. Grazing has lower labour requirements, gives higher animal production than most cut-and-remove systems and can be made feasible in many farm systems.

In this chapter some critical studies of plant relations in intercropping which shed light on the interplay of processes controlling the outcome of competition for resources have been presented. There still has to be created for other crop combinations and specific regional conditions the body of knowledge which can form the basis for predicting crop husbandry effects on radiation level, the leaf area of component crops and their interception of radiation, the efficiency of conversion of intercepted radiation to carbohydrate and the influence these have on the sequence of processes leading to the end-products of cropping. These are assessed not only in terms of grain or tuber yield, but also in relation to N fixation and OM accretion, and in the value of crop residues to livestock. It is from these studies that farmers working with scientists can access the information systems which help them choose between different suites of options in farm practice, knowing their riskiness and the value of inputs and outputs, in order to meet their farming goals.

CHAPTER 12

Conclusions

12.1 Central issues for the development of appropriate technology

This book has described the principles upon which mixed farming and its environment may be sustained. It has provided from tropical and subtropical regions of four continents illustrations of technology which use these principles, of the processes that determine whether land is being degraded and whether the goals of farmers are met. It is not a handbook of information which gives recipes for improved farm practice in differing local conditions; rather it attempts to provide the basic approaches that can be adapted in the farm situation and from which new solutions will evolve with the assistance of ongoing research as learning processes progress amongst farmers and scientists.

The successful integration of forages in cropping systems depends upon the development of appropriate technology directed to the following central issues:

- conservation of the soil,
- biological N fixation,
- cycling of nutrients,
- efficient use of resources by élite germplasm,
- opportunities for diversified production.

Conservation of the soil

The first requirement for any system is the preservation of the soil resource *in situ* and the retention or improvement of the physical and chemical fertility upon which the level of continued crop yields depends. Each system is judged in the first instance by its capacity to meet this requirement, and the simplest measure of degradation is the capacity of plants to grow in response to rainfall (Walker 1993).

Reliance on mechanical structures as a defence against erosion has proved a failed technology, since inter-bank erosion is uncontrolled, fertile soil is buried or displaced, and bank failure leads to accelerated gullying. On sloping land farmers are looking to contour grass strips or to shrub legume plantings and plantation crops, whilst the stability of areas cropped to annuals requires minimum or zero-tillage systems, crop residue retention or the growing of live mulch, intercropping with forage legumes, or the use of a ley pasture phase to minimize the duration of erosion risk and to promote OM accretion.

Biological N fixation

The predominant soil deficiency is N, and this is the nutrient over which farmers have greatest control through the farming practice they adopt and which is most influential in determining production. There are some highly acid tropical soils in which P supply may be of equal importance, and this also bears on the level of biological N fixation which is feasible. The use of artificial fertilizers to promote legume growth and N fixation is often essential if productivity and OM accretion are to be enhanced; this can be managed with no detrimental effects on the environment or the quality of product output. This issue is related to the third issue of nutrient cycling; if soil fertility is low there are few nutrients to be recycled.

This book has provided notable examples of the successful incorporation of forage legumes in farm practice, but it would be foolish to claim that a robust legume technology acceptable to farmers is available in all situations in the humid and subhumid tropics and subtropics. A greater concentration of research and interaction with farmers is indicated in the generation of élite germplasm, especially in respect to pest and disease resistance, in the development of appropriate techniques of legume management, particularly with regard to synchronizing release of legume N and crop demand, and in the understanding of N transfer processes. An alternative dependence upon fertilizer N has unfortunate consequences in terms of the depletion of fossil fuel, environmental pollution and the drain on foreign currency in tropical countries which are not rich in oil reserves.

Cycling of nutrients

Emphasis has been given to the role of ruminants in converting forages and crop residues high in long-chain carbohydrates to sources of nutrients which are more readily available to the crops, and to the function of OM in conserving nutrients and in deactivating harmful compounds. Two key areas of concern relate to the spatial transfer of nutrients consumed by livestock.

The first is the significant loss of nutrients from excreta where animals are housed rather than grazed. Special attention to the development of new techniques for the conservation of urine is indicated, and current practice results in substantial losses of N from the system which can otherwise benefit crop production, as demonstrated by Thomas and Addy (1977). The second, which follows, is the severe environmental pollution of streams and of the local atmosphere through the concentration of animals in housed or corralled conditions.

Efficient use of resources by élite germplasm

The theme that farming systems might be designed to effect greater capture of natural resources is a continuing thread in this book. Mixed farming involving livestock can convert more of the primary resources of light, air and water to carbohydrate than monocropping systems which provide less continuous leaf cover, and this also results in the greater accessibility of soil nutrients to plants, the opportunity to recycle the nutrients in forage and crop products and the incorporation of 'waste' lands in the production system. Mixed farming also provides better resilience in response to abiotic disturbance.

Species diversity is a key element in maintaining the structure of the system and its continuing output (Hadley 1993). This applies both to plant and to animal adaptation to the farm environment. The incorporation of élite germplasm in farm practice is the innovation most simply introduced and readily adopted by farmers, and plant improvement is rightly the focus of many agricultural research programmes. The genetic advances made in the past two decades are substantial, but a better marriage of the objectives of farmer and scientist is needed, for example in the attention paid to the volume and feeding value of crop residues in crop improvement programmes (Reed *et al.* 1988).

Opportunities for diversified production

The case for the diversification of output of farm products has been argued in terms of greater resource capture and of minimization of risk, which contribute to food and income security. The capacity of smallholders to grow new crops or to produce new animal products in response to market demand and opportunity is not to be underestimated. Changing needs alter the possibilities for diversification; ley pastures directed to forage seed production may make the rotation of pastures with crops more economically attractive than a system directed to meat production (Manidool 1987). The balance of forest/pasture/crop land is not fixed and changed land use can be made sustainable in most subhumid and humid regions.

However, government interventions which distort resource allocation in ways which are out of synchrony with market opportunity or which promote cultivation of marginal lands through subsidies for grain or for fertilizer N threaten environmental stability. The continual search for innovative crop or animal products whose production is especially suited to particular farm environments and farmer skills and their marketing overseas might receive more attention than the focus many institutions have on traditional cereal crops.

12.2 Facilitating farm practice which sustains tropical cropping systems

This book would be incomplete without some reference to the ways in which farm practice which sustains tropical cropping systems may be facilitated. The cry is often heard that farmers need to use the great reservoir of existing knowledge which might improve their cropping systems; it is less often appreciated that much of this knowledge is inapplicable in its current form to meet the current goals of farmers or is deficient in some aspects that require further research, development and adaptation. The central issues include:

- involvement of farmers in research,
- learning and the development of decision support systems,
- institutional policies.

Involvement of farmers in research

A common model of change in farming depicts a linear series of processes in which research generates knowledge and innovations; these are communicated by 'extension' workers to farmers and adopted by the farmers most receptive to them; there is a further diffusion as other farmers copy the innovators. This 'trickle down' process has been criticized as ineffective; especially is this so if research, extension and regulatory institutions act as if they were closed systems, formulating problems in isolation from the external farm environment, and little attention is paid to the quality and nature of the relationship between the system and its environment (Holt and Schoorl 1990).

A preferred model (Figure 12.1; Ison and Ampt 1992) is cyclical and dynamic. The first three steps require that farmers (especially including women), local extension workers, subject matter specialists and research workers interact for 'the purpose of working synergistically to support decision making, problem solving and innovation' (Roling 1990) related to the farm situation and its environment, which are perceived quite differently by the participants according to their varying backgrounds of social and cultural experience. The validity of farmer observation is given equal authority with that of scientists' perceptions, although the latter may contribute more to the understanding of the underlying biological processes which are operating. A problem may be defined when there is agreement among these participants about its characteristics, which leads the group through the rational/preliminary conceptual stage (Knipscheer et al. 1983) to the more developed stage (step 4, Figure 12.1) where specific agronomic research incorporating ecological principles may

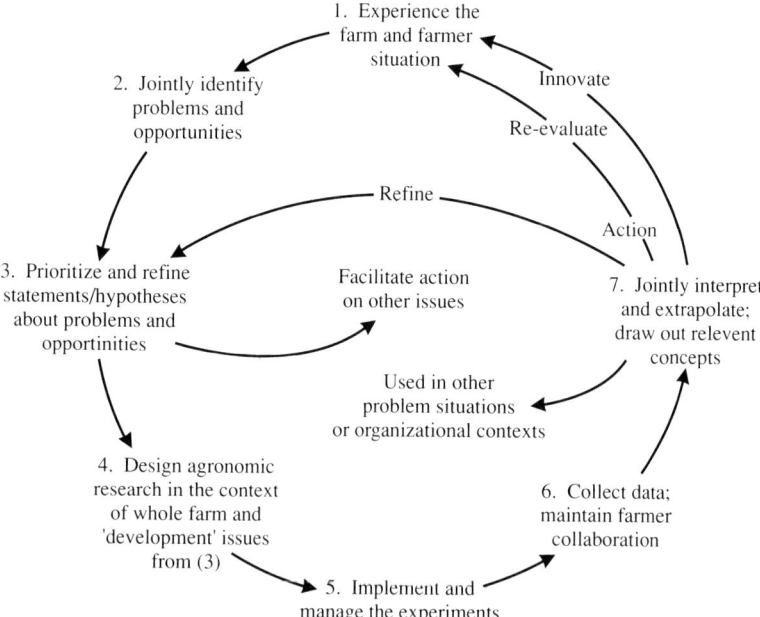

Figure 12.1 A conceptual model for future agronomic research based on the problem-solving or learning model (from Ison and Ampt 1992).

be designed through participatory rural appraisal. Where feasible the research will be carried out on farmer properties and with the participation of the landholder. The outcomes of research are then jointly evaluated, leading to innovative action on the farm and/or to the redefinition of the problem which forms the basis for continuing research. There is now considerable evidence (for example, Farrington and Martin 1988) of the value of this model.

It should be recognized that the development of innovations often arises from farmer initiative but depends in the long term on advances in reductionist science which are effectively integrated into farming systems, and 'display greater coherence with the expressed needs of the day-to-day lives of the people involved' (Russell and Ison 1992).

Learning and the development of decision support systems

The belief that knowledge is directly transferable from one person to another has currency among extension workers, but the evidence is quite contrary (Sless 1986). Rather in 'conversation' people share communication of the worlds they experience and may thereby develop a shared meaning or their worlds may fail to interact. Collaboration between scientists and farmers emerges as shared enthusiasms for action develop (Russell and Ison 1992) which lead to changed perceptions of how farming practice may be modified to fulfil the multi-purpose goals which exist amongst farmers, and which enhance the knowledge base and learning of both farmers and scientists.

These activities are most effective if the end-point is not one saleable extension recipe but rather a basket of technologies (Hadley 1993) from which farmers are able to select modifications which are in tune with their farm resources and objectives. The role of the scientist is to describe the components of the farm system and their interaction, to quantify the expected inputs and outputs involved and to provide an assessment of the risks involved; the scientist acts as an honest broker rather than an evangelist for some current scientific fashion which may have little validation in farming practice. Concurrently the interaction helps to develop the problem solving skills of the farmer which provide a continuing basis for sustainable farming.

The construction of decision support systems which enable farmers to understand the implications of the choices that are available to them have utility if they are not directed to formulating a universal solution but which provide specific solutions in the varying circumstances of the user profile, as defined by farm ecological niche, availability of skills, labour and capital resources, market infrastructure, local culture and personal needs. Such support systems have a different character according to the degree of subsistence farming or cash economy involved (Stuth *et al.* 1993), and the success of all such systems depends upon human resource development.

Institutional policies

There are few short-term solutions available from research in farming systems directed to developing the technologies enumerated in the first half of this chapter that facilitate the integration of forages in cropping systems in ways which sustain output and protect the environment. Effective institutions are therefore committed to investment in long-term research and the development of a research infrastructure which attracts, retains and supports the best minds in their investigations and interaction with the rural community.

Institutional programmes may be focused on the producer but these need also to be compatible with community objectives in land use and to deal with inter-regional externalities (Nores and Vera 1993), especially in respect of trade. The best research and development is led by community demand and takes account of changing community needs and expectations. The burgeoning dairy industry of Thailand is an example of how an emerging demand which has benefited farm income, environmental protection and community health has been assisted by the intervention of government with the political will to control the dumping of external dairy products.

Long-term land tenure or sustained usufruct appear to be essential features of sustainability in most societies; the agrosilvopastoral solutions developed in Chapters 8 and 9 and the rotation of crops with pastures in the ley systems of Chapter 10 are not feasible unless the farmer is able to benefit from activities and investments with delayed returns. Subsistence farmers having temporary occupancy of land display short-term perspectives on land use (Toledo and Formosa 1993).

Attachment to the preservation of the primitive would condemn the forest-dwelling hunter-gatherers and their families to a life of continued poverty. Better solutions are available in which the continued growing of trees is guaranteed by the investment farmers make in agrosilvopastoral systems. The sequestration of carbon from the atmosphere is ensured where trees are grown or where forages accumulate soil OM. As resistances to herbicides and pesticides develop farmers may be forced to adopt ley pastures to control weeds and pests and to detoxify their soils. The excess nitrate which drains into the subsoil in zero-tillage systems may be recovered by deep rooted forages.

These final examples illustrate the marriage of production incentives and of strategies of environmental protection which are available in mixed farming but which require to be developed in local contexts.

The main pathway to enhanced and sustained farm income and food security in the humid and subhumid zones of the tropics and subtropics lies in the combination of ruminant livestock with crop production of many forms. The new technologies of tropical forage production play a significant role in the fulfillment of these global objectives.

References

Agamuthu, P, Chan Y K, Jesinger R, Khoo K M, Broughton W J (1981) Effect of differently managed legumes on the early development of oil palms (*Elaeis guineensis* Jacq.) *Agro-Ecosystems* **6**: 315–23

Agamuthu P, Furtado J I, Broughton W J (1980) Establishment and subsequent effects of legume covercrops on the development of young oil palms (*Elaeis guineensis* Jacq.) In *Proceedings fifth international symposium of tropical ecology*, 16–21 April 1979. Kuala Lumpur, Malaysia: 561–8

Agarwal M L, Tribhuwan Singh, Singh N P, Matambar (1986) Effect of intercropping and nitrogen on the yield of spring planted sugarcane. *Indian Journal of Agronomy* **31**: 209–10

Agyemang K, Abiye Astatke, Anderson F M, Wolde-Ab Wolde-Mariam (1991b) Effects of work on reproductive and productive performance of crossbred dairy cows in the Ethiopian highlands. *Tropical Animal Health and Production* **23**: 241–9

Agyemang K, Dwinger R H, Grieve A S, Bah M L (1991a) Milk production characteristics and productivity of N'Dama cattle kept under village management in the Gambia. *Journal of Dairy Science* **74**: 1599–608

Ahn J H, Robertson B M, Elliott R, Gutteridge R C, Ford C V (1989) Quality assessment of tropical browse legumes: tannin content and protein degradation. *Animal Feed Science and Technology* **27**: 147–56

Aken'Ova M E, Atta-Krah A N (1986) Control of spear grass (*Imperata cylindrica* (L.) Beauv.) in an alley cropping fallow. *Nitrogen Fixing Tree Research Reports* **4**: 27–8

Akobundu I O (1982) Live mulch crop production in the tropics. *World Crops* **34**: 125–45

Alazard D, Ndoye I, Dreyfus B (1988) *Sesbania rostrata* and other stem-nodulated legumes. In Bothe H, Bruijin F J de, Newton W E (eds) *Nitrogen fixation: Hundred years after. Proceedings of the 7th International Congress on Nitrogen Fixation*, March 1988. Stuttgart, Gustav Fischer Verlag 765–9

Allen B J (1985) Dynamics of fallow successions and introduction of robusta coffee in shifting cultivation areas in the lowlands of Papua New Guinea. *Agroforestry Systems* **3**: 227–38

References 359

Allmaras R R, Burwell R E, Holt R F (1967) Plow-layer porosity and surface roughness from tillage as affected by initial porosity and soil moisture at tillage time. *Soil Science Society America Proceedings* **31**: 550–3

Altieri M A, Liebman M (1986) Insect, weed and plant disease management in multiple cropping systems. In Francis C A (ed.) *Multiple cropping systems*. New York, Macmillan Publishing Company: 183–218

Alva A K, Blamey F P C, Edwards D G, Asher C J (1986) An evaluation of aluminium indices to predict aluminium toxicity to plants grown in nutrient solutions. *Communications in Soil Science and Plant Analysis* **17**: 1271–80

Anderson H A, Vaughan D (1985) Soil nitrogen: its extraction, distribution and dynamics. In Vaughan D, Malcolm R E (eds) *Soil organic matter and biological activity*. Dordrecht, Martinus Nijhoff/Dr W Junk, Developments in Plant and Soil Science **16**: 290–327

Anderson L S, Sinclair F L (1993) Ecological interactions in agroforestry systems. *Agroforestry Abstracts* **6**: 57–91

Andrews A C (1981) Mixed cropping of pigeonpea and upland rice. Final Report, Thai-Australian Highland Agricultural Project, Department of Agriculture, University of Queensland: 23–5

Andrews A C, Kwaengsopha S (1982) The effect of cattle grazing and fertilizer on the growth of *Pinus* and *Eucalyptus* species in the highlands of northern Thailand. *Thai Journal of Agricultural Science* **15**: 267–74

Anecksamphant C, Boonchee S, Sajjapongse A (1990) Management of sloping lands for sustainable agriculture in northern Thailand. *Transactions of the 14th International Congress of Soil Science* **vi**: 198:203

Anoka U A, Akobundu I O, Okonkwo S N C (1991) Effects of *Gliricidia sepium* (Jacq.) Steud and *Leucaena leucocephala* (Lam.) de Wit on growth and development of *Imperata cylindrica* (L.) Raeuschel. *Agroforestry Systems* **16**: 1–12

Anosike N, Coughenour C M (1990) The socioeconomic basis of farm enterprise diversification decisions. *Rural Sociology* **55**: 1–24

Applegate G (1991) Caribbean pine and pasture on the Atherton Tableland. *Agricultural Science* NS **4**: 33–5

Applegate G B, Nicholson D I (1988) Caribbean pine in an agroforestry system on the Atherton Tableland in north east Australia. *Agroforestry Systems* **7**: 3–15

Araya Sanchez J F (1987) [Effect of *Gliricidia sepium* as a mulch in a maize (*Zea mays*)–beans (*Phaseolus vulgaris*) system on a hillside at Acosta-Puriscal, San José, Costa Rica]. Thesis, Universidad de Costa Rica/Centro Agronómico Tropical de Investigación y Enseñanza

Arias I, López G, Aurrecoecha P (1980) Weight gain of cattle continuously grazing sorghum stubble and regrowth in eastern Gúarico. *Agronomía Tropical* **30**: 269–78

Armbruster T, Peters K J, Hadji-Thomas A, Lamizana P (1991) Sheep production in the humid zone of West Africa. I Reproductive performance in improved production systems in Côte d'Ivoire. *Journal of Animal Breeding and Genetics* **108**: 203–9

Ashokan P K, Nair R V, Sudhakara K (1985) Studies on cassava–legume intercropping systems for the Oxisols of Kerala State, India. *Tropical Agriculture*, UK **62**: 313–8

Asiedu F H K, Oppong E N W, Opoku A A (1978) Utilisation by sheep of herbage under tree crops in Ghana. *Tropical Animal Health and Production* **10**: 1–10

Atta-Krah A N (1990) Alley farming with *Leucaena*: effect of short grazed fallows on soil fertility and crop yields. *Experimental Agriculture* **26**: 1–10

Atta-Krah A N (1993) Trees and shrubs as a secondary component of pasture. In *Proceedings XVII International Grassland Congress*, Palmerston North NZ: 2045-52

Atta-Krah A N, Sumberg J E (1988) Studies with *Gliricidia sepium* for crop/livestock production systems in West Africa. *Agroforestry Systems* **6**: 97–118

Avila M (1988) Integrating livestock with cropping systems. In *Cropping in the semiarid areas of Zimbabwe*. Harare, Zimbabwe, Department of Agricultural Technical and Extension Services: 404–16

Ayanaba A, Tuckwell S B, Jenkinson D S (1976) The effects of clearing and cropping on the organic reserves and biomass of tropical forest soils. *Soil Biology and Biochemistry* **8**: 519–25

Baker E F I (1981) Population, time, and crop mixtures. In *Proceedings of the international workshop on intercropping, Hyderabad, India, 10–13 January 1979*. Patancheru, Hyderabad, India; ICRISAT (1981): 52–60

Bakrie B (1991) Power output in wet-soil tillage operations at three villages in Subang, West Java. *Draught Animal Bulletin* (**1**): 9–12

Barnes D L (1961) Residual effects of cutting-frequency and fertilizing with nitrogen on root and shoot growth, and the available carbohydrate and nitrogen content of the roots of Sabi panicum. *Rhodesia Agricultural Journal* **58**: 365–9

Barnes D L (1981a) Residual effects of grass leys on the productivity of sandy granite-derived soils. 1. Harvested leys. *Zimbabwe Journal of Agricultural Research* **19**: 51–67

Barnes D L (1981b) Residual effect of grass leys on the productivity of sandy granite-derived soils. 2. Grazed leys. *Zimbabwe Journal of Agricultural Research* **19**: 69–82

Barrow N J (1987) Return of nutrients by animals. In Snaydon R W (ed.) *Managed grasslands. Analytical studies*. Amsterdam, Elsevier: 181–6

Baum E (1984) Opportunities and constraints of small scale farms to adopt agroforestry methods in the western Usambaras of Tanzania. *Tropenlandwirt* **85** (April): 67–76

Bawden R J, Ison R L (1992) The purpose of field-crop ecosystems: social and economic aspects. In Pearson C J (ed.) *Field crop ecosystems of the world* Amsterdam, Elsevier: 11–35

Beaty E R, Sampaio E V S B, Ashley D A, Brown R H (1974) Partitioning and translocation of ^{14}C photosynthate by Bahiagrass (*Paspalum notatum* Flugge). In *Proceedings XII International Grassland Congress*, Moscow **1**: 259–67

Beckwith R S, Butler J H A (1983) Aspects of the chemistry of organic matter. In *Soils: an Australian viewpoint*. Melbourne, CSIRO/London, Academic Press: 565–82

Beets W C (1979) Relevant cropping systems research for the Asian farmer. *Malaysian Agricultural Journal* **52**: 58–64

Beets W C (1982) *Multiple cropping and tropical farming systems*. Boulder, Colorado, Westview
Belsky A J, Amundson R G, Duxbury J M, Riha S J, Ali A R, Mwonga S M (1989) The effects of trees on their physical, chemical and biological environments in a semi-arid situation in Kenya. *Journal of Applied Ecology* **26**: 1005–24
Benjamin A, Shelton H M, Gutteridge R C (1991) Shade tolerance of some tree legumes. In Shelton H M, Stür W W (eds) *Forages for plantation crops*. Canberra, ACIAR Proceedings **32**: 75–6
Berndt R D, White B J (1976) A simulation-based evaluation of three cropping systems on cracking-clay soils in a summer-rainfall environment. *Agricultural Meteorology* **16**: 211–29
Bessho T, Bell L C (1992) Soil solid and solution phase changes and mung bean response during amelioration of aluminum toxicity with organic matter. *Plant and Soil* **140**: 183–96
Birch H F (1958) The effect of soil drying on humus decomposition and nitrogen availability. *Plant and Soil* **10**: 9–31
Bird R (1991) Tree and shelter effects on agricultural production in southern Australia. *Agricultural Science* NS **4**: 37–9
Bishop C E, Toussaint W D (1958) *Introduction to agricultural economic analysis*. New York, Wiley and Sons
Bishop J P (1983) Tropical forest sheep on legume forage/fuelwood fallows. *Agroforestry Systems* **1**: 79–84
Black A S, Waring S A (1972) Ammonium fixation and availability in some cereal producing soils in Queensland. *Australian Journal of Soil Research* **10**: 197–207
Blamey F P C, Edwards D G, Asher C J (1983) Effects of aluminium, OH:Al and P:Al molar ratios, and ionic strength on soybean root elongation in solution culture. *Soil Science* **136**: 197–207
Blunt C G, Humphreys L R (1970) Phosphate response of mixed swards at Mt. Cotton, south-eastern Queensland. *Australian Journal of Experimental Agriculture and Animal Husbandry* **10**: 431–41
Boardman N K (1977) Comparative photosynthesis of sun and shade plants. *Annual Review of Plant Physiology* **28**: 355–77
Boddey R M, Döbereiner J (1988) Nitrogen fixation associated with grasses and cereals: recent results and perspectives for future research. *Plant and Soil* **108**: 53–65
Boddey R M, Victoria R L (1986) Estimation of biological nitrogen fixation associated with *Brachiaria* and *Paspalum* grasses using ^{15}N labelled organic matter and fertilizer. *Plant and Soil* **90**: 265–92
Böhnert E, Lascano C, Weniger J H (1985) Botanical and chemical composition of the diet selected by fistulated steers under grazing on improved grass–legume pastures in the tropical savannas of Colombia. 1. Botanical composition of forage available and selected. *Zeitschrift für Tierzüchtung und Züchtungsbiologie* **102**: 385–94
Boonkird S A, Fernandes E C M, Nair P K R (1984) Forest villages: an agroforestry approach to rehabilitating forest land degraded by shifting cultivation in Thailand. *Agroforestry Systems* **2**: 87–102
Bourke R M (1974) The role of legumes in soil fertility maintenance in the

lowlands of Papua New Guinea; a legume evaluation study at Keravat, New Britain; and notes on the use of legumes as food in Papua New Guinea. *Science in New Guinea* **2**: 63–9

Bouton J H, Smith R L, Schank S C, Burton G W, Tyler M E, Littell R C, Gallaher R N, Quesenberry K H (1979) Response of pearl millet inbreds and hybrids to inoculation with *Azospirillum brasilense*. *Crop Science* **19**: 12–16

Bradfield R (1974) Intensive multiple cropping. *Tropical Agriculture* **51**: 91–3

Brady N C (1990) *The nature and properties of soils*, 10th edn. New York, Macmillan

Brandon N J (1993) Establishment requirements of *Leucaena leucocephala* (Lam.) de Wit cv. Cunningham. PhD thesis, University of Queensland

Braunack M V, Walker J (1985) Recovery of some surface soil properties of ecological interest after sheep grazing in a semi-arid woodland. *Australian Journal of Ecology* **10**: 451–60

Bray R A, D P A Sands (1987) Arrival of the *Leucaena* psyllid in Australia: impact, dispersal, and natural enemies. *Leucaena Research Reports*. **7**: 61–5

Bray R A, Woodroffe T D (1991) Effect of the leucaena psyllid on yield of *Leucaena leucocephala* cv. Cunningham in south-east Queensland. *Tropical Grasslands* **25**: 356–7

Bridge B J, Mott J J, Winter W H, Hartigan, R J (1983a) The formation of degraded areas in the dry savanna woodlands of northern Australia. *Australian Journal of Soil Research* **21**: 91–104

Bridge B J, Mott J J, Winter W H, Hartigan, R J (1983b) Improvement in soil structure resulting from sown pastures on degraded areas in the dry savanna woodlands of northern Australia. *Australian Journal of Soil Research* **21**: 83–90

Briscoe C B (1983) Integrated forestry–agriculture–livestock land use at Jari Florestal e Agropecuária. In Huxley, P A (ed.) *Plant research and agroforestry*. Nairobi, Kenya, International Council for Research in Agroforestry: 63–8

Brockington N R, Stobbs T H, Newhouse P W, Wadsworth G A (1965) The effect of leys on soil fertility in the annual cropping areas of Uganda. *African Soils* **10**: 473

Bromfield S M, Cumming R W, David D J, Williams C H (1983) Change in soil pH, manganese and aluminium under subterranean clover pasture. *Australian Journal of Experimental Agriculture and Animal Husbandry* **23**: 181–91

Broughton W J (1977) Effect of various covers on soil fertility under *Hevea brasiliensis* Muell. Arg. and on growth of the tree. *Agro-ecosystems* **3**: 147–70

Brown L R (1981) Eroding the base of civilization. *Journal of Soil and Water Conservation* **39**: 162–5

Bruce R C (1965) Effect of *Centrosema pubescens* Benth. on soil fertility in the humid tropics. *Queensland Journal of Agricultural and Animal Sciences* **22**: 221–6

Bryan W W, Evans T R (1973) Effects of soils, fertilizers and stocking rates on pastures and beef production on the Wallum of southeast

Queensland. 1. Botanical composition and chemical effects on plants and soils. *Australian Journal of Experimental Agriculture and Animal Husbandry* **13**: 516–29

Budelman A (1988a) Leaf dry matter productivity of three selected perennial leguminous species in humid tropical Ivory Coast. *Agroforestry Systems* **7**: 47–62

Budelman A (1988b) The decomposition of the leaf mulches of *Leucaena leucocephala, Gliricidia sepium* and *Flemingia macrophylla* under humid tropical conditions. *Agroforestry Systems* **7**: 33–45

Budelman A (1989) Nutrient composition of the leaf biomass of three selected woody leguminous species. *Agroforestry Systems* **8**: 39–51

Burrows W H, Frost P G H (1993) Deforestation in the savanna context: problems and benefits for pastoralism. *Proceedings XVII International Grassland Congress*, Palmerston North NZ: 2223–30

Burrows W H, Scanlan J C, Rutherford M T (1988) *Native pastures in Queensland: the resources and their management*. Brisbane, Queensland Department of Primary Industries

Burrows W H, Carter J O, Scanlan J C, Anderson E R (1990) Management of savannas for livestock production in north-east Australia: contrasts across the tree–grass continuum. *Journal of Biogeography* **17**: 503–12

Burton G W, Prine G M, Jackson J E (1957) Studies of drouth tolerance and water use of several southern grasses. *Agronomy Journal* **49**: 498–503

Bushby H V A (1982) Rhizosphere populations of *Rhizobium* strains and nodulation of *Leucaena leucocephala*. *Australian Journal of Experimental Agriculture and Animal Husbandry* **22**: 293–8

Cadish G, Sylvester-Bradley R, Nösberger J (1989) ^{15}N-based estimation of nitrogen fixation by eight tropical forage-legumes at two levels of P:K supply. *Field Crops Research* **22**: 181–94

Cameron D F, Miller C P, Edye L A, Miles J W (1993) Advances in research and development with *Stylosanthes* and other tropical pasture legumes. *Proceedings XVII International Grassland Congress*. Palmerston North, NZ: 2109–14

Cameron D M, Rance S J, Jones R M, Charles-Edwards D A, Barnes A (1989) Project Stag: an experimental study in agroforestry. *Australian Journal of Agricultural Research* **40**: 699–714

Cameron D M, Gutteridge R C, Rance S J (1991) Sustaining multiple production systems. 1. Forest and fodder trees in multiple use systems in the tropics. *Tropical Grasslands* **25**: 165–72

Campbell C A, Myers R J K, Weir K L (1981) Potentially mineralizable nitrogen: decomposition rates and their relationship to temperature for five Queensland soils. *Australian Journal of Soil Research* **19**: 323–32

Carangal V R, Calub A D (1987) Crop residues and fodder crops in rice-based farming systems. In Dixon R M (ed.) *Ruminant feeding systems utilizing fibrous agricultural residues* – 1986. Canberra, IDP: 3–23

Carvalho M M de, Edwards D G, Andrew C S, Asher C J (1981) Aluminium toxicity, nodulation, and growth of *Stylosanthes* species. *Agronomy Journal* **73**: 261–5

Casenave A, Valentin C (1989) *Les états de surface de la zone Sahèliènne influence sur l'infiltration*. Paris, ORSTOM, Collection Didactiques

Castro T F, Yoshida T (1974) Effect of organic matter on the biodegradation of some organochlorine insecticides in submerged soils. *Soil Science and Plant Nutrition* **20**: 363–70

Catchpole D W, Blair G J (1990a) Forage tree legumes. I Productivity and N economy of *Leucaena*, *Gliricidia*, *Calliandra* and *Sesbania* and tree/green panic mixtures. *Australian Journal of Agricultural Research* **41**: 521–30

Catchpole D W, Blair G J (1990b) Forage tree legumes. III Release of nitrogen from leaf, faeces and urine derived from *Leucaena* and *Gliricidia* leaf. *Australian Journal of Agricultural Research* **41**: 539–47

Cenpukdee U, Fukai S (1992a) Cassava/legume intercropping with contrasting cassava cultivars. 1. Competition between component crops under three intercropping conditions. *Field Crops Research* **29**: 113–33

Cenpukdee U, Fukai S (1992b) Cassava/legume intercropping with contrasting cassava cultivars. 2. Selection criteria for cassava genotypes in intercropping with two contrasting legume crops. *Field Crops Research* **29**: 135–49

Cenpukdee V, Fukai S (1992c) Agronomic modification of competition between cassava and pigeonpea in intercropping. *Field Crops Research* **30**: 131–46

Cerri C C, Volkoff B, Andreaux F (1991) Nature and behaviour of organic matter in soils under natural forest, and after deforestation, burning and cultivation, near Manaus. *Forest Ecology and Management* **38**: 247–57

Chadhokar P A (1983) The effect of *Gliricidia* supplemented dry season forage on the milk yield and composition of MRY (Netherland) cows in Sri Lanka. *Tropical Grasslands* **17**: 39–41

Chadhokar P A, Kantharaju H R (1980) Effect of *Gliricidia maculata* on growth and breeding of Bannur ewes. *Tropical Grasslands* **14**: 78–82

Chakraborty S, Pettitt A N, Boland R M, Lowchoy S, Cameron D F, Irwin J A G, Davis R D (1993) Stylo host heterogeneity for anthracnose management. *Proceedings XVII International Grassland Congress*. Palmerston North, NZ: 2137–40

Chamala S, Coughenour M C (1986) Constraints and incentives to stubble mulching among Queensland grain growers. *Journal of Soil Conservation, New South Wales* **42**: 92–7

Chamberlin R J, the late Peake D C I, McCown R L, Vallis I, Jones R K (1986) Competition for nitrogen between a maize crop and forage legume intercrops in a wet-dry tropical environment. In Hague I, Jutzi S and Neate P J H (eds) *Potentials of forage legumes in farming systems of sub-Sahara Africa*. Addis Ababa, ILCA: 82–99

Chandler M R, Date R A, Roughley R J (1982) Infection and root-nodule development in *Stylosanthes* species by *Rhizobium*. *Journal of Experimental Botany* **33**: 47–57

Chantalakhana C (1990) *Small farm animal production and sustainable agriculture*. Taipei, ASPAC Food and Fertilizer Technology Center Extension Bulletin 309

Chapman A L, Myers R J K (1987) Nitrogen contributed by grain legumes in rotation with rice on the Kununurra soils of the Ord Irrigation area, Western Australia. *Australian Journal of Experimental Agriculture* **27**: 155–63

Chee Y K, Ahmad Faiz (1991) Forage resources in Malaysia rubber estates.

In Shelton H M, Stür W W (eds) *Forages for plantation crops*. Canberra, ACIAR Proceedings 32: 32–5

Chen C P (1991) Cattle productivity under oil palm in Malaysia. In Shelton H M, Stür W W (eds) *Forages for plantation crops*. Canberra, ACIAR Proceedings 32: 97–101

Chen C P (1993) Pastures as the secondary component in tree–pasture systems. *Proceedings XVII International Grassland Congress*. Palmerston North, NZ: 2037–43

Chen C P, Bong Julita I (1983) Performance of tropical forages under the closed canopy of the oil palm. 1. Grasses. *MARDI Research Bulletin* **11**: 248–63

Chen C P, Othman A (1984) Performance of tropical forages under the closed canopy of the oil palm. 2. Legumes. *MARDI Research Bulletin* **12**: 21–37

Chen Y, Tarchitzky J, Brouwer J, Morinj, Banin A (1980) Scanning electron microscope observations on soil crusts and their formation. *Soil Science* **130**: 45–55

Cheshire M V (1985) Carbohydrates in relation to soil fertility. In Vaughan D, Malcolm R G (eds) *Soil organic matter and biological activity*. Dordrecht, Martinus Nijhoff/Dr W Junk, Developments in Plant and Soil Science **16**: 263–88

Cheva-Isarakul B (1987) Integration of small ruminants and mixed deciduous forest in northern Thailand. In Devendra C (ed.) *Small ruminant production systems in south and southeast Asia*. Ottawa, Canada, IDRC

Chiyenda S S, Materechera S A (1989) Effect of incorporating prunings of *Leucaena leucocephala*, *Cassia siamea* and *Cajanus cajan* on yield of maize in alley cropping system. In Heide J van der (ed.) *Nutrient management for food crop production in tropical farming systems*. Haren, Netherlands, Institute for Soil Fertility

Chong D T, Tajuddin I, Abd. Samat M S (1991) Stocking rate effect on sheep and forage productivity under rubber in Malaysia. In Shelton H M, Stür W W (eds) *Forages for plantation crops*. Canberra, ACIAR Proceedings 32: 102–6

Choonluchanon S, Boonkerd N, Sawatdee P (1988) Adaption of exotic *Azolla* to tropical environment of Thailand. *Plant and Soil* **108**: 67–70

CIAT (1989) Legumes: the key to productive pastures. Cali, Colombia, CIAT Report 1989: 55–7

CIAT (1990) Rice–pastures association. *Tropical Pastures Annual Report for 1989*. Cali, Colombia, CIAT: 24-1-14

CIAT (1991a) Litter decomposition. *Tropical Pastures Program 1987–1991 Annual Report*. Cali, Colombia, CIAT: 12-17-25

CIAT (1991b) Pastures and nutrient recycling. In *CIAT Report: Highlights for 1990 and early 1991*. Cali, Colombia, CIAT: 81–4

CIAT (1991c) Associated pastures and erosion control. In *CIAT Report 1991, highlights for 1990 and early 1991*. Cali, Colombia, CIAT: 93–5

CIAT (1991d) 20. Livestock production systems. *Tropical Pastures Program 1987–1991 Annual Report*. Cali, Colombia, CIAT: 20-1-75

CIAT (1991e) *Tropical pastures program: 1987–1991 Annual Report*. Cali, Colombia, CIAT

CIAT (1992) *Pastures for the tropical lowlands. CIAT's contribution*. Cali, Colombia, CIAT

Ciesiolka, C (1987) *Catchment management in the Nogoa watershed.* Australian Water Resources Council Research Project Report 80/128. Canberra, Department of Resources and Energy

Clarke A L, Greenland D J, Quirk J P (1967) Changes in some physical properties of the surface of an impoverished red-brown earth under pasture. *Australian Journal of Soil Research* **5**: 59–68

Clatworthy J N (1986) Establishment and yields of pasture legumes under cutting in Zimbabwe. 2. Legumes for ley pasture. *Zimbabwe Journal of Agricultural Research* **24**: 149–66

Clayton E (1983) *Agriculture, poverty and freedom in developing countries.* London, Macmillan

Clements R J (1989) Rates of destruction of growing points of pasture legumes by grazing cattle. In *Proceedings XVI International Grassland Conference.* Versailles, Association Française pour la Production Fourragère **2**: 1027–8

Coaldrake J E (1967) Depth of ploughing in relation to depth of suckering and soil type in the control of root suckers of brigalow (*Acacia harpophylla*). *Australian Journal of Experimental Agriculture and Animal Husbandry* **7**: 523

Cobbina J, Atta-Krah A N (1992) Forage productivity of *Gliricidia* accessions on a tropical alfisol soil in Nigeria. *Tropical Grasslands* **26**: 248–54

Cobbina J, Atta-Krah A N, Kang B T (1989) Leguminous browse supplementation effect on the agronomic value of sheep and goat manure. *Biological Agriculture and Horticulture* **6**: 115–21

Cogle A L, Strong W M, Saffigna P G, Ladd J N, Amato M (1987) Wheat straw decomposition in subtropical Australia. II Effect of straw placement on decomposition and recovery of added ^{15}N-urea. *Australian Journal of Soil Research* **25**: 481–90

Colbran, R C (1969) Cover crops and nematode control in pineapples. *Queensland Agricultural Journal* **95**: 658–61

Combe J (1983) Agroforestry techniques in tropical countries: Potentials and limitations. *Agroforestry Systems* **1**: 13–27

Conway G R (1987) The properties of agro-ecosystems. *Agricultural Systems* **24**: 95–118

Cook B G, Grimes R F (1977) Multiple land use of open forest in south-eastern Queensland for timber and improved pasture: Establishment and early growth. *Tropical Grasslands* **11**: 239–46

Cook B G, Garthe R J, Grimes R F (1984) Tropical pastures in eucalypt forest near Gympie. *Queensland Agricultural Journal* **110**: 45–6

Cook C C, Grut M (1989) *Agroforestry in sub-Saharan Africa: a farmer's perspective.* Washington, World Bank Technical Paper 112

Cook G D, So H B, Dalal R C (1992) Structural degradation of two Vertisols under continuous cultivation. *Soil and Tillage Research* **24**: 47–64

Cook S J (1985) Effect of nutrient application and herbicides on root competition between green panic seedlings and a *Heteropogon* grassland sward. *Grass and Forage Science* **40**: 171–5

Copland J W (ed.) (1985) *Draught animal power for production.* Canberra, ACIAR Proceedings 10

Corby H D L (1974) Systematic implications of nodulation among Rhodesian legumes. *Kirkia* **9**: 301–29

Corby H D L (1988) Types of rhizobial nodules and their distribution amongst the Leguminoseae. *Kirkia* **13**: 53–123

Corby H D L (1990) The incidence of rhizobial nodulation among legumes dominant in the Flora Zambesiaca area of Africa. *Kirkia* **13**: 365–75

Corlett J E, Ong C K, Black C R, Monteith J L (1992a) Above- and below-ground interactions in a leucaena/millet alley cropping system. I Experimental design, instrumentation and diurnal trends. *Agricultural and Forest Meteorology* **60**: 53–72

Corlett J E, Black C R, Ong C K, Monteith J L (1992b) Above- and below-ground interactions in a leucaena/millet alley cropping system. II Light interception and dry matter production. *Agricultural and Forest Meteorology* **60**: 73–91

Cowan R T, O'Grady P, Moss R J, Byford I J R (1974) Milk and fat yields of Jersey and Friesian cows grazing tropical grass–legume pastures. *Tropical Grasslands* **8**: 117–20

Crack B J (1972) Changes in soil nitrogen following different establishment procedures for Townsville stylo on a solodic soil in north-eastern Queensland. *Australian Journal of Experimental Agriculture and Animal Husbandry* **12**: 274–80

Cramb R A (1989) The use and productivity of labour in shifting cultivation: an east Malaysian case study. *Agricultural Systems* **29**: 97–115

Craswell E T, Pushparajah E (1991) Soil management and crop technologies for sustainable agriculture in marginal upland areas of southeast Asia. In Blair G, Lefroy R (eds) *Technologies for sustainable agriculture on marginal uplands in southeast Asia*. Canberra, ACIAR Proceedings No. 33: 93–100

Crosby D G (1970) The nonbiological degradation of pesticides in soils. In *International symposium on pesticides in soil*. East Lansing USA, Michigan State University: 86–94

Crowder L V, Chheda H R (1982) *Tropical grassland husbandry*. London, Longman

Dalal R C (1974) Effects of intercropping maize with pigeonpeas on grain yield and nutrient uptake. *Experimental Agriculture* **10**: 219–24

Dalal R C, Mayer R J (1986a) Long term trends in fertility of soils under continuous cultivation and cereal cropping in southern Queensland. I Overall changes in soil properties and trends in winter cereal yields. *Australian Journal of Soil Research* **24**: 265–79

Dalal R C, Mayer R J (1986b) Long term trends in fertility of soils under continuous cultivation and cereal cropping in southern Queensland. II Total organic carbon and its rate of loss from the soil profile. *Australian Journal of Soil Research* **24**: 281–92

Dalal R C, Mayer R J (1986c) Long term trends in fertility of soils under continuous cultivation and cereal cropping in southern Queensland. III Distribution and kinetics of soil organic matter in particle size fractions. *Australian Journal of Soil Research* **24**: 293–300

Dalal R C, Mayer R J (1986d) Long term trends in fertility of soils under continuous cultivation and cereal cropping in southern Queensland. IV Loss of organic carbon from different density fractions. *Australian Journal of Soil Research* **24**: 301–9

Dalal R C, Mayer R J (1986e) Long term trends in fertility of soils under continuous cultivation and cereal cropping in southern Queensland. V Rate of loss of total nitrogen from the soil profile and changes in C/N ratios. *Australian Journal of Soil Research* **24**: 493–504

Dalal R C, Mayer R J (1987a) Long-term trends in fertility of soils under continuous cultivation and cereal cropping in southern Queensland. VI Loss of total nitrogen from different particle-size and density fractions. *Australian Journal of Soil Research* **25**: 83–93

Dalal R C, Mayer R J (1987b) Long-term trends in fertility of soils under continuous cultivation and cereal cropping in southern Queensland. VII Dynamics of nitrogen mineralization potentials and microbial biomass. *Australian Journal of Soil Research* **25**: 461–72

Dalal R C, Mayer R J (1990) Long-term trends in fertility of soils under continuous cultivation and cereal cropping in southern Queensland. VIII Available N indices and their relationships to crop yield. *Australian Journal of Soil Research* **28**: 563–75

Dalal R C, Strong W M, Weston E J, Gaffney J (1991) Sustaining multiple production systems. 2. Soil fertility decline and restoration of cropping lands in sub-tropical Queensland. *Tropical Grasslands* **25**: 173–80

Darnhofer I (1992) Economic implications of a labour study on pruning in agroforestry systems. Thesis, University for Agriculture and Forestry, Vienna, Austria

Date R A (1991) Nitrogen fixation in *Desmanthus*: strain specificity of *Rhizobium* and responses to inoculation in acidic and alkaline soil. *Tropical Grasslands* **25**: 47–55

Date R A, Norris D O (1979) *Rhizobium* screening of *Styloanthes* species for effectiveness in nitrogen fixation. *Australian Journal of Agricultural Research* **30**: 85–104

Date R A, Burt R L, Williams W T (1979) Affinities between various *Stylosanthes* species as shown by rhizobial, soil pH and geographic relationships. *Agro-Ecosystems* **5**: 57–61

Day J M, Neves M C P, Döbereiner J (1975) Nitrogenase activity on the roots of tropical forage grasses. *Soil Biology and Biochemistry* **7**: 107–12

DeBoer A J (1973) Selected measures of bovine performance in three Thai villages. *Thai Journal of Agricultural Science* **6**: 177–90

De la Cruz R E, Manalo M Q, Aggargan N S, Tambalo J D (1988) Growth of three legume trees inoculated with VA mycorrhizal fungi and *Rhizobium*. *Plant and Soil* **108**: 111–15

Deregibus V A, Sanchez R H, Casal J J, Trlica M J (1985) Tillering responses to enrichment of red light beneath the canopy in a humid natural grassland. *Journal of Applied Ecology* **22**: 199–206

Devendra C (1987) Herbivores in the arid and wet tropics. In Hacker J B, Ternouth J H (eds) *The nutrition of herbivores*. Sydney, Academic Press: 23–46

Devendra C (1992) Nutritional potential of fodder trees and shrubs as protein sources in ruminant nutrition. In Speedy A, Pugliese P-L (eds) *Legume trees and other fodder trees as protein sources for livestock*. Rome, FAO Animal Production and Health Paper 102: 95–113

Devendra C, McLeroy G B (1982) *Goat and sheep production in the tropics*. Harlow, UK, Longman

Diatloff A (1974) Factors involved in the amelioration of retarded symbiosis in Tinaroo glycine. *Australian Journal of Agricultural Research* **25**: 577–82

Dixit R S, Jain N K (1979) Studies on ley farming – effect of two year ley on soil fertility and crop yields. *Indian Journal of Agronomy* **24**: 282–90

Dixon R M (ed.) (1987) *Ruminant feeding systems utilizing fibrous agricultural residues – 1986.* Canberra, IDP

D'Mello J P F (1992) Nutritional potentialities of fodder trees and fodder shrubs as protein sources in monogastric nutrition. In Speedy A, Pugliese P-L (eds) *Legume trees and other fodder trees as protein sources for livestock.* Rome, FAO Animal Production and Health Paper 102: 115–27

Doll J P, Orazem F (1984) *Economics. Theory with applications*, 2nd edn. New York, John Wiley & Sons

Dommergues Y R (1963) Evaluation du toux de fixation de l'azote dans un sol dunaire reboise en filao (*Casuarina equisetifolia*). *Agrochimica* **7**: 335–40

Doorn J van, Eijk-Bos C van, Leguizamo B A, Moreno V L A (1987) [Agroforestry techniques as an alternative to the use of burning in the cultivation of maize in Urabá, Colombia]. Serie Técnica, Corporatión Nacional de Investigación y Fomento Forestal, Colombia 20

Dove M R (1983) Theories of swidden agriculture, and the political economy of ignorance. *Agroforestry Systems* **1**: 85–99

Dreyfus B L, Diem H G, Freire J, Keya S O, Dommergues Y R (1987) Nitrogen fixation in tropical agriculture and forestry. In DaSilva E J, Dommergues Y R, Nyns E J, Ratledge C J (eds) *Microbial technology in the developing world.* Oxford, Oxford University Press: 7–50

Duguma B, Kang B T, Okali D U U (1988) Effect of pruning intensities of three woody leguminous species grown in alley cropping with maize and cowpea on an alfisol. *Agroforestry Systems* **6**: 19–35

Dunne T (1979) Sediment yield and land use in tropical catchments. *Journal of Hydrology* **42**: 281–300

Eaglesham A R J, Ayanaba A, Rao V R, Eskew D L (1981) Improving the nitrogen nutrition of maize by intercropping with cowpea. *Soil Biology and Biochemistry* **13**: 169–71

Edwards C A, Lal R, Madden P, Miller R H, House G (1990) *Sustainable agricultural systems in humid tropics of South America.* Ankeny, Iowa, Soil and Water Conservation Society

Edwards G (1991) Balancing cost–benefit analysis and ecological considerations in developing priorities in R and D in upland agriculture. In Blair G, Lefroy R (eds) *Technologies for sustainable agriculture in marginal uplands in southeast Asia.* Canberra, ACIAR Proceedings 33: 15–24

Edwards P (1991) The potential of integrated Asian farming systems. In Speedy A W (ed.) *Developing world agriculture.* London, Grosvenor Press International: 144–51

Edye L A, Burt R L, Norris D O, Williams W T (1974) The symbiotic effectiveness and geographic origin of morphological–agronomic groups of *Stylosanthes* accessions. *Australian Journal of Experimental Agriculture and Animal Husbandry* **14**: 349–57

Eldridge D J, Rothon J (1992) Runoff and sediment yield from a semi-arid woodland in eastern Australia. 1. The effect of pasture type. *The Rangeland Journal* **14**: 26–39

Ella A, Jacobsen C, Stür W W, Blair G (1989) Effect of plant density and cutting frequency on the productivity of four tree legumes. *Tropical Grasslands* **23**: 28–34

Ella A, Blair G J, Stür W W (1991) Effect of age of forage tree legumes at the first cutting on subsequent production. *Tropical Grasslands* **25**: 275–80

El-Swaify S A, Dangler E W (1982) Rainfall erosion in the tropics: a state of the art. In *Soil Erosion and Conservation in the Tropics*. ASA Spec. Publ. 43, Madison, Wisconsin, American Society of Agronomy: 1–25

Emerson W W (1977) Physical properties and structure. In Russell J S, Greacen E L (eds) *Soil factors in crop production in a semi-arid environment*. Brisbane, University of Queensland Press: 78–104

Eng P K, Kerridge P C, Mannetje L't (1978) Effects of stocking rate on pasture and animal production from a guinea-grass legume pasture in Johore, Malaysia. 1. Dry matter yields, botanical and chemical composition. *Tropical Grasslands* **12**: 188–97

Enyi B A C (1973) An analysis of the effect of weed competition on growth and yield attributes in sorghum (*Sorghum vulgare*), cowpeas (*Vigna unguiculata*) and green gram (*Vigna aureus*). *Journal of Agricultural Science* **81**: 449–53

Epstein E, Grant W J (1967) Soil losses and crust formation as related to some soil physical properties. *Soil Science Society of America Proceedings* **31**: 547–50

Eriksen F I, Whitney A S (1981) Effect of light intensity on growth of some tropical forage species. 1. Interaction of light intensity and nitrogen fertilization on six forage grasses. *Agronomy Journal* **73**: 427–33

Eriksen F I, Whitney A S (1982) Growth and N fixation of some tropical forage legumes as influenced by solar radiation regimes. *Agronomy Journal* **74**: 703–9

Escalante C L (1985) Promising agroforestry systems in Venezuela. *Agroforestry Systems* **3**: 209–21

Evans D O, Rotar P P (1987) *Sesbania in agriculture*. Boulder, Colorado, Westview Press

Evans P T (1988) Designing agroforestry innovations to increase their adoptability: a case study from Paraguay. *Journal of Rural Studies* **4**: 45–55

Falvey J L (1986) *An introduction to working animals*. Melbourne, MPW Australia

Faris M A, Burity H A, Dos Reis O V, Mafra R C (1983a) Intercropping of sorghum or maize with cowpeas or common beans under two fertility regimes in northeastern Brazil. *Experimental Agriculture* **19**: 251–61

Faris M A, Araújo M R A de, Lira M de A, Arcovere A S S (1983b) Yield stability in intercropping studies of sorghum or maize with cowpea or common bean under different fertility levels in northeastern Brazil. *Canadian Journal of Plant Science* **63**: 789–99

Farrington J, Martin A M (1988) Farm participatory research: A review of concepts and recent field work. *Agricultural Administration and Extension* **29**: 247–64

Felker R P, Chamala R K, Glumac E L, Wiesman C, Greenstein M (1991) Mechanized forage production of *Leucaena leucocephala* and *L. pulverulenta*. *Tropical Grasslands* **25**: 342–8

Fenner R J, Rattray A G H (1966) The residual effect of nitrogen applied to a *Panicum coloratum* (Makarikari) ley. *Rhodesia Agricultural Journal* **63**: 93–6

Ferdinandez D E F (1972) Intercropping with coconut. *Ceylon Coconut Quarterly* **23**: 51–3

Ferraris R (1979) Productivity of *Leucaena leucocephala* in the wet tropics of north Queensland. *Tropical Grasslands* **13**: 20–7

Filius A M (1982) Economic aspects of agroforestry. *Agroforestry Systems* **1**: 29–39

Floate M J S (1987) Nitrogen cycling in managed grasslands. In Snaydon R W (ed.) *Managed grasslands. Analytical studies*. Amsterdam, Elsevier: 163–72

Ford R (1991) The effects of agricultural activities on water catchments. *Agricultural Science NS* **4**: 35–9

Fosset G W, Venamore P C (1993) Evaluation of animal productivity and pasture yield following treatment of brigalow (*Acacia harpophylla*) regrowth with tebuthiuron. *Proceedings XVII International Grassland Congress* Palmerston North NZ: 2233–4

Foster H L (1971) Crop yields after different elephant grass ley treatments at Kawanda Research Station, Uganda. *East African Agricultural and Forestry Journal* **37**: 63–72

Francis C A (ed.) (1986) *Multiple cropping systems*. New York, Macmillan

Francis P A (1987) Land tenure systems and agricultural innovation. The case of alley farming in Nigeria. *Land Use Policy* **4**: 305–19

Freebairn D M, Loch R J (1991) How does organic farming perform in relation to soil conservation? In Thompson J P, Thomas G A (eds) *Organic farming in field crop production*. Brisbane, Queensland Department of Primary Industries Conference and Workshop Series QC91001: 49–55

Freebairn D M, Wockner G H (1986a) A study of soil erosion on vertisols of the eastern Darling Downs, Queensland. I Effects of surface conditions on soil movement within contour bay catchments. *Australian Journal of Soil Research* **24**: 135–58

Freebairn D M, Wockner G H (1986b) A study of soil erosion on vertisols of the eastern Darling Downs, Queensland. II The effect of soil, rainfall, and flow conditions on suspended sediment losses. *Australian Journal of Soil Research* **24**: 159–72

Fresco L O, Westphal E (1988) A hierarchical classification of farm systems. *Experimental Agriculture* **24**: 399–419

Fujita H, Humphreys L R (1992) Variation in seasonal stocking rate and the dynamics of *Lotonis bainesii* in *Digitaria decumbens* pastures. *Journal of Agricultural Science* **118**: 47–53

Gaskins M H, Hubbell D H, Albrecht S L (1983) Interactions between grasses and rhizosphere nitrogen-fixing bacteria. In Smith A J, Hays V

W (eds) *Proceedings of the XIV International Grassland Congress, June 1981*. Boulder, Colorado, Westview Press: 324–9

Gates C T (1974) Nodule and plant development in *Stylosanthes humilis* H.B.K.: symbiotic response to phosphorus and sulphur. *Australian Journal of Botany* **22**: 45–56

Gauthier D, Diem H G, Dormmergues Y R, Ganry F (1985) Assessment of N_2 fixation by *Casuarina equisetifolia* inoculated with *Frankia* ORS 021001 using ^{15}N methods. *Soil Biology and Biochemistry* **17**: 375–9

Gianluppi V, Camargo A A H, Serräo E A de S (1983) System of sequential production of rice with forage crops, in cerrado soils of Roraima. II *Brachiaria humidicola* and pigeon pea. Pesquisa em Andamento de Boa Vista (1983) No 5. Unidade de Execucao de Pesquisa de Ambito Territorial, Vista, RR, Brazil

Gibson T (1983) Toward a stable low-input highland agricultural system. Ley farming in *Imperata cylindrica* grasslands of northern Thailand. *Mountain Research and Development* **3**: 378–85

Gibson T (1987) Northeast Thailand. A ley farming system using dairy cattle in the infertile uplands. *World Animal Review* **61**: 36–43

Gibson T A (1988) The agronomic and soil fertility effects of leguminous pastures in sandy, upland soils of north-east Thailand. PhD thesis, University of Queensland

Gibson T A, Humphreys L R (1973) The influence of nitrogen nutrition of *Desmodium uncinatum* on seed production. *Australian Journal of Agricultural Research* **24**: 667–76

Gichuru M P, Kang B T (1989) *Calliandra calothyrsus* (Melssn.) in an alley cropping system with sequentially cropped maize and cowpea in southwestern Nigeria. *Agroforestry Systems* **9**: 191–203

Gifford G F (1978) Infiltrometer studies in rangeland plant communities of the Northern Territory. *Australian Rangeland Journal* **1**: 142–9

Gill A S, Patil B D (1985) Agroforestry studies of *Leucaena* with sugarcane. *Leucaena Research Reports* **6**: 35

Gill M P S, Dhillon N S, Dev G (1987) Phosphorus nutrition of groundnut–wheat and greengram–wheat crop sequences in Tulewal sandy loam. *Journal of the Indian Society of Soil Science* **35**: 426–31

Giller K E, Wilson K J (1991) *Nitrogen fixation in tropical cropping systems*. Wallingford, UK, CAB International

Gillespie A R (1989) Modelling nutrient flux and interspecies root competition in agroforestry interplantings. *Agroforestry Systems* **8**: 257–65

Gillingham A G (1987) Phosphorus cycling in managed grasslands. In Snaydon R W (ed.) *Managed grasslands. Analytical studies*. Amsterdam, Elsevier: 173–80

Giri G, De R (1979) Effect of preceding grain legumes on dryland pearl millet in north-west India. *Experimental Agriculture* **15**: 169–72

Giri G, De R (1980) Effect of preceding grain legumes on growth and nitrogen uptake of dryland pearl millet. *Plant Soil* **54**: 459–64

Glatzle A, Ramirez E (1993) Potential suitability of spontaneously reseeding *Styloanthes* spp. for ley farming in the Central Choco of Paraguay. *Proceedings XVII International Grassland Congress*. Palmerston North NZ: 2191–2

Gliessman S R, Garcia E R, Amador A M (1981) The ecological basis for the application of traditional agricultural technology in the management of tropical agro-ecosystems. *Agro-Ecosystems* **7**: 173–85

Glover N, Beer J (1986) Nutrient cycling in two traditional Central American agroforestry systems. *Agroforestry Systems* **4**: 77–87

Goldson J R (1967) Undersowing as a means of establishing the ley in western Kenya. *East African Agriculture and Forestry Journal* **32**: 274

Gomez A A, Gomez K A (1983) *Multiple cropping in the humid tropics of Asia.* Ottawa, IDRC

Good R (1964) *The geography of the flowering plants*, 3rd edn. London, Longman, Green & Co.

Graham T W G, Webb A A, Wearing S A (1981) Soil nitrogen status and pasture productivity after clearing of brigalow (*Acacia harpophylla*). *Australian Journal of Experimental Agriculture and Animal Husbandry* **21**: 109–18

Grainger A (1980) The development of tree crops and agroforestry systems. *International Tree Crops Journal* **1**: 3–14

Gray S G (1970) The place of trees and shrubs as sources of forage in tropical and subtropical pastures. *Tropical Grasslands* **4**: 57–62

Greenland D J (1971) Changes in the nitrogen status and physical condition of soils under pastures, with special reference to the maintenance of the fertility of Australian soils used for growing wheat. *Soils and Fertilisers* **34**: 237–51

Greenland D J, Ford G W (1964) Separation of partially humified organic materials from soils by ultrasonic dispersion. *Transactions 8th International Congress of Soil Science* **3**: 137–48

Gryseels G, Asamenew G (1985) Links between livestock and crop production in the Ethiopian highlands. *ILCA Newsletter* **4**: 5–6

Guerrero M R (1975) Soils of the eastern region of Colombia. In Bornemisza E and Alvarado A (eds) *Soil management in tropical America*. Raleigh NC, Soil Science Department North Carolina State University: 61–91

Gutteridge R C (1983) Productivity of forage legumes on rice-paddy walls in north-east Thailand. *Proceedings XIV International Grassland Congress*: 226–9

Gutteridge R C (1988a) Alley cropping. An alternative farming practice. *Agricultural Science* **1**(7): 18–21

Gutteridge R C (1988b) Alley cropping kenaf (*Hibiscus cannabinus*) with leucaena (*Leucaena leucocephala*) in south-eastern Queensland. *Australian Journal of Experimental Agriculture* **28**: 481–4

Gutteridge R C (1992) Evaluation of the leaf of a range of tree legumes as a source of nitrogen for crop growth. *Experimental Agriculture* **28**: 195–202

Gutteridge R C, Boonklinkajorn P (1979) Improved pasture under coconuts – an important concept for southern Thailand. *Thai Journal of Agricultural Science* **12**: 323–34

Gutteridge R C, MacArthur S (1988) Productivity of *Gliricidia sepium* in a subtropical environment. *Tropical Agriculture* **65**: 275–6

Gutteridge R C, Shelton H M (1993) *Forage tree legumes in tropical agriculture.* Wallingford, UK, CAB International

Gutteridge R C, Sorensson C T (1992) Frost tolerance of a *Leucaena diversifolia*×*Leucaena leucocephala* hybrid in Queensland, Australia.*Leucaena Research Reports* **13**: 3–5

Gutteridge R C, Humphreys L R, Topark-Ngarm A (1984) Legumes in north-east Thailand. *ILCA Newsletter* **7**: 3–5

Hadley M (1993) Grasslands for sustainable ecosystems. In Proceedings XVII International Grassland Congress, Palmerston North, NZ: 21–7

Haggar J P, Beer J W (1993) Effect on maize growth of the interaction between increased nitrogen availability and competition with trees in alley cropping. *Agroforestry Systems* **21**: 239–49

Haggar R J (1971) The production and management of *Stylosanthes gracilis* at Shika, Nigeria. 1. In sown pastures. *Journal of Agricultural Science* **77**: 427–36

Haizel K A (1974) The agronomic significance of mixed cropping. 1. Maize interplanted with cowpea. *Ghana Journal of Agricultural Science* **7**: 169–78

Hall I J, Walduck G D, Walker R W (1993) Digitaria milanjiana cv. Jarra ley pasture for banana cropping in tropical north Queensland. Proceedings XVII International Grassland Congress. Palmerston North, New Zealand, 2202–3

Hartley C W S (1988) *The oil palm* (Elaeis guineensis *Jacq.*) 3rd ed. Harlow, UK, Longman Scientific & Technical

Havel A K, Lakhani D A, Ndunguru B J (1980) Intercropping of maize and cowpea: effect of plant populations on insect pests and seed yield. In Keswani C L, Ndunguru B J (eds). *Proceedings of the second symposium on intercropping in semi-arid areas, 4–7 August 1980*. Ottawa, Canada, IDRC **186e**: 102–9

Haynes R J (1983) Soil acidification induced by leguminous crops. *Grass and Forage Science* **38**: 1–11

Haynes R J, Williams P H (1993) Nutrient cycling and soil fertility in the grazed pasture ecosystem. *Advances in Agronomy* **49**: 119–99

Hazra C R, Tripathi S B (1986) Effect of tree canopy and nitrogen application on forage production of oats. *Journal of the Indian Society of Soil Science* **34**: 520–3

Heady E O (1952) Economics of agricultural production and resource use. Eagley Cliffs, New Jersey, Prentice-Hall

Heide J van der, Kruijs A C B M van der, Kang B T, Vlek P L (1985) Nitrogen management in multiple cropping systems. In Kang B T, Heide J van der (eds) *Nitrogen management in farming systems in humid and subhumid tropics*. Haren, Netherlands, Institute for Soil Fertility: 291–306

Hendricksen R E, Minson D J (1985) Growth, canopy structure and chemical composition of *Lablab purpureus* cv. Rongai at Samford, S.E. Queensland. *Tropical Grasslands* **19**: 81–7

Henzell E F (1988) The role of biological nitrogen fixation research in solving problems in tropical agriculture. *Plant and Soil* **108**: 15–21

Herridge D F, Brockwell J (1988) Contributions of fixed nitrogen and soil nitrate to the nitrogen economy of irrigated soybean. *Soil Biology and Biochemistry* **20**: 711–1

Hill D H (1988) Cattle and buffalo meat production in the tropics. Harlow, UK, Longman

Hirata M, Sugimoto Y, Ueno M (1986) Energy and matter flows in bahiagrass pasture. II Net primary production and efficiency for solar energy utilisation. *Journal of Japanese Society of Grassland Science* **31**: 387–96

Hocking D, Rao D G (1990) Canopy management possibilities for arboreal *Leucaena* in mixed sorghum and livestock small farm production systems in semi-arid India. *Agroforestry Systems* **10**: 135–52

Hoffmann D, Nari J, Petheram R J (eds) (1989) *Draught animals in rural development*. Canberra, ACIAR Proceedings 27

Hofstad O (1978) Preliminary evaluation of the taungya system for combined wood and food production in north-eastern Tanzania. Record, Division of Forestry, University of Dar es Salaam 2

Holt J A, Easey J F (1984) Biomass of mound-building termites in a red and yellow earth landscape, north Queensland. In *Proceedings National Soil Conference*. Brisbane, Australia, Australian Society of Soil Science Incorporated: 363

Holt J E, Schoorl D (1990) The application of open and closed systems theory to change in agricultural institutions. *Agricultural Systems* **34**: 123–32

Homchan J, Date R A, Roughley R J (1989a) Responses to inoculation with root-nodule bacteria by *Leucaena leucocephala* in soils of N.E. Thailand. *Tropical Grasslands* **23**: 92–7

Homchan J, Date R A, Roughley R J (1989b) Responses to inoculation with root-nodule bacteria by *Stylosanthes humilis* and *S. hamata* in soils of N.E. Thailand. *Tropical Grasslands* **23**: 98–104

Hong A (1978) Evaluation on the use of vegetative covers for soil conservation in FELDA. *Malaysian Agricultural Journal* **51**: 335–42

Hoogmoed W B, Stroosnijder L (1984) Crust formation on sandy soils in the Sahel. 1. Rainfall and infiltration. *Soil and Tillage Research* **4**: 5–23

Hosmani S A, Hunshal C S, Rangaswamy K T (1986) Intercropping of fodder legumes in sorghum in the transitional belt of North Karnataka. *Journal of Farming Systems* **2(3–4)**: 4–9

Hudson N (1981) *Soil conservation*, 2nd edn. Ithaca NY, Cornell University Press

Hudson N W (1957) Soil erosion and tobacco growing. *Rhodesia Agricultural Journal* **54**: 547–55

Hue N V, Amien I (1989) Aluminium detoxification with green manures. *Communications in Soil Science and Plant Analysis* **20**: 1499–511

Hughes C E (1993) Leucaena *genetic resources*. Oxford, UK, Oxford Forestry Institute

Hulugalle N R (1989) Effect of tied ridges and undersown *Styloanthes hamata* (L.) on soil properties and growth of maize in the Sudan savannah of Burkina Faso. *Agriculture, Ecosystems and Environment* **25**: 39–51

Humphreys L R (1966) Sub-tropical grass growth. III Effects of stage of defoliation and inflorescence removal. *Queensland Journal of Agricultural and Animal Sciences* **23**: 499–531

Humphreys L R (1980) *A guide to better pastures for the tropics and sub-tropics*, 4th edn. Silverwater NSW, Wright Stephenson

Humphreys L R (1981) *Environmental adaptation of tropical pasture plants*. London, Macmillan

Humphreys L R (1986) The improved integration of forage production with rice culture in south-east Asia. *International Rice Commission Newsletter* **34**: 275-96

Humphreys L R (1987) *Tropical pastures and fodder crops*, 2nd edn. Harlow UK, Longman

Humphreys L R (1991) *Tropical pasture utilisation*. Cambridge, Cambridge University Press

Humphreys L R, Riveros F (1986) *Tropical pasture seed production*. FAO Plant Production and Protection Paper 8. Rome, Food and Agriculture Organization of United Nations: 111-21

Humphreys L R, Robinson A R (1966) Subtropical grass growth. 1. Relationship between carbohydrate accumulation and leaf area in growth. *Queensland Journal of Agriculture and Animal Science* **23**: 211-59

Hutton E M, Beattie W M (1976) Yield characteristics in three bred lines of the legume *Leucaena leucocephala*. *Tropical Grasslands* **10**: 187-94

Hutton E M, Bonner I A (1960) Dry matter and protein yields in four strains of *Leucaena glauca* Benth. *Journal of the Australian Institute of Agricultural Science* **26**: 276-7

Hutton E M, Williams W T, Beall L B (1972) Reactions of lines of *Phaseolus atropurpureus* to four species of root-knot nematode. *Australian Journal of Agricultural Research* **23**: 623-32

Ikombo B M (1984) Effects of farmyard manure and fertilizers on maize in semi-arid areas of eastern Kenya. *East African Agricultural and Forestry Journal* **4**: 266-74

ILCA (1988a) Evaluation of forage legumes in management systems – highlands. Annual Report 1988. Addis Ababa, ILCA: 99-103

ILCA (1988b) Evaluation of multipurpose trees in management systems – alley farming. Annual Report 1988. Addis Ababa, ILCA: 94-5

ILCA (1989) Agronomic studies on selected forage legumes – subhumid zone. Annual Report 1988. Addis Ababa, Ethiopia, ILCA: 98-9

IRRI (1983) Selection and testing of cultivars for rice-based cropping systems. Annual Report for 1982. Los Baños, Philippines, IRRI: 420-4

Ismail T (1986) Integration of animals in rubber plantations. *Agroforestry Systems* **4**: 55-66

Ison R L (1990) Teaching threatens sustainable agriculture. London, International Institute for Environment and Development, Gatekeeper Series 21

Ison R L, Ampt P R (1992) Rapid rural appraisal: a participatory problem formulation method relevant to Australian agriculture. *Agricultural Systems* **38**: 363-86

Ive J R, Rose C W, Wall B H, Torssell B W R (1976) Estimation and simulation of sheet run-off. *Australian Journal of Soil Research* **14**: 129-38

Jabbar M A, Cobbina J, Reynolds L (1992) Optimum fodder-mulch allocation of tree foliage under alley farming in southwest Nigeria. *Agroforestry Systems* **20**: 187-98

Jain N C (1993) Productivity of silvipastoral areas and management needs in arid regions of India. *Proceedings XVII International Grassland Congress*. Palmerston North, NZ: 2069-70

Jaiyebo E O, Moore A W (1964) Soil fertility and nutrient storage in different soil–vegetation systems in a tropical rain-forest environment. *Tropical Agriculture* **41**: 129–39

Jama B, Getahun A (1991) Fuelwood production from *Leucaena leucocephala* established in fodder crops at Mtwapa, Coast Province, Kenya. *Agroforestry Systems* **16**: 119–28

Jama B, Getahun A, Nguigi D W (1991) Shading effects of alley cropped *Leucaena leucocephala* on weed biomass and maize yield at Mtwapa, Coast Province, Kenya. *Agroforestry Systems* **13**: 1–12

Jenkinson D S, Ayanabe A (1977) Decomposition of carbon-14 labelled plant material under tropical conditions. *Soil Science Society of America Journal* **41**: 912–5

Jenkinson D S, Fox R H, Rayner J H (1985) Interaction between fertilizer nitrogen and soil nitrogen – the so-called 'priming' effect. *Journal of Soil Science* **36**: 425–44

Jennings P G, Holme W (1985) Supplementary feeding to dairy cows grazing typical pasture: a review. *Tropical Agriculture* **62**: 266–72

Johansen C, Kerridge P C (1979) Nitrogen fixation and transfer in tropical legume-grass swards in south-eastern Queensland. *Tropical Grasslands* **13**: 165–70

Johansen C, Kerridge P C, Luck P E, Cook B G, Lowe K F, Ostrowski H (1977) The residual effect of molydenum fertilizer on growth of tropical pasture legumes in a sub-tropical environment. *Australian Journal of Experimental Agriculture and Animal Husbandry* **17**: 961–8

Johnson D V, Nair P K R (1985) Perennial crop-based agroforestry systems in northeast Brazil. *Agroforestry Systems* **2**: 281–92

Johnson R W, Back P V (1977) Combination of cropping and spraying to control brigalow (*Acacia harpophylla*) suckers. *Queensland Journal of Agriculture and Animal Science* **34**: 197–204

Jones E (1972) Principles for using fertilizers to improve red ferrallitic soils in Uganda. *Experimental Agriculture* **8**: 315–32

Jones, G H G (1942) The effect of a leguminous cover crop in building up soil fertility. *East African Agricultural Journal* **8**: 48–52

Jones M B, Long S P, Roberts M J (1992) Synthesis and conclusions. In Long S P, Jones M B, Roberts M J (eds) *Primary productivity of grass ecosystems of the tropics and sub-tropics*. London, Chapman & Hall: 212–55

Jones M J (1971) The maintenance of soil organic matter under continuous cultivation at Samaru, Nigeria. *Journal of Agricultural Science* (Cambridge) **77**: 473–82

Jones R J (1967) The effects of some grazed tropical grass–legume mixtures and nitrogen fertilized grass on total soil nitrogen, organic carbon, and subsequent yields of *Sorghum vulgaris*. *Australian Journal of Experimental Agriculture and Animal Husbandry* **7**: 66–71

Jones R J (1988) The future for the grazing herbivore. *Tropical Grasslands* **22**: 97–115

Jones R J, Megarrity R G (1986) Successful transfer of DHP-degrading bacteria from Hawaiian goats to Australian ruminants. *Australian Veterinary Journal* **63**: 259–62

Jones R J, Davies J Griffiths, Waite R B (1967) The contribution of some tropical legumes to pasture yields of dry matter and nitrogen at Samford, south-eastern Queensland. *Australian Journal of Experimental Agriculture and Animal Husbandry* **7**: 57–65

Jones R K, Dalgliesh N P, Dimes J P, McCown R L (1991) Sustaining multiple production systems. 4. Ley pastures in crop–livestock systems in the semi-arid tropics. *Tropical Grasslands* **25**: 189–96

Jones R M, Harrison R E (1980) Note on the survival of individual plants of *Leucaena leucocephala* in grazed stands. *Tropical Agriculture* **57**: 265–6

Jones R M, Ratcliff D (1983) Patchy grazing and its relation to deposition of cattle dung pats in coastal sub-tropical Queensland. *Journal of the Australian Institute of Agricultural Science* **49**: 109–11

Jonsson K, Fidjeland L, Maghembe J A, Högberg P (1988) The vertical distribution of fine roots of five tree species and maize in Morogoro, Tanzania. *Agroforestry Systems* **6**: 63–9

Jordan C F (ed.) (1981) *Tropical ecology*. Stroudsberg, Hutchinson Ross

Juo A S R, Lal R (1977) The effect of fallow and continuous cultivation on the chemical and physical properties of an alfisol in western Nigeria. *Plant and Soil* **47**: 567–84

Jutzi S, Anderson F M, Astatke Abiye (1987) Low-cost modifications of the traditional Ethiopian tine plough for land sloping and surface drainage of heavy clay soils: Preliminary results from on-farm verification trials. *ILCA Bulletin* **27**: 28–31

Kahn S U (1978) The interaction of organic matter with pesticides. In Schnitzer M, Kahn S U (eds) *Soil organic matter*. Oxford, Elsevier Press

Kahurananga J (1991) Intercropping Ethiopian *Trifolium* species with wheat. *Experimental Agriculture* **27**: 385–90

Kaligis D A, Mamonto S (1991) Intake and digestibility of some forages for shaded environments. In Shelton H M, Stür W W (eds) *Forages for plantation crops*. Canberra, ACIAR Proceedings 32: 89–91

Kamara C A, Haque I (1992) *Faidherbia albida* and its effects on Ethiopian highland vertisols. *Agroforestry Systems* **18**: 17–29

Kamnalrut A, Evenson J P (1992) Monsoon grassland in Thailand. In Long S P, Jones M B, Roberts M J (eds) *Primary productivity of grass ecosystems of the tropics and sub-tropics*. London, Chapman & Hall: 100–26

Kang B T, Wilson G F, Sipkens L (1981) Alley cropping maize (*Zea mays* L.) and Leucaena (*Leucaena leucocephala* Lam.) in southern Nigeria. *Plant and Soil* **63**: 165–79

Kang B T, Wilson G F, Lawson T L (1984) *Alley cropping. A stable alternative to shifting cultivation*. Ibadan, IITA

Kang B T, Grimme H, Lawson T L (1985) Alley cropping sequentially cropped maize and cowpea with *Leucaena* on a sandy soil in southern Nigeria. *Plant and Soil* **85**: 267–77

Kang B T, Reynolds L, Atta-Krah A N (1990) Alley farming. *Advances in Agronomy* **43**: 315–59

Kannegieter A (1967) Zero cultivation and other methods of reclaiming *Pueraria* fallowed land for foodcrop cultivation in the forest zone of Ghana. *Tropical Agriculturalist* **123**: 51–73

Kanyama G Y, Edje O T (1976) *Effects of undersowing maize with stylo on seed and dry matter yields*. Research Bulletin Bunda Agricultural College, Lilongwe, Malawi

Karim A B, Savill P S (1991) Effect of spacing on growth and biomass production of *Gliricidia sepium* (Jacq.) Walp. in an alley cropping system in Sierra Leone. *Agroforestry Systems* **16**: 213–22

Kass D L, Jimenez H M, Camacho H Y (1987) Second year results of alley cropping with *Gliricidia sepium* (Jacq.) Steud. on an oxic dystropept in San Carlos, Costa Rica. *Nitrogen Fixing Tree Research Reports* **5**: 42–3

Kaufman D D (1970) Pesticides in soil. In *International symposium on pesticides in soil*. East Lansing USA, Michigan State University: 73–86

Kearney P C, Harris C I, Kaufman D, Sheets T J (1965) Behaviour and fate of chlorinated aliphatic acids in soil. *Advances in Pest Control Research* **6**: 1–30

Kellman M (1979) Soil enrichment by neotropical savanna trees. *Journal of Ecology* **67**: 565–77

Kellman M C (1969) Some environmental components of shifting cultivation in upland Mindanao. *Journal of Tropical Geography* **28**: 40–56

Kelly R D, Walker B H (1976) The effects of different forms of land use on the ecology of a semi-arid region in south-eastern Rhodesia. *Journal of Ecology* **64**: 553–76

Kerkhof P (1990) *Agroforestry in Africa – a survey of project experience*. London, Panos Publications

Kerven G L, Edwards D G, Asher C J, Hallman P S, Kokot S (1989) Aluminum determination in soil solution. II Short term colorimetric procedures for the measurement of inorganic aluminium in the presence of organic acid ligands. *Australian Journal of Soil Research* **27**: 91–102

Keswani C L, Mreta R A D (1980) Effect of intercropping on the severity of powdery mildew on green-grass. In Keswani C L, Ndunguru B J (eds) *Proceedings of the second symposium on intercropping in semi-arid areas, 4–7 August 1980*. Ottawa, Canada, IDRC **186e**: 910–4

Kessler C D J, Shelton H M (1980) Dry season legume forages to follow paddy rice in N.E. Thailand. III Influence of time and intensity of cutting on *Crotolaria juncea*. *Experimental Agriculture* **16**: 207–14

Keya N C O (1974) Grass/legume pastures in Western Kenya. II Legume performance at Kitale, Kisii and Kakamega. *East African Agriculture and Forestry Journal* **39**: 247–57

Kharwara P C, Sharma P K, Singh L N (1987) Rice-based cropping systems under irrigation in North India. *International Rice Research Newsletter* **12**: 50

Kimber A J (1974) Crop rotations, legumes and more productive arable farming in the highlands of Papua New Guinea. *Science in New Guinea* **2**: 70–9

King K F S (1979) Agroforestry and the utilisation of fragile ecosystems. *Forest Ecology and Management* **2**: 161–8

Kirkegaard J A, So H B, Troedson R J, Wallis E S (1992a) The effect of compaction on the growth of pigeon pea on clay soils. I Mechanisms of crop response and seasonal effects on a vertisol in a sub-humid environment. *Soil and Tillage Research* **24**: 107–27

Kirkegaard J A, Troedson R J, So H B, Kushwaha B L (1992b) The effect of compaction on the growth of pigeon pea on clay soils. II Mechanisms of crop response and seasonal effects on an oxisol in a humid coastal environment. *Soil and Tillage Research* **24**: 129–47

Knight F H (1971) *Risk, uncertainty and profit*. Chicago, University of Chicago Press

Knipscheer H C, Menz K M, Verinumbe I (1983) The evaluation of preliminary farming systems technologies: zero-tillage systems in West Africa. *Agricultural Systems* **11**: 95–103

Koch B L (1977) Associative nitrogenase activity by some Hawaiian grass roots. *Plant and Soil* **47**: 703–6

Kowal J M, Kassam A H (1976) Energy and instantaneous intensity of rainstorms at Samaru, northern Nigeria. *Tropical Agriculture* **53**: 185–98

Kruger A J, Van Den Berg M (1993) Potential of subtropical forage legume maize intercrops in comparison with monocropped maize. *Proceedings XVII International Grassland Congress*, Palmerston North, NZ

Kumar R, Vaithiyanathan S (1990) Occurrence, nutritional significance and effect on animal productivity of tannins in tree leaves. *Animal Feed Science and Technology* **30**: 21–38

Ladd J N, Parsons J W, Amato M (1977) Studies of N immobilization and mineralization in calcareous soils. 2. Mineralization of immobilized N from soil fractions of different particle size and density. *Soil Biology and Biochemistry* **9**: 312–25

Lal R (1976) No-tillage effects on soil properties under different crops in Western Nigeria. *Soil Science Society of America Journal* **40**: 762–8

Lal R (1982) Effective conservation farming systems for the humid tropics. In Kussow W, El-Swaify S A, Mannering J (eds) *Soil erosion and conservation in the tropics*. ASA Publication 43, Madison, Wisconsin, American Society of Agronomy: 57–96

Lal R (1985) Soil surface management. In Muchow R C (ed.) *Agro-research for the semi-arid tropics*. St Lucia, University of Queensland Press: 273–300

Lal R (1989a) Agroforestry systems and soil surface management of a tropical alfisol. 1. Soil moisture and crop growth. *Agroforestry Systems* **8**: 7–29

Lal R (1989b) Agroforestry systems and soil surface management of a tropical alfisol: IV Effects on soil physical and mechanical properties. *Agroforestry Systems* **8**: 197–215

Lal R (1989c) Agroforestry systems and soil surface management of a tropical alfisol: V Water infiltrability, transmissivity and soil water sorptivity. *Agroforestry Systems* **8**: 217–38

Lal R (1990) *Soil erosion in the tropics. Principles and management*. New York, McGraw-Hill

Lal R, Kang B T (1982) Management of organic matter in soils of the tropics and subtropics. In *Transactions of 12th International Soil Science Congress*, Delhi: 152–77

Lal R, Wilson G F, Okigbo B N (1978) No-tillage farming after various grasses and leguminous cover crops in a tropical alfisol. I Crop performance. *Field Crops Research* **1**: 71–84

Lal R, Wilson G F, Okigbo B N (1979) Changes in properties of an alfisol produced by various crop covers. *Soil Science* **127**: 377–82

Lamotte M (1982) Consumption and decomposition in tropical grassland ecosystems at Lamto, Ivory Coast. In Huntley B J, Walker B H (eds) *Ecology of tropical savannas*. New York, Springer-Verlag: 415–29

Lamotte M, Bourlière F (1983) Energy flow and nutrient cycling in tropical savannas. In Bourlière F (ed.) *Ecosystems of the world: 13 tropical savannas*. Amsterdam, Elsevier: 583–603

Lascano C E, Avila P (1991) Potencial de producción de leche en pasturas solas y asociadas con leguminosas adaptadas a suelos ácidos. *Pasturas Tropicales* 13(3): 2–10

Lathwell D J, Bouldin D R (1981) Soil organic matter and soil nitrogen behaviour in cropped soils. *Tropical Agriculture (Trinidad)* 58: 341–8

Lawes J B, Gilbert J H, Pugh E 1861 On the sources of nitrogen of vegetation. In Lawes J B, Gilbert J H *Agricultural Chemistry*, Volume II. London, Clowes and Sons: 1–88

Lawn R J, Brun W A (1974) Symbiotic nitrogen fixation in soybeans. 1. Effect of photosynthetic source-sink manipulations. *Crop Science* 14: 11–16

Lawson T L, Kang B T (1980) Yield of maize and cowpea in an alley cropping system in relation to available light. *Agriculture and Forest Metereology* 39: 177–84

Le Hourou A (1980) *Browse in Africa*. Addis Ababa, Ethopia, ILCA

Lee K-K, Dibereiner J (1982) Effect of excessive temperatures on rhizobia growth, nodulation and nitrogen fixing activity in symbiosis with Siratro. *Pesquisa Agropecuária Brasileira* 17: 181–4

Lee Y S, Bartlett R J (1976) Stimulation of plant growth by humic substances. *Journal of Soil Science Society of America* 40: 876–9

Lefroy E C, Dann P R, Wildin J H, Wesley-Smith R W, McGowan A A (1992) Trees and shrubs as sources of fodder in Australia. *Agroforestry Systems* 20: 117–39

Lekchom C, Witayanuparpyunyong K, Sukpituskal P, Watkin B R (1989) The use of improved pastures by grazing dairy cows for economic milk production in Thailand. In *Proceedings XVI International Grassland Congress*. Versailles, Association Française pour la Production Fourragère 2: 1163–4

Leng R A (1990) Ruminant nutrition in the tropics. In Speedy A W (ed.) *Developing world agriculture*. London, Grosvenor Press: 221–8

Leng R A, Choo B S, Arreaza C (1992) Practical technologies to optimize feed utilisation by ruminants. In Speedy A, Pugliese P-L (eds) *Legume trees and other fodder trees as protein sources for livestock*. Rome, FAO Animal Production and Health Paper 102: 75–93

Lenné J M (1981) Control of anthracnose of the tropical forage legume Stylosanthes capitata by burning. *14th International Grassland Congress Summaries of Papers*, 321

Liem C, Morawali H, Kanahebi A, Tonga U, Petheram R J (1988) Traction, trampling, tractors and toil in Tarus village, W. Timor NTT. *DAP Project Bulletin* (5): 11–15

Lieth H F H (ed.) (1978) *Patterns of primary productivity in the biosphere*. Stroudsberg, Hutchinson Ross

Litscher T, Whiteman P C (1982) Light transmission and pasture composition under smallholder coconut plantations in Malaita, Solomon Islands. *Experimental Agriculture* **18**: 383–91

Little D A, Said A N (eds) (1987) *Utilization of agricultural by-products as livestock feeds in Africa.* Addis Ababa, ILCA

Littler J W (1984) Effect of pasture on subsequent wheat crops on a black earth soil of the Darling Downs. 1. The overall experiment. *Queensland Journal of Agriculture and Animal Science* **41**: 1–12

Littler J W, Whitehouse M J (1987) Effect of pasture on subsequent wheat crops on a black earth soil of the Darling Downs. III Comparison of nitrogen from pasture and fertiliser sources. *Queensland Journal of Agricultural and Animal Sciences* **44**: 1–8

Liyanage L V K (1991) Forages for plantation crops in Sri Lanka. In Shelton H M, Stür W W (eds) *Forages for plantation crops.* Canberra, ACIAR Proceedings 32: 157–61

Lloyd D L, Smith K P, Clarkson N M, Weston E J, Johnson B (1991) Sustaining multiple production systems. 3. Ley pastures in the subtropics. *Tropical Grasslands* **25**: 181–8

Loch R J, Coughlan K J (1984) Effects of zero tillage and stubble retention on some properties of a cracking clay. *Australian Journal of Soil Research* **22**: 91–8

Long S P, Garcia Moya E, Imbamba S K, Kamnalrut A, Picdade M T F, Scurlock J M O, Shen Y K, Hall D O (1989) Primary productivity of natural grass ecosystems of the tropics. A reappraisal. *Plant and Soil* **115**: 155–66

Loomis R S, Connor D J (1992) *Crop ecology. Productivity and management in agricultural systems.* Melbourne, Cambridge University Press

Low F (1967) Estimating potential erosion in developing countries. *Journal of Soil and Water Conservation* **22**: 147–8

Lowe K F, Hamilton B A (1986) Dairy pastures in the Australian tropics and subtropics. *Tropical Grassland Society of Australia Occasional Publication* **3**: 68–79

Lowry J B (1989) Agronomy and forage quality of *Albizia lebbeck* in the semi-arid tropics. *Tropical Grasslands* **23**: 84–91

Luchiari A, Resende M, Ritchey K D, de Freitas Jr E, de Souza P I M (1985) Manejo do solo e aproveitamento de agua. In Goedert W (ed.) *Solos dos cerrados.* Brasilia, EMBRAPA: 285–322

Ludlow M M, Charles-Edwards D A (1980) Analysis of the regrowth of a tropical grass/legume sward subjected to different frequencies and intensities of defoliation. *Australian Journal of Agricultural Research* **32**: 673–92

Ludlow M M, Wilson G L (1971) Photosynthesis of tropical pasture plants. 2. Temperature and illuminance history. *Australian Journal of Biological Sciences* **24**: 1065–75

Lulandala L L L, Hall J B (1987) Fodder and wood production from *Leucaena leucocephala* intercropped with maize and beans at Mafiga, Morogoro, Tanzania. *Forest Ecology and Management* **21**: 109–17

Lumpkin T A, Plucknett D L (1980) *Azolla*: botany, physiology, and use as a green manure. *Economic Botany* **34**: 111–53

Lynam J K, Herdt R W (1989) Sense and sustainability: sustainability as an objective in international agricultural research. *Agricultural Economics* **3**: 381–98

Macfarlane D, Shelton H M (1986) *Pastures in Vanuatu.* Canberra, ACIAR Technical Reports Series 24

Macfarlane D C, Whiteman P C (1983) Grazing under *Eucalyptus deglupta* reafforestation on Kolombangara Island, Solomon Islands. Final Report (1977–1982). Brisbane, Department of Agriculture, University of Queensland

Macklin B, Jama B, Reshid Kedir, Getahun A (1988) Results of alley cropping experiments with *Leucaena leucocephala* and *Zea mays* at the Kenya coast. *Leucaena Research Reports* **9**: 61–4

Maclean R H, Litsinger J A, Moody K, Watson A K (1992a) Increasing *Gliricidia sepium* and *Cassia spectabilis* biomass production. *Agroforestry Systems* **20**: 199–212

Maclean R H, Litsinger J A, Moody K, Watson A K (1992b) The impact of alley cropping *Gliricidia sepium* and *Cassia spectabilis* on upland rice and maize production. *Agroforestry Systems* **20**: 213–28

Macqueen D J (1993) Calliandra *series Racemosae: taxonomic information; OFI seed collections; trial design.* Oxford, Oxford Forestry Institute

Maghembe J A, Kaoneka A R S, Lulandala L L L (1986) Intercropping, weeding and spacing effects on growth and nutrient content in *Leucaena leucocephala* at Morogoro, Tanzania. *Forest Ecology and Management* **16**(1–4): 269–79

Mahyuddin P, Little D A, Lowry J B (1988) Drying treatments drastically affects feed evaluation and feed quality with certain tropical forage species. *Animal Feed Science and Technology* **22**: 69–78

Mangan J L (1988) Nutritional effects of tannins in animal feeds. *Nutrition Research Reviews* **1**: 209–31

Manidool C (1983) Pastures under coconuts in Thailand. In Juang T C (ed.) *Asian pastures.* Taipei, FFTC Book Series 25

Manidool C (1987) *Livestock based farming systems in Thailand.* Taipei, ASPAC Food and Fertilizer Technology Center Extension Bulletin 266

Mann J S, Badrul Hasan, Beniwal R K (1989) Adopt silvipasture system for more fodder in arid and semi-arid regions of Rajasthan. *Indian Farming* **39**: 33–5

Mannetje L't, Jones R M (1990) Pasture and animal productivity of buffel grass with Siratro, lucerne or nitrogen fertilizer. *Tropical Grasslands* **24**: 269–81

Martin W S (1944) Grass covers in their relation to soil structure. *Empire Journal of Experimental Agriculture* **12**: 21–32

Martinez M R, Roger P A, Mercado B L (eds) (1986) Studies on nitrogen-fixing blue-green algae and their symbiotic forms in the Philippines. *The Philippine Agriculturalist* **69**: 547–676

Mason S C, Leihner D E, Vorst J J (1986a) Cassava–cowpea and cassava–peanut intercropping. 1. Yield and land use efficiency. *Agronomy Journal* **78**: 43–6

Mason S C, Leihner D E, Vorst J J (1986b) Cassava–cowpea and cassava–peanut intercropping. 2. Leaf area index and dry matter accumulation. *Agronomy Journal* **78**: 47–53

Mason S C, Leihner D E, Vorst J J (1986c) Cassava-cowpea and cassava-peanut intercropping. 3. Nutrient concentrations and removal. *Agronomy Journal* **78**: 441-4

Matthews R B, Lungu S, Volk J, Holden S T, Solberg K (1992a) The potential of alley cropping in improvement of cultivation systems in the high rainfall areas of Zambia. II Maize production. *Agroforestry Systems* **17**: 241-61

Matthews R B, Holden S T, Volk J, Lungu S (1992b) The potential of alley cropping in improvement of cultivation systems in the high rainfall areas of Zambia. I *Chitemene* and *Fundikila*. *Agroforestry Systems* **17**: 219-40

May P H, Anderson A B, Frazo J M F, Balick J M (1985) Babassu palm in the agroforestry systems in Brazil's mid-north region. *Agroforestry Systems* **3**: 275-95

McCarl B A, Knight T O, Wilson J R, Hastie J B (1987) Stochastic dominance over potential portfolios: caution regarding covariance. *American Journal of Agricultural Economics* **69**: 804-12

McCown R L, Jones R K, Peake D C I (1985) Evaluation of a no-till, tropical legume ley farming strategy. In Muchow R C (ed.) *Agro-Research for the semi-arid tropics*. St Lucia, University of Queensland Press: 451-69

McCown R L, Winter W H, Andrew M H, Jones R K, the late Peake D C I (1986) A preliminary evaluation of legume ley farming in the Australian semi-arid tropics. In Haque I, Jutzi S, Neate P J H (eds) *Potentials of forage legumes in farming systems of Sub-Saharan Africa*. Addis Ababa, ILCA: 397-419

McDowell L R, Conrad J H, Ellis G L, Loosli J K (1983) *Minerals for grazing ruminants in tropical regions*. Gainesville, University of Florida

McDowell R E (1988) Importance of crop residues for feeding livestock in smallholder farming systems. In Reed, J D, Capper B S, Neate P J H (eds) *Plant breeding and the nutritive value of crop residues*. Addis Adaba, ILCA: 3-27

McIntyre D S (1958a) Permeability measurements of soil crusts formed by raindrop impact. *Soil Science* **85**: 185-9

McIntyre D S (1958b) Soil splash and the formation of surface crusts by raindrop impact. *Soil Science* **85**: 261-6

McIvor J G (1984) Leaf growth and senescence in *Urochloa mosambicensis* and *U. oligotricha* in a seasonally dry tropical environment. *Australian Journal of Agricultural Research* **35**: 177-87

McManmon M, Crawford R M M (1971) A metabolic theory of flooding tolerance: the significance of enzyme distribution and behaviour. *New Phytologist* **70**: 299-306

Mead R, Willey R W (1980) The concept of a 'land equivalent ratio' and advantages in yield from intercropping. *Experimental Agriculture* **16**: 217-8

Mears P T, Humphreys L R (1974) Nitrogen response and stocking rate of *Pennisetum clandestinum* pastures. I Pasture nitrogen requirement and concentration, distribution of dry matter and botanical composition. *Journal of Agricultural Science* **83**: 451-68

Medina O A, Sylvia D M, Kretschmer Jr A E (1987) Growth response of tropical forage legumes to inoculation with *Glomus intraradices*. *Tropical Grasslands* **24**: 24-7

Medina O A, Kretschmer Jr A E, Sylvia D M (1988) The occurrence of vesicular-arbuscular mycorrhizal fungi on tropical forage legumes in south Florida. *Tropical Grasslands* **22**, 73–8

Melillo J M, Aber J D, Linkins A E, Ricca A, Fry B, Nadelhoffer K J (1989) Carbon and nitrogen dynamics along the decay continuum: plant litter to soil organic matter. *Plant and Soil* **115**: 53–62

Mello F de A F de (1980) Amounts of nitrogen fixed by some legumes in the state of Sao Paulo. *Revista de Agricultura Piracicaba, Brazil* **55**: 41–2

Menz K (1992) No easy answers to hilly land erosion. *ACIAR Newsletter* (22): 5

Middleton C H, Mellor W (1982) Grazing assessment of the tropical legume *Calopogonium caeruleum*. *Tropical Grasslands* **16**: 213–6

Miller C P, Winter W H, Coates D B, Kerridge P C (1990) Phosphorus and beef production in northern Australia. 10 Strategies for phosphorus use. *Tropical Grasslands* **24**: 239–49

Miller I L, Williams W T (1981) Tolerance of some tropical legumes to six months of simulated waterlogging. *Tropical Grasslands* **15**: 39–43

Mills J (1759) *A practical treatise of husbandry*. London, J Whiston and B White, R Baldwin, W Johnston, P Davey, B Law: 367

Mittal J P (1993) Proper utilisation of silvipastoral resources of arid tropics by raising sheep and goats together. *Proceedings XVII International Grassland Congress*. Palmerston North, NZ

Mittal S P, Singh P (1989) Intercropping field crops between rows of *Leucaena leucocephala* under rainfed conditions in northern India. *Agroforestry Systems* **8**: 165–72

Moffett M M (1973) Seed transmission of *Pseudomonas phaseolicola* in *Macroptilium atropurpureum* cv. Siratro. *Tropical Grasslands* **7**: 195–9

Mohamed Saleem M A (1985) Effect of sowing time on the grain yield and fodder potential of sorghum undersown with stylo in the subhumid zone of Nigeria. *Tropical Agriculture* **62**: 151–3

Mohamed Saleem M A, Otsyina R M (1986) Grain yield of maize and the nitrogen contribution following *Stylosanthes* pasture in the Nigerian Subhumid Zone. *Experimental Agriculture* **22**: 207–14

Mokwunye V (1980) Interaction between farmyard manure and N P K fertilizers in savanna soils. In *Organic recycling in Africa*. Rome, FAO Soils Bulletin **43**: 192–200

Monsoon Asia Agroforestry Joint Research Team (1986) *Comparative studies on the utilization and conservation of the natural environment by agroforestry systems*. Kyoto, Japan, Faculty of Agriculture, Kyoto University

Moog F A (1980) Backyard cattle raising and its feeding system in a Batangas barrio. Thesis, University of the Philippines, Los Baños

Moog F A, Faylon P S (1991) Integrated forage–livestock systems under coconuts in the Philippines. In Shelton H M, Stür W W (eds) *Forages for plantation crops*. Canberra, ACIAR Proceedings 32: 144–6

Moore A W (1962) The influence of a legume on soil fertility under a grazed tropical pasture. *Empire Journal of Experimental Agriculture* **30**: 239–48

Moore A W (1974) Availability to rhodes grass (*Chloris gayana*) of nitrogen in tops and roots added to soil. *Soil Biology and Biochemistry* **6**: 249–55

Morgan R P C (1979) *Soil erosion*. London, Longman

Morris J W, Bezuidenbout J J, Furniss, P R (1982) Litter decomposition. In Huntley B J, Walker B H (eds) *Ecology of tropical savannas. Ecological Studies 42*. New York, Springer-Verlag: 535–53

Mott J J, Bridge B J, Arndt W (1979) Soil seals in tropical tall grass pastures in northern Australia. *Australian Journal of Soil Research* **17**: 483–94

Mugabe N R, Sinje M E, Sibuga K P (1980) A study of crop/weed competition in intercropping. In Keswani C L, Ndunguru B J (eds) *Proceedings of second symposium on intercropping in semi-arid areas, 4–7 August 1980*. Ottawa, Canada, IDRC 186e: 96–101

Mugwira L M, Shumba E M (1986) Rate of manure applied in some communal areas and the effect on plant growth and maize grain yields. *Zimbabwe Agricultural Journal* **83**: 99–104

Mulder E G, Lie T A, Woldendorp J W (1969) Biology and soil fertility. In *Soil Biology, Reviews of Research*. Paris, UNESCO: 163–208

Mulongoy K (1986) Microbial biomass and maize nitrogen uptake under a *Psophocarpus palustris* live mulch grown on a tropical alfisol. *Soil Biology and Biochemistry* **18**: 395–8

Murphy P M, Sherwood M T (1989) Nitrogen fixation, cycling, and utilisation. *Proceedings of the XVI International Grassland Congress, 1989*. Versailles, France, Association Française pour la Production Fourragère: **3**: 1805–10

Myers R J K (1976) Nitrogen accretion and other soil changes in Tindall clay loam under Townsville stylo/grass pastures. *Australian Journal of Experimental Agriculture and Animal Husbandry* **16**: 94–8

Myers R J K, Vallis I, McGill W B, Henzell E F (1986) Nitrogen in grass-dominant, unfertilized pasture systems. In *Proceedings of the XIII Congress of the International Society of Soil Science*. Hamburg **6**: 761–71

Nair K P P, Patel V K, Singh R P, Kauslik M K (1979) Evaluation of legume intercropping in conservation of fertilizer nitrogen in maize culture. *Journal of Agricultural Science* **93**: 189–94

Nair P K R (1985) Classification of agroforestry systems. *Agroforestry Systems* **3**: 97–128

Nair P K R (1987) Soil productivity under agroforestry. In Gholz H L (ed.) *Agroforestry: realities, possibilities and potentials*. Dordrecht, Netherlands, Martinus Nijhoff: 21–30

Nair P K R (1989a) Agroforestry defined. In Nair P K R (ed.) *Agroforestry systems in the tropics*, Dordrecht, Netherlands, Kluwer Academic Publishers: 13–18

Nair P K R (ed.) (1989b) *Agroforestry systems in the tropics*. Dordrecht, Netherlands, Kluwer Academic Publishers

Nair P K R, Fernandes E C M, Wambugu P (1984) Multipurpose trees and shrubs for agroforestry. *Agroforestry Systems* **2**: 145–63

Natarajan M, Willey R W (1980) Sorghum–pigeonpea intercropping and the effects of plant population density. 1. Growth and yield. *Journal of Agricultural Science* **95**: 51–8

National Research Council (1993) *Vetiver grass: a thin green line against erosion*. Washington DC, National Academy Press

Nautiyal C S, Hegde S V, Berkum P van (1988) Nodulation, nitrogen fixation, and hydrogen oxidation by pigeon pea *Bradyrhizobium* spp. in symbiotic association with pigeon pea, cowpea, and soybean. *Applied and Environmental Microbiology* **54**: 94–7

Nelder J H (1962) New kinds of systematic designs for spacing experiments. *Biometrics* **18**: 283–307

Netsinghe V H (1966) On fertilizer placement using radioisotopes. *Ceylon Coconut Planters' Review* **4**: 55–60

Newton K, Jamieson G I (1968) Cropping and soil fertility studies at Keravat, New Britain 1954–1962. *Papua New Guinea Agricultural Journal* **20**: 25–51

Nitis I M (1977) Stylosanthes *as companion crop to cassava*. IFS Final Report, Udayana University, Bali

Nitis I M (1985) Present state of grassland production and utilisation and future perspectives for grassland farming in humid tropical Asia. *Proceedings XV International Grassland Congress*: 39–44

Nitis I M, Lara K, Suarra M, Sukanten W, Putra S, Arya W (1989) *Three strata system for cattle feeds and feeding in dryland farming area in Bali*. Denpasar, Indonesia, Udayana University

Nnadi L A, Haque I (1988) Forage legumes in African crop–livestock systems. *ILCA Bulletin* 30: 19

Nores G A, Vera R R (1993) Science and information for our grasslands. In *Proceedings XVII International Grassland Congress*. Palmerston North, NZ: 3–7

Norton B W (1993) The nutritive value of tree legumes. In Gutteridge R C, Shelton H M (eds) *Forage tree legumes in tropical agriculture*. Wallingford, UK, CAB International: 177–91

Norton B W, Wilson J R, Shelton H M, Hill K D (1991) The effect of shade on forage quality. In Shelton H M, Stür W W (eds) *Forages for plantation crops*. Canberra, ACIAR Proceedings 32: 83–8

Noy-Meir I, Walker B H (1986) Stability and resilience in rangelands. In Joss P J, Lynch P W, Williams O B (eds) *Rangelands: a resource under siege*. Canberra, Australian Academy of Science: 21–5

Nurhayati D P, Ivory D A, Stür W W (1989) The effectiveness and competitiveness of some Indonesian *Rhizobium* strains on tropical legumes grown in four soil types of Java. *Plant and Soil* **117**: 146–50

Nye P H, Greenland D J (1960) *The soil under shifting cultivation*. Harpenden, UK, Commonwealth Agriculture Bureaux

Obi A O (1989) Long-term effects of the continuous cultivation of a tropical Ultisol in southwestern Nigeria. *Experimental Agriculture* **25**: 207–15

Ofori C S (1973) Decline in fertility status of a tropical forest ochrosol under continuous cropping. *Experimental Agriculture* **9**: 15–22

O'Hara G W, Boonkerd N, Dilworth M J (1988) Mineral constraints to nitrogen fixation. *Plant and Soil* **108**: 93–110

Okafor J C, Fernandes E C M (1987) Compound farms of southeastern Nigeria: a predominant agroforestry homegarden system with crops and small livestock. *Agroforestry Systems* **5**: 153–68

Okali C, Sumberg J E (1985) Sheep and goats, men and women: household relations and small ruminant development in southwest Nigeria. *Agricultural Systems* **18**: 39–59

Okon Y, Heytler P B, Hardy R W F (1983) N_2 fixation by *Azospirillum brasilense* and its incorporation into host *Setaria italica*. *Applied and Environmental Microbiology* **46**: 694–7

Oliveira P R P de, Humphreys L R (1986) Influence of level and timing of shading on seed production in *Panicum maximum* cv. Gatton. *Australian Journal of Agricultural Research* **37**: 417–24

Ong C K, Odongo J C W, Marshall F, Black C R (1991) Water use by trees and crops: five hypotheses. *Agroforestry Today* **3**(2): 7–10

Onim J F M, Mathuva M, Otieno K, Fitzhugh H A (1990) Soil fertility changes and response of maize and beans to green manures of leucaena, sesbania and pigeonpea. *Agroforestry Systems* **12**: 197–215

Othman W M W, Asher C J (1987) The effects of height and frequency of previous defoliation on nodulation, nitrogen fixation and regrowth of phasey bean. *Pertanika* **10**: 1–10

Othman W M W, Asher C J, Wilson G L (1988) ^{14}C-labelled assimilate supply to root nodules and nitrogen fixation of phasey bean plants following defoliation and flower removal. In Shamsuddin Z H, Othman W M N, Marziah M, Sundram J (eds) *Biotechnology of nitrogen fixation in the tropics*. Kuala Lumpur, Universiti Pertanian Malaysia: 217–24

Owino F (1992) Improving multipurpose tree and shrub species for agroforestry systems. *Agroforestry Systems* **19**: 131–7

Palada M C, Kang B T, Claassen S L (1992) Effect of alley cropping with *Leucaena leucocephala* and fertilizer application on yield of vegetable crops. *Agroforestry Systems* **19**: 139–47

Palmer B, Schlink A C (1992) The effect of drying on the intake and rate of digestion of the shrub legume *Calliandra calothyrsus*. *Tropical Grasslands* **26**: 89–93

Paningbatan E P (1986) Alley cropping in the Philippines. In Latham M J (ed.) *Soil management under humid conditions in Asia and the Pacific (Asialand). IBSRAM Proceedings No 5*. Bangkok, Thailand, International Board for Soil Research and Management Inc.

Paningbatan E P (1990) Alley cropping for managing soil erosion in sloping lands. *Transactions of the 14th International Congress of Soil Science* **VII**: 376–7

Papendick R I (ed.) (1977) *Multiple cropping*. Madison, Wisconsin, American Society of Agronomy

Parsons J W (1984) Green manuring. *Outlook on Agriculture* **13**: 20–3

Paterson R T (1993) *Use of trees by livestock 4: anti-nutritive factors*. Chatham, UK, Natural Resources Institute

Patil B P, Pal M (1985) Studies on nitrogen and energy relationship in intercropped pearl millet–wheat–greengram cropping system. *Annals of Agricultural Research* **6**: 196–202

Payne C B (1968) Wind and fruit trees. *Hortus Rhodesia* (9): 24–7

Payne W J A (1985) A review of the possibilities for integrating cattle and tree crop production systems in the tropics. *Forest Ecology and Management* **12**: 1–36

Pearce D W, Turner R K (1990) *Economics of natural resources and the environment*. New York, Harvester Wheatsheaf

Pearcy R W (1987) Photosynthetic responses of tropical forest trees. In *Proceedings International Conference on Tropical Plant Ecophysiology*. Bogor, Indonesia. BIOTROP Special Publication 31. SEAMEO-BIOTROP, Bogor, Indonesia: 49–66

Pearson C J, Ison R L (1987) *Agronomy of grassland systems*. Cambridge, Cambridge University Press
Pellek R (1992) Contour hedgerows and other soil conservation interventions for hilly terrain. *Agroforestry Systems* **17**: 135–52
Peltier R, Eyog-Matig O (1990) [Agroforestry trials in northern Cameroon]. *Bois et Forêts des Tropiques* (217): 3–31
Peoples M B, Herridge D F (1990) Nitrogen fixation by legumes in tropical and subtropical agriculture. *Advances in Agronomy* **44**: 155–223
Pera A, Vallini G, Sireno I, Bianchin M L, Bertoldi M de (1983) Effects of organic matter on rhizosphere organisms and root development of *Sorghum* plants in two different soils. *Plant and Soil* **74**: 3–18
Pereira H C, Chenery E M, Mills W R (1954) The transient effects of grasses on the structure of tropical soils. *Empire Journal of Experimental Agriculture* **22**: 148–60
Pereira H C, Hosegood P H, Dagg M (1967) Effect of tied ridges, terraces and grass leys on a lateritic soil in Kenya. *Experimental Agriculture* **3**: 89–98
Philip J R (1957) The theory of infiltration. 4. Sorptivity and algebraic infiltration equations. *Soil Science* **84**: 257–64
Phiri D M, Coulman B, Steppler H A, Kamara C S, Kwesiga F (1992) The effect of browse supplementation on maize husk utilization by goats. *Agroforestry Systems* **17**: 153–8
Plazinski J, Franche C, Liu C-C, Lin T, Shaw N, Gunning B E S, Rolfe B G (1988) Taxonomic status of *Anabaena azollae*: An overview. *Plant and Soil* **108**: 185–90
Plucknett D L (1979) *Managing pastures and cattle under coconuts*. Tropical Agriculture Series 2. Boulder, Colorado, Westview
Pongskul V, Wilaipon B, Gutteridge R C (1982) Undersowing upland crops with pasture legumes. II Kenaf with *Stylosanthes hamata* cv. Verano. *Thai Journal of Agricultural Science* **15**: 7–10
Poschen P (1986) An evaluation of the *Acacia albida* agroforestry practices in the Hararghe highlands of eastern Ethiopia. *Agroforestry Systems* **4**: 129–43
Posner J L (1982) Cropping systems and soil conservation in the hill areas of tropical America. *Turrialba* **32**: 287–99
Pott A, Humphreys L R (1983) Persistence and growth of *Lotononis bainesii–Digitaria decumbens* pastures. 1. Sheep stocking rate. *Journal of Agricultural Science* **101**: 1–7
Pottier D (1984) Running cattle under trees: an experiment in agroforestry. *Unasylva* **36**: 23–7
Poultney R G (1963) A comparison of direct seeding and undersowing on the establishment of grass and the effect on the cover crop. *East African Agriculture and Forestry Journal* **29**: 26–30
Powell J M (1986) Manure for cropping: a case study from central Nigeria. *Experimental Agriculture* **22**: 15–24
Powell J M, Waters-Bayer A (1985) Interactions between livestock husbandry and cropping in a West African savanna. In Tothill J C, Mott J C (eds) *Ecology and management of the world's savannas*. Canberra, Australia, Australian Academy of Science: 252–5
Pressland A J (1982) Fire in the management of grazing lands in Queensland. *Tropical Grasslands* **16**: 104–12

Pressland A J, Cowan D C (1987) Responses of plant growth to removal of surface soil of the rangelands of western Queensland. *Australian Rangelands Journal* **9**: 74–8

Preston T R (1992) The role of multi-purpose trees in integrated farming systems for the wet tropics. In Speedy A, Pugliese P-L (eds) *Legume trees and other fodder trees as protein sources for livestock*. Rome, FAO Animal Production Protection Paper 102: 193–209

Prinsen J H (1987) Potential of *Albizia lebbeck* (Mimosaceae) as a tropical fodder tree. A review of literature. *Tropical Grasslands* **20**: 78–83

Probert M E (1982) Interactions of nitrogen with phosphorus and sulfur with emphasis on tropical pasture legumes. In Galbally I E, Freney J R (eds) *The cycling of carbon, nitrogen, sulfur and phosphorus in terrestrial and aquatic ecosystems*. Canberra, Australian Academy of Science: 47–50

Probert M E, Williams J (1986) The nitrogen status of red and yellow earths in the semi-arid tropics as influenced by Caribbean Stylo (*Stylosanthes hamata*) grown at various rates of applied phosphorus. *Australian Journal of Soil Research* **24**: 405–21

Probert M E, Okalebo J R, Simpson J R, Jones R K (1992) The role of boma manure for improving soil fertility. In Probert M E (ed.) *A search for strategies for sustainable dryland cropping in semi-arid eastern Kenya*. Canberra, ACIAR Proceedings 41: 63–70

Punj V, Gupta R P (1988) VA-mycorrhizal fungi and *Rhizobium* as biological fertilizers for *Leucaena leucocephala*. *Acta Microbiologica Polonica* **37**: 327–36

Purcino A A C, Lynd J Q (1985) Tripartite symbiosis of *Stylosanthes scabra* Vog. influenced by soil fertility treatments of a Typic Eutrustox. *Agronomy Journal* **77**: 455–8

Quirk M F, Bushell J J, Jones R J, Megarrity R C, Butler K L (1988) Liveweight gains on leucaena and native grass pastures after dosing cattle with rumen bacteria capable of degrading DHP, a ruminal metabolite from leucaena. *Journal of Agricultural Science* **111**: 165–70

Rai R S V, Suresh K K (1988) Agrosilvicultural studies – optimum species combination. *International Tree Crops Journal* **5**: 1–8

Ranga Rao V (1977) Effect of root temperature on the infection process and nodulation in *Lotus* and *Stylosanthes*. *Journal of Experimental Botany* **28**: 241–59

Rao M M, Edmunds J E (1984) Intercropping of banana with food crops: cowpeas, maize and sweet potato. *Tropical Agriculture* **61**: 9–11

Rao M R, Willey R W (1983) Effects of genotype in cereal/pigeonpea intercropping on the Alfisols of the semi-arid tropics of India. *Experimental Agriculture* **19**: 67–78

Rao M R, Sharma M M, Ong C K (1990) A study of the potential of hedgerow intercropping in semi-arid India using a two-way systematic design. *Agroforestry Systems* **11**: 243–58

Rapp A, Murray-Rust D H, Christiansson C, Berry L (1972) Soil erosion and sedimentation in four catchments near Dodoma, Tanzania. *Geografiska Annaler* **54A**: 255–318

Rattray A G H (1961) The value of maize stover. *Rhodesia Agricultural Journal* **58**: 350–3

Reddell P, Rosbrook P A, Bowen G D, Gwaze D (1988) Growth responses in *Casuarina cunninghamiana* plantings to inoculation with *Frankia*. *Plant and Soil* **108**: 79–86

Reed J D, Capper B S, Neale P J H (eds) (1988) *Plant breeding and the nutritive value of crop residues*. Addis Ababa, ILCA

Rees R G, Platz G J (1979) The occurrence and control of yellow leaf spot of wheat in north-eastern Australia. *Australian Journal of Experimental Agriculture and Animal Husbandry* **19**: 369–72

Rerkasem K, Rerkasem B (1988) Yields and nitrogen nutrition of intercropped maize and ricebean (*Vigna umbellata*). *Plant and Soil* **108**: 151–62

Rerkasem B, Rerkasem R, Peoples M B, Herridge D F, Bergersen F J (1988) Measurement of N_2 fixation in maize (*Zea mays* L.)–ricebean (*Vigna umbellata* (Thumb) Ohwi and Ohashi) intercrops. *Plant and Soil* **108**: 125–35

Reynolds L, Atta-Krah A N, Francis P A (1988) *Alley farming with livestock – guidelines*. Ibadan, Nigeria, ILCA

Reynolds S G (1981) Grazing trials under coconuts in Western Samoa. *Tropical Grasslands* **15**: 3–10

Reynolds S G (1982) Contributions to yield, nitrogen fixation and transfer by local and exotic legumes in tropical grass–legume mixtures in Western Samoa. *Tropical Grasslands* **16**: 76–83

Reynolds S G (1988) *Pastures and cattle under coconuts*. Rome, Food and Agriculture Organization of the United Nations, Plant Production and Protection Paper 91

Richards B N, Bevege D I (1967) The productivity and nitrogen economy of artificial ecosystems comprising various combinations of perennial legumes and coniferous tree species. *Australian Journal of Botany* **15**: 467–80

Richards P W (1977) Tropical forests and woodlands: an overview. *Agro-Ecosystems* **3**: 225–38

Rickert K G, Humphreys L R (1970) Effects of variation in density and phosphate application on growth and chemical composition of Townsville stylo (*Stylosanthes humilis*). *Australian Journal of Experimental Agriculture and Animal Husbandry* **10**: 442–9

Rika I K, Nitis I M, Humphreys L R (1981) Effects of stocking rate on cattle growth, pasture production and coconut yield in Bali. *Tropical Grasslands* **15**: 149–57

Riveros F (1992) The genus *Prosopis* and its potential to improve livestock production in arid and semi-arid regions. In Speedy A, Pugliese P-L (eds) *Legume trees and other fodder trees as protein sources for livestock*. Rome, FAO Animal Production and Health Paper 102: 257–76

Robbins G B (1984) Relationships between productivity and age since establishment of pastures of *Panicum maximum* var. *trichoglume*. PhD thesis, University of Queensland

Robbins G B, Bushell J J, Butler K L (1987) Decline in plant and animal production from ageing pastures of green panic (*Panicum maximum* var. *trichoglume*). *Journal of Agricultural Science* **108**: 407–17

Robbins G B, Bushell J J, McKeon G M (1989) Nitrogen immobilization in decomposing litter contributes to productivity decline in ageing pastures of green panic (*Panicum maximum* var. *trichoglume*). *Journal of Agricultural Science* **113**: 401–6

Robertson A D, Humphreys L R (1976) Effects of frequency of heavy grazing and of phosphorus supply on an *Arundinaria ciliata* association oversown with *Stylosanthes humilis*. *Thai Journal of Agricultural Science* **9**: 181–8

Rodel M G W, Boultwood J N, Scheerhorn G (1981) The effects of applied nitrogen on the yields of maize following heavily fertilized and intensively grazed grass. *Rhodesia Agricultural Journal* **78**: 15–19

Roling W (1990) The agricultural research–technology transfer interface: a knowledge systems perspective. In *Making the link: agricultural research and technology transfer in developing countries*. Boulder, Colorado, Westview Press

Roose E J (1977) Application of the Universal Soil Loss Equation of Wischmeier and Smith in West Africa. In Greenland D J, Lal R (eds) *Conservation and soil management in the humid tropics*. Chichester, UK, Wiley: 177–87

Rose C W (1988) Research progress on soil erosion processes and a basis for soil conservation practices. In Lal R (ed.) *Soil erosion research methods*. Iowa, Soil and Water Conservation Society: 119–39

Rosecrance R C, Brewbaker J C, Fownes J H (1992a) Alley cropping of maize with nine leguminous trees. *Agroforestry Systems* **17**: 159–68

Rosecrance R C, Rogers S, Tofinga M (1992b) Effects of alley cropped *Calliandra calothyrsus* and *Gliricidia sepium* hedges on weed growth, soil properties and taro yields in Western Samoa. *Agroforestry Systems* **19**: 57–66

Ruaysoongnern S (1987) A study of seedling growth of *Leucaena leucocephala* (Lam.) de Wit cv. Cunningham with special reference to its nitrogen and phosphorus nutrition in acid soils. PhD thesis, University of Queensland

Ruaysoongnern S, Aitken R L (1980) Nitrogen fixation by cultivars of *Stylosanthes hamata* in two upland soils from northeastern Thailand. *Thai Journal of Agricultural Science* **13**: 291–301

Ruaysoongnern S, Shelton H M, Edwards D G (1989) The nutrition of *Leucaena leucocephala* de Wit cv. Cunningham seedlings. I External requirements and critical concentrations in index leaves of nitrogen, phosphorus, potassium, calcium, sulphur and manganese. *Australian Journal of Agricultural Research* **40**: 1241–51

Rudder T H, Burrow H, Seifert G W, Maynard P J (1982) The effects of year of birth, damage, breeding and dam reproductive efficiency on liveweights and age at sale of commercially managed steers in central Queensland. *Proceedings of the Australian Society of Animal Production* **14**: 281–4

Ruhigwa B A, Gichuru M P, Mambari B, Tarich N M (1992) Root distribution of *Acioa barteri*, *Alchornia cordifolia*, *Cassia siamea* and *Gmelina arborea* in an acid Ultisol. *Agroforestry Systems* **19**: 67–78

Russell D B, Ison R L (1992) The research–development relationship in rangelands: an opportunity for contextual science. In *Proceedings IV International Rangeland Congress*, Montpellier, France: 1047–54

Russell J S (1991) Likely climatic changes and their impact on the northern pastoral industry. *Tropical Grasslands* **25**: 211–8

Russell J S, Williams C H (1982) Biogeochemical interactions of carbon, nitrogen, sulfur and phosphorus in Australian agroecosystems. In Galbally I E, Freney J R (eds) *Cycling of carbon, nitrogen, sulfur and phosphorus in terrestrial and aquatic ecosystems*. Canberra, Australian Academy of Science: 61–75

Russo S L (1986) Introduction of *Leucaena leucocephala* as a forage and alley crop in The Gambia. *Leucaena Research Reports* **7**: 105

Ruthenberg H (1980) *Farming systems in the tropics*. Oxford, Clarendon Press

Said A N, Dzowela B H (1989) *Overcoming constraints to the efficient utilization of agricultural by-products as animal feed*. Addis Ababa, ILCA

Sajjapongse A (1991) Asialand – Management of sloping lands. *IBSRAM Newsletter* (19): 4–6

Samarakoon S P, Shelton H M, Wilson J R (1990a) Voluntary feed intake by sheep and digestibility of shaded *Stenotophrum secundatum* and *Pennisetum clandestinum* herbage. *Journal of Agricultural Science* **114**: 143–50

Samarakoon S P, Wilson J R, Shelton H M (1990b) Growth, morphology and nutritive value of shaded *Stenotophrum secundatum*, *Ascoropus compressus* and *Pennisetum clandestinum*. *Journal of Agricultural Science* **114**: 161–9

Sánchez M D (1991) Supplementation of sheep under rubber in Indonesia. In Shelton H M, Stür W W (eds) *Forages for plantation crops*. Canberra, ACIAR Proceedings 32: 107–11

Sánchez P A (1976) *Properties and management of soils in the tropics*. New York, John Wiley

Sánchez P A (1982) Nitrogen in shifting cultivation systems of Latin America. *Plant and Soil* **67**: 91–103

Sánchez P A, Benites J R (1987) Low-input cropping for arid soils of the humid tropics. *Science* **238**: 1521–7

Sánchez P A, Bandy D E, Villachica J H, Nicholaides J J (1982) Amazon Basin soils: management for continuous crop production. *Science* **216**: 821–7

Sandanam S, Somaratne A, Amarasekera A R, Yatawatte, S T, Samarajeewa S, Ananthacumaraswamy A (1982) An assessment of the suitability of five graminaceous species for soil reconditioning before replanting tea. I Effect of species on enrichment of organic matter status by top and root residues. *Tea Quarterly* **51**: 99–107

Sangakkara V R (1989) Forage legumes as a component of smallholder rice farming systems, a case study. In *Proceedings of the XVI International Grassland Congress* **2**: 1303–4

Sanginga N, Zapata F, Danso S K A, Bowen G D (1990) Effect of successive cuttings on uptake and partitioning of ^{15}N among plant parts of *Leucaena leucocephala*. *Biology and Fertility of Soils* **9**: 37–42

Sanoria C L, Singh K L, Ramamurthy, K, Maurya B R (1982) Field trials with *Azospirillum brasilense* in an Indo-Gangetic alluvium. *Journal of the Indian Society of Soil Science* **30**: 208–9

Sarwatt S V (1993) The potential role of *Crotalaria ochroleuca* ('mareja') in traditional crop and livestock systems in Tanzania. *Proceedings XVII International Grassland Congress*, Palmerston North, NZ: 2190–1

Scaife M A (1971) The long-term effects of fertilizers, farmyard manure and leys at Mwanhala, Western Tanzania. *East African Agricultural and Forestry Journal* **37**: 8–14

Scanlan J C (1984) Herbicidal control of woody weeds in central Queensland. 1. Brigalow (*Acacia harpophylla*). *Tropical Grasslands* **18**: 26–33

Scanlan J C, McKeon G M (1993) Competitive effects of trees on pasture are a function of rainfall distribution and soil depth. *Proceedings XVII International Grassland Congress*, Palmerston North, NZ: 2231–2

Schank S C, Day J M, Delgado de Lucas E (1977) Nitrogenose activity, nitrogen content, *in vitro* digestibility and yield of 30 tropical forage grasses in Brazil. *Tropical Agriculture Trinidad* **54**: 119–26

Schofield J L (1941) Introduced legumes in north Queensland. *Queensland Agricultural Journal* **56**: 378

Schroder V C (1970) Soil temperature effects on shoot and root growth of pangolagrass, slenderstem digitgrass, coastal bermudagrass, and Pensacola bahiagrass. *Proceedings Soil and Crop Science Society of Florida* **30**: 241

Schultze-Kraft R, Clements R J (eds) (1990) Centrosema: *Biology, agronomy, and utilisation*. Cali, Colombia, CIAT

Scott N M (1985) Sulphur in soils and plants. In Vaughan D, Malcolm R E (eds) *Soil organic matter and biological activity*. Dordrecht, Martinus Nijhoff/Dr W Junk. Developments in Plant and Soil Science **16**: 380–401

Seguin B, Gignoux N (1974) [Experimental study on the effect of shelterbelts on the vertical profile of wind speed]. *Agricultural Meteorology* **13**: 15–23

Serrão E A, Toledo, J M (1991) The search for sustainability in Amazonia pastures. In Anderson A B (ed.) *Steps towards sustainable use of the Amazon rain forest*. New York, Columbia University Press: 195–214

Serrão E A, Uhl C, Nepstad D (1993) Deforestation for pasture in the humid tropics: Is it economically and environmentally sound in the long term? *Proceedings XVII International Grassland Congress*, Palmerston North, NZ: 2215–21

Shamoot S, Donald L, Bartholemew V (1968) Rhizodeposition of organic debris in soil. *Soil Science Society America Proceedings* **32**: 817–20

Sharpley A N (1980) The enrichment of soil phosphorus in runoff sediments. *Journal of Environmental Quality* **9**: 521–6

Shaw N H (1978) Superphosphate and stocking rate effects on a native pasture oversown with *Stylosanthes humilis* in central coastal Queensland. 2. Animal production. *Australian Journal of Experimental Agriculture and Animal Husbandry* **18**: 800–7

Shea G M (1991) Towards the development of an agroforestry strategy in Queensland. Position paper, Brisbane, Queensland Forest Service, QDPI

Shekhawat J S, Sen N L, Somani L L (1988) Evaluation of agro-forestry systems under semi-arid conditions of Rajasthan. *Indian Forester* **114**: 98–101

Shelton H M (1980) Dry-season legume forages to follow paddy rice in northeast Thailand. 1. Species evaluation and effectiveness of native *Rhizobium* for nitrogen fixation. *Experimental Agriculture* **11**: 89–95

Shelton H M (1991) Productivity of cattle under coconuts. In Shelton H M, Stür W W (eds) *Forages for plantation crops*. Canberra, ACIAR Proceedings 32: 92–6

Shelton H M (1993) Environmental adaptation of forage tree legumes. In Gutteridge R C, Shelton H M (eds) *Forage tree legumes in tropical agriculture*. Wallingford, UK, CAB International: 120–31

Shelton H M, Humphreys L R (1972) Pasture establishment in upland rice crops at Na Pheng, Central Laos. *Tropical Grasslands* **6**: 223–8

Shelton H M, Humphreys L R (1975a) Undersowing of *Oryza sativa* with *Stylosanthes guyanensis*. I Plant density. *Experimental Agriculture* **11**: 89–95
Shelton H M, Humphreys L R (1975b) Undersowing of *Oryza sativa* with *Stylosanthes guyanensis*. II Delayed sowing time and crop variety. *Experimental Agriculture* **11**: 97–101
Shelton H M, Humphreys L R (1975c) Undersowing of *Oryza sativa* with *Stylosanthes guyanensis*. III Nitrogen supply. *Experimental Agriculture* **11**: 103–11
Shelton H M, Stür W W (eds) (1991) *Forages for plantation crops*. Canberra, ACIAR Proceedings 32
Shelton H M, Humphreys L R, Batello C (1987) Pastures in the plantations of Asia and the Pacific: performance and prospect. *Tropical Grasslands* **21**: 159–68
Shepherd J A, Coombs R F (1979) The effect of four *Meloidogyne* species (Nematoda: Meloidogynidae) on *Panicum maximum* cv. Umtali. *Rhodesia Journal of Agricultural Research* **17**: 155–6
Sheriff D W, Ludlow M M (1984) Physiological reactions to an imposed drought by *Macroptilium atropurpureum* and *Cenchrus ciliaris* in a mixed sward. *Australian Journal of Plant Physiology* **11**: 23–34
Siaw D E K A, Kang B T, Okali D U U (1991) Alley cropping with *Leucaena leucocephala* (Lam.) de Wit and *Acioa barteri* (Hook. f.) Engl. *Agroforestry Systems* **14**: 219–31
Silver B A (1987) Shade is important for milk production. *Queensland Agricultural Journal* **113**: 25–6
Simons A J, Dunsdon A J (1992) *Evaluation of the potential for genetic improvement of* Gliricidia sepium. Oxford, Oxford Forestry Institute
Simons A J, Stewart J L (1993) *Gliricidia sepium* – a multipurpose forage tree legume. In Gutteridge R C, Shelton H M (eds) *Forage tree legumes in tropical agriculture*. Wallingford, UK, CAB International
Simpson K (1983) *Soil*. London, Longman
Sinclair R, Wedge L, Romero A (1991) Utilización de rastrojos en la alimentación de animales. *Pastures Tropicales* **13**: 20–2
Singh D, Kohli R K (1992) Impact of *Eucalyptus tereticornis* shelter belt on crops. *Agroforestry Systems* **20**: 253–66
Singh N B, Singh P P, Nair K P P (1986) Effect of legume intercropping of enrichment of soil nitrogen, bacterial activity and productivity of associated maize crops. *Experimental Agriculture* **22**: 339–44
Singleton P W, Tavares J W (1986) Inoculation response of legumes in relation to the number and effectiveness of indigenous *Rhizobium* populations. *Applied and Environmental Microbiology* **51**: 1013–8
Sinha A K, Chatterjee B N (1966) Fertility building under grasses. *Journal of British Grassland Society* **21**: 153–61
Siregar M E, Semali A (1982) The effect of fodder legumes intercropped with corn on production. *Ilmu dan Peternakan* **1**: 35–8
Skerman P J (1977) *Tropical forage legumes*. Rome, FAO Plant Production and Protection Series No. 2
Skerman P J, Riveros F (1990) *Tropical grasses*. Rome, FAO
Skerman P J, Cameron D G, Riveros F (1988) *Tropical forage legumes*, 2nd edn. Rome, FAO

Skjemstad U O, Vallis I, Myers R J K (1988) Decomposition of soil organic nitrogen. In Wilson J R (ed.) *Advances in nitrogen cycling in agricultural ecosystems*. Wallingford, UK, CAB International: 134–44

Sless D (1986) *In search of semantics*. London, Croom Helm

Smith G W (1966) The relation between rainfall, soil water and yield of copra on a coconut estate in Trinidad. *Journal of Applied Ecology* **3**: 117–25

Smith M A, Whiteman P C (1983) Evaluation of tropical grasses in increasing shade under coconut canopies. *Experimental Agriculture* **19**: 153–61

Smith M A, Whiteman P C (1985) Animal production from rotationally-grazed natural and sown pastures under coconuts at three stocking rates in the Solomon Islands. *Journal of Agricultural Science* **104**: 173–80

Smith O B (1992a) Alley farming and protein banks for tropical Africa. In Speedy A, Pugliese P-L (eds) *Legume trees and other fodder trees as protein sources for livestock*. Rome, FAO Animal Production and Health Paper 102: 245–56

Smith O B (1992b) Fodder trees and shrubs in range and farming systems in tropical humid Africa. In Speedy A, Pugliese P-L (eds) *Legume trees and other fodder trees as protein sources for livestock*. Rome, FAO Animal Production and Health Paper 102: 43–59

Smith R L, Boulton J H, Schank S C, Quesenberry K H, Tyler M E, Milam J R, Gaskins M H, Littell R C (1976) Nitrogen fixation in grasses inoculated with *Spirillum lipoferum*. *Science* **193**: 1003–5

Smith R L, Schank S C, Milam J R, Baltensperger A A (1984) Responses of *Sorghum* and *Pennisetum* species to the N_2-fixing bacterium *Azospirillum brasilense*. *Applied and Environmental Microbiology* **47**: 1331–6

Smith R M, Stamey W L (1965) Determining the range of tolerable erosion. *Soil Science* **100**: 414–24

So H B, Cook G D, Dalal R C (1988) Structural degradation of vertisols associated with continuous cultivation. *International Soil Tillage Research Organisation, XI International Conference*, Edinburgh, 1988: 123–8

Sollenberger L E, Jones Jr C S (1989) Beef production from nitrogen-fertilized Mott Dwarf Elephantgrass and Pensacola Bahiagrass pastures. *Tropical Grasslands* **23**: 129–34

Sophanodora P (1989) Productivity and nitrogen nutrition of some tropical pasture species under low radiation environments. PhD thesis, University of Queensland

Sophanodora P (1991) Compatibility of grass–legume swards under shade. In Shelton H M, Stür W W (eds) *Forages for plantation crops*. Canberra, ACIAR Proceedings 32: 117–19

Sophanodora P, Tudsri S (1991) Integration of forages for cattle and goats into plantation systems in Thailand. In Shelton H M, Stür W W (eds) *Forages for plantation crops*. Canberra, ACIAR Proceedings 32: 147–50

Sparling G P (1985) The soil biomass. In Vaughan D, Malcolm R E (eds) *Soil organic matter and biological activity*. Dordrecht, Martinus Nijhoff/Dr W Junk. Developments in Plant and Soil Science **16**: 223–62·

Spedding C R W (1971) *Grassland ecology*. London, Oxford University Press

Speedy A, Pugliese P-L (1992) *Legume trees and other fodder trees as protein sources for livestock*. Rome, FAO Animal Production and Health Paper 102

Ssekabembe C (1985) Perspectives of hedgerow intercropping. *Agroforestry Systems* **3**: 339–56
Stace H M, Edye L A (eds) (1984) *The biology and agronomy of* Stylosanthes. Sydney, Academic Press
Standing Committee on Agriculture (1991) *Sustainable agriculture*. Melbourne, CSIRO, Australian Agricultural Council, Working Group on Sustainable Agriculture, SCA technical report series 36
Steel R J H, Humphreys L R (1974) Growth and phosphorus response of some pasture legumes sown under coconuts in Bali. *Tropical Grasslands* **8**: 171–8
Steele K W, Vallis I (1988) The nitrogen cycle in pastures. In Wilson J R (ed.) *Advances in nitrogen cycling in agricultural ecosystems*. Wallingford, UK, CAB International: 274–91
Stephens D (1967) Effects of grass fallow treatments in restoring fertility of Bugarda clay loam in South Uganda. *Journal of Agricultural Science* **68**: 391–403
Stevenson F J (1982) *Humus chemistry – genesis, composition, reactions*. New York, John Wiley and Sons
Stobbs T H (1969a) The value of *Centrosema pubescens* (Benth.) for increasing animal production and improving soil fertility in northern Uganda. *East African Agricultural and Forestry Journal* **35**: 197–202
Stobbs T H (1969b) The effect of grazing resting land upon subsequent arable crop yields. *East African Agricultural and Forestry Journal* **35**: 28–32
Stocking M A, Elwell H A (1973a) Soil erosion hazard in Rhodesia. *Rhodesian Agricultural Journal* **70**: 93–101
Stocking M A, Elwell H A (1973b) Prediction of subtropical storm losses from field plant studies. *Agricultural Meteorology* **12**: 193–201
Stolzy L H (1972) Soil aeration and gas exchange in relation to grasses. In Youngner V B, McKell C M (eds) *The biology and utilization of grasses*. New York, Academic Press: 251
Stoop W A (1981) Cereal-based intercropping systems for the West African semi-arid tropics, particularly Upper Volta. In *Proceedings of the international workshop on intercropping, Hyderabad, India, 10–13 January 1979*. Patancheru, Hyderabad, India, ICRISAT: 61–8
Stür W W, Humphreys L R (1987) Seed production in *Brachiaria decumbens* and *Paspalum plicatulum* as influenced by system of residue disposal. *Australian Journal of Agricultural Research* **38**: 869–80
Stür W W, Shelton H M (1991a) Response of forage resources in plantation crops of southeast Asia and the Pacific. In Shelton H M, Stür W W (eds) *Forages for plantation crops*. Canberra, ACIAR Proceedings 32: 25–31
Stür W W, Shelton H M (1991b) Compatibility of forages and livestock with plantation crops. In Shelton H M, Stür W W (eds) *Forages for plantation crops*. Canberra, ACIAR Proceedings 32: 112–16
Stür W W, Shelton H M, Gutteridge R C (1993) Defoliation management of forage tree legumes. In Gutteridge R C, Shelton H M (eds) *Forage tree legumes in tropical agriculture*. Wallingford, UK, CAB International: 158–67
Sturmheit P (1990) Agroforestry and soil conservation needs of smallholders in southern Zambia. *Agroforestry Systems* **10**: 265–89

Stuth J W, Hamilton W T, Conner J C, Sheehy DP (1993) Decision support systems (DSS) in the transfer of grassland technology. In *Proceedings XVII International Grassland Congress*, Palmerston North, NZ: 749–57

Sugimoto Y, Hirata M, Ueno M (1987) Energy and matter flows in bahiagrass pasture. V Excreting behaviour of Holstein heifers. *Journal of Japanese Society of Grassland Science* **32**: 8–14

Sumberg J E (1983) 'Leuca-fence': a living fence for sheep using *Leucaena leucocephala*. *ILCA Newsletter* (2): 5

Sumberg J E, Atta-Krah A N (1988) The potential of alley farming in humid West Africa: A re-evaluation. *Agroforestry Systems* **6**: 163–68

Swasdiphanich S (1993) Environmental influences on forage yields of shrub legumes. PhD thesis, University of Queensland

Swift M J, Sanchez P A (1984) Biological management of tropical soil fertility for sustained productivity. *Nature and Resources* **20**: 2–10

Swift M J, Heal O W, Anderson J M (1979) *Decomposition in terrestrial ecosystems*. Oxford, Blackwell Scientific Publications

Swift M J, Frost P G H, Campbell B M, Hatton J C, Wilson K B (1989) Nitrogen cycling in farming systems derived from savanna: perspectives and challenges. *Plant and Soil* **115**: 63–76

Szott L T, Palm C A, Sanchez P A (1991) Agroforestry in acid soils of the humid tropics. *Advances in Agronomy* **45**: 275–301

Talineau J C, Hainnaux G, Bonzon B, Fillonneau C, Picard D, Sicot M (1976) [Agronomic aspects of the inclusion of a forage crop in a crop rotation under humid tropical conditions in the Ivory Coast]. *Cahiers ORSTOM, Biologie* **11**: 277–90

Tang M, Menéndez J (1988) Evaluation of *Rhizobium* strains in undisturbed soil cores of three tropical legumes. *Pastos y Forrajes* **11**: 37–42

Tanner P D, Mugwira L (1984) Effectiveness of communal area manures as sources of nutrients for young maize plants. *Zimbabwe Agricultural Journal* **81**: 31–5

Tarawali, G (1991) The residual effect of *Stylosanthes* fodder banks on maize yield at several locations in Nigeria. *Tropical Grasslands* **25**: 26–31

Tate K R (1985) Soil phosphorus. In Vaughan D, Malcolm R E (eds) *Soil organic matter and biological activity*. Dordrecht, Martinus Nijhoff/Dr W Junk. Developments in Plant and Soil Science **16**: 330–77

Taylorson R B, Borthwick H A (1969) Light filtration by foliar canopies: significance for light-controlled weed seed germination. *Weed Science* **17**: 48–51

Teitzel J K (1992) Sustainable pasture systems in the humid tropics of Queensland. *Tropical Grasslands* **26**: 196–205

Thomas D (1978). Pastures and livestock under tree crops in the humid tropics. *Tropical Agriculture (Trinidad)* **55**: 39–44

Thomas D, Bennett A J (1975a) Establishing a mixed pasture under maize in Malawi. I Time of sowing. *Experimental Agriculture* **11**: 257–63

Thomas D, Bennett A J (1975b) Establishing a mixed pasture under maize in Malawi. II Method of sowing. *Experimental Agriculture* **11**: 273–6

Thomas D, Addy B L (1977) Stall-fed beef production in Malawi. *World Review of Animal Production* **13**: 23–30

Thomas D, Schulze-Kraft R (1990) Evaluation of five shrubby legumes in comparison with *Centrosema acutifolium* Carimagua, Colombia. *Tropical Grasslands* **24**: 87–92

Thomas R, Humphreys L R (1970) Pasture improvement at Na Pheng, Central Laos. *Tropical Grasslands* **4**: 229–36

Thomas R J (1992) The role of the legume in the nitrogen cycle of productive and sustainable pastures. *Grass and Forage Science* **47**: 133–42

Thompson J P (1987) Decline of vesicular-arbuscular mycorrhizae in long fallow disorder of field crops and its expression in phosphorus deficiency of sunflower. *Australian Journal of Agricultural Research* **38**: 846–62

Thompson J P (1991) How does organic farming perform in relation to soil biology? In Thompson J P, Thomas G A (eds) *Organic farming in field crop production*. Brisbane, Queensland Department of Primary Industries Conference and Workshop Series QC91001: 23–30

Thompson J P, Mackenzie J P, Amos R (1987) Fallow management for winter cereal production – Results from the Hermitage trial 1968–87. In Coughlan K J, Truong P N (eds) *Effects of management practices on soil physical properties*. Brisbane, Queensland Department of Primary Industries Conference and Workshop Series QC87006: 218–9

Thorburn P J, Cowie B A, Lawrence P A (1991) Effect of land development on groundwater recharge determined from non-steady chloride profiles. *Journal of Hydrology* **124**: 43–58

Tisdall J M, Oades J M (1982) Organic matter and water-stable aggregates in soils. *Journal of Soil Science* **33**: 141–63

Toky O P, Ramakrishnan P S (1981) Cropping and yields in agricultural systems of the north-eastern hill region of India. *Agro-Ecosystems* **7**: 11–25

Toledo J M (1985) Pasture development for cattle production in the major ecosystems of the tropical American lowlands. In *Proceedings XV International Grassland Congress*: 74–81

Toledo J M, Formosa D (1993) Sustainability of sown pastures in the tropics and subtropics. In *Proceedings of the XVII International Grassland Congress*, Palmerston North, NZ: 1891–6

Toledo J M, Serrão E A (1982) Pasture and animal production in Amazonia. In Hecht G (ed.) *Amazonia – agriculture and land use research*. Cali, Columbia, CIAT: 281–309

Torres F (1983) Role of woody perennials in animal agroforestry. *Agroforestry Systems* **1**: 131–63

Tothill J C, Dzowela B H, Diallo A K (1989) Present and future role of grasslands in inter-tropical countries with special reference to ecological and sociological constraints. *Proceedings XVI International Grassland Congress*. Versailles, Association Française pour la Production Fourragère **3**: 1719–24

Traoré N'G, Breman H (1993) New wine in old bags? External inputs could save pastoralism. *ILEIA Newsletter* **9**(1): 10–12

Turenne J F (1977) Culture itinérante et jachère forestière en Guyane. Évolution de la matière organique. *Cahiers ORSTOM, Pédologie* **15**: 449–61

Uhl C, Murphy P (1981) A comparison of productivities and energy values between slash and burn agriculture and secondary succession in the upper Rio Negro region of the Amazon Basin. *Agro-Ecosystems* **7**: 63–83

Umali-Garcia M, Hubbell D H, Gaskins M H, Dazzo F B (1980) Association of *Azospirillum* with grass roots. *Applied and Environmental Microbiology* **39**: 219–26

Uren N (1991) The management of soil organic matter for sustainable agriculture. *Agricultural Science NS* **4**: 45–8

Vallis I (1972) Soil nitrogen changes under continuously grazed legume–grass pastures in subtropical coastal Queensland. *Australian Journal of Experimental Agriculture and Animal Husbandry* **12**: 495–501

Vallis I (1979) The effect of climatic, edaphic and animal factors on nitrogen transformations in permanent pastures. In *Proceedings Workshop on Nitrogen Relations in Pasture Systems of Southern Queensland*. Brisbane, Queensland Department of Primary Industries: 3–26

Vallis I, Gardener C J (1984) Short-term nitrogen balance in urine-treated areas of pasture on a yellow earth in the subhumid tropics of Queensland. *Australian Journal of Experimental Agriculture and Animal Husbandry* **24**: 522–8

Vallis I, Jones R J (1973) Net mineralisation of nitrogen in leaves and leaf litter of *Desmodium intortum* and *Phaseolus atropurpureus* mixed with soil. *Soil Biology and Biochemistry* **5**: 391–8

Vallis I, Henzell E F, Evans T R (1977) Uptake of soil nitrogen by legumes in mixed swards. *Australian Journal of Agricultural Research* **28**: 413–25

Vallis I, Peake (the late) D C I, Jones R K, McCown R L (1985) Fate of urea-nitrogen from cattle urine in a pasture–crop sequence in a seasonally dry tropical environment. *Australian Journal of Agricultural Research* **36**: 809–17

Vandermeer J (1989) *The ecology of intercropping*. Cambridge, Cambridge University Press

Van Veen J A, Ladd J N, Frissel M J (1984) Modelling C and N turnover through the microbial biomass in soil. *Plant and Soil* **76**: 257–74

Vargas M A T, Suhet A R (1981) Efficiency of commercial inoculants and native *Rhizobium* strains on six forage legumes on cerrado soil. *Pesquisa Agropecuária Brasileira* **16**: 357–62

Vaughan D, Malcolm R E (1985) Influence of humic substances on growth and physiological processes. In Vaughan D, Malcolm R E (eds) *Soil organic matter and biological activity*. Dordrecht, Martinus Nijhoff/Dr W Junk. Developments in Plant and Soil Science **16**: 37–75

Vaughan D, Ord B G (1985) Soil organic matter – a perspective of its nature, extraction, turnover and role in soil fertility. In Vaughan D, Malcolm R E (eds) *Soil organic matter and biological activity*. Dordrecht, Martinus Nijhoff/Dr W Junk. Developments in Plant and Soil Science **16**: 1–35

Velásquez G J A, González J E (1972) The nutritive value of groundnut (*Arachis hypogea*) straw. *Agronomía Tropical* **22**: 287–90

Vergara N T, Nair P K R (1985) Agroforestry in the South Pacific region – an overview. *Agroforestry Systems* **3**: 363–79

Vieweg B, Wilms W (1974) Problems associated with a change from shifting to permanent cultivation on a light soil in the Kilombero Valley, Tanzania. *Soils Bulletin, FAO* No. 24: 228–9

Von Carlowitz P G, Wolf G V (1991) Potential and limitations of natural repellents against early destructive browsing by livestock and game. *Agroforestry Systems* **16**: 33–40

Waidyanatha U P de S, Wijesinghe D S, Stauss R (1984) Zero-grazed pasture under immature *Hevea* rubber: productivity of some grasses and grass–legume mixtures and their competition with *Hevea*. *Tropical Grasslands* **18**: 21–6

Walker B (1980) Effects of stocking rate on perennial tropical legume grass pastures. PhD thesis, University of Queensland

Walker B, Weston E J (1990) Pasture development in Queensland – a success story. *Tropical Grasslands* **24**: 257–68

Walker B H (1993) Stability in rangelands: ecology and economics. *Proceedings XVII International Grassland Congress*, Palmerston North, NZ: 1885–90

Wanapat M (1990) *Nutritional aspects of ruminant production in southeast Asia with special reference to Thailand*. Khon Kaen, Department of Animal Science, Khon Kaen University

Wandera F P, Dzowela B H, Karachi M K (1991) Production and nutritive value of browse species in semi-arid Kenya. *Tropical Grasslands* **25**: 349–55

Wannawony S, Belt G H, McKetta C W (1991) Benefit–cost analysis of selected agroforestry systems in Northeastern Thailand. *Agroforestry Systems* **16**: 83–94

Warembourg F R, Roumet C (1989) Why and how to estimate the cost of symbiotic N_2 fixation. A progressive approach based on the use of ^{14}C and ^{15}N isotopes. In Clarholm M, Bergström (eds) *Ecology of arable land*. Dordrecht, Kluwer Academic Publishers: 31–41

Waring S A (1991) Nutrients from organic matter. In Thompson J P, Thomas G A (eds) *Organic farming in field crop production*. Brisbane, Queensland Department of Primary Industries Conference and Workshop Series QC91001: 1–8

Watson G A (1983) Development of mixed tree and food crop systems in the humid tropics: a response to population pressure and deforestation. *Experimental Agriculture* **19**: 311–32

Watson G A, Wong P W, Narayanan R (1964a) Effect of cover plants on soil nutrient status and on growth of *Hevea*. III A comparison of leguminous creepers with grasses and *Mikania cordata*. *Journal of Rubber Research Institute of Malaya* **18**: 80–95

Watson G A, Wong P W, Narayanan R (1964b) Effects of cover plants on soil nutrient status and growth of *Hevea*. IV Leguminous creepers compared with grasses, *Mikania cordata* and mixed indigenous covers on four soil types. *Journal of Rubber Research Institute of Malaya* **18**: 123–34

Watson S E, Whiteman P C (1981) Animal production from naturalized and sown pastures at three stocking rates under coconuts in the Solomon Islands. *Journal of Agricultural Science* **97**: 669–76

Weaver P (1980) Agri-silviculture in tropical America. *Unasylva* **31**(126): 2–12

Weaver R W, Wright S F, Varanka M W, Smith O E, Holt E C (1980)

Dinitrogen fixation (C_2H_2) by established forage grasses in Texas. *Agronomy Journal* **72**: 965–8

Weier K L, MacRae I C, Whittle J (1981) Seasonal variation in the nitrogenase activity of a *Panicum maximum* var. *trichoglume* pasture and identification of associated bacteria. *Plant and Soil* **63**: 189–97

Wetselaar R (1967) Estimation of nitrogen fixation by four legumes in a dry monsoonal area of north-western Australia. *Australian Journal of Experimental Agriculture and Animal Husbandry* **7**: 518–22

Whitehouse M J, Littler J W (1984) Effect of pasture on subsequent wheat crops on a black earth soil of the Darling Downs. II Organic C, nitrogen and pH changes. *Queensland Journal of Agricultural and Animal Sciences* **41**: 31–40

Whiteman P C (1970a) Seasonal changes in growth and nodulation of perennial tropical pasture legumes in the field. II Effects of controlled defoliation levels on nodulation of *Desmodium intortum* and *Phaseolus atropurpureus*. *Australian Journal of Agricultural Research* **21**: 207–14

Whiteman P C (1970b) Seasonal changes in growth and nodulation of perennial tropical pasture legumes in the field. III Effects of flowering on nodulation of three *Desmodium* species. *Australian Journal of Agricultural Research* **21**: 215–22

Whiteman P C (1972) The effects of inoculation and nitrogen application on seedling growth and nodulation of *Glycine wightii* and *Phaseolus atropurpureus* in the field. *Tropical Grasslands* **6**: 11–6

Whiteman P C, Seitlhekom, Siregar M E, Chudasama A K, Javier R R (1984) Short-term flooding tolerance of seventeen commercial tropical pasture legumes. *Tropical Grasslands* **18**: 91–6

Whitney A S, Green R E (1969) Legume contributions to yields and compositions of *Desmodium* spp.–pangolagrass mixtures. *Agronomy Journal* **61**: 741–6

Wieder R K, Lang G E (1982) A critique of the analytical methods used in examining decomposition data from litter bags. *Ecology* **63**: 1636–42

Wiersum K F (1982) Tree gardening and taungya on Java: examples of agroforestry techniques in the humid tropics. *Agroforestry Systems* **1**: 53–70

Wiersum K F (1985) Trees in agricultural and livestock development. *Netherlands Journal of Agricultural Science* **33**: 105–14

Wilaipon B, Gutteridge R C, Chutikul K (1981) Undersowing upland crops with pasture legumes. 1. Cassava with *Stylosanthes hamata* cv. Verano. *Thai Journal of Agricultural Science* **14**: 333–7

Wildin J H (1993) Beef production from broadacre leucaena in central Queensland. In Gutteridge R C, Shelton H M (eds) *Forage tree legumes in tropical agriculture*. Wallingford, UK, CAB International: 352–6

Wilkins R J (1982) Improving forage quality by processing. In Hacker J B (ed.) *Nutritional limits to animal production from pastures*. Farnham Royal, UK, Commonwealth Agricultural Bureaux: 389–408

Wilkinson G E (1975a) Rainfall characteristics and soil erosion in the rainforest area of western Nigeria. *Experimental Agriculture* **11**: 247–55

Wilkinson G E (1975b) Effect of grass fallow rotations on the infiltration of water into a savanna zone soil of Northern Nigeria. *Tropical Agriculture* **52**: 97–103

Wilkinson G E, Aina P O (1976) Infiltration of water into two Nigerian soils under secondary forest and subsequent arable cropping. *Geoderma* **15**: 51–9

Willey R W, Singh R P, Reddy M S (1989) Cropping systems for Vertisols in different rainfall regimes in the semi-arid tropics. In *Management of Vertisols for improved agricultural production*. Patancheru, India, ICRISAT: 119–31

Williams J (1991) Search for sustainability: agriculture and its place in the natural ecosystem. *Agricultural Science NS* **4**: 32–9

Williams J, Chartres C J (1991) Sustaining productive pastures in the tropics. 1. Managing the soil resource. *Tropical Grasslands* **25**: 73–84

Wilson J R, Ludlow M M (1991) The environment and potential growth of herbage under plantations. In Shelton H M, Stür W W (eds) *Forages for plantation crops*. Canberra, ACIAR Proceedings 32: 10–24

Wilson J R, Mannetje L't (1978) Senescence, digestibility and carbohydrate content of buffel grass and panic leaves in swards. *Australian Journal of Agricultural Research* **29**: 503–16

Wilson J R, Wild D W M (1991) Improvement of nitrogen nutrition and grass growth under shading. In Shelton H M, Stür W W (eds) *Forages for plantation crops*. Canberra, ACIAR Proceedings 32: 77–82

Wilson J R, Wong C C (1982) Effects of shade on some factors influencing nutritive quality of green panic and Siratro pastures. *Australian Journal of Agricultural Research* **33**: 937–49

Wilson J R, Catchpoole V R, Weir K L (1986) Stimulation of growth and nitrogen uptake by shading a rundown green panic pasture on brigalow clay soil. *Tropical Grasslands* **20**: 134–43

Wilson J R, Hill K, Cameron D M, Shelton H M (1990) The growth of *Paspalum notatum* under the shade of a *Eucalyptus grandis* plantation canopy or in full sun. *Tropical Grasslands* **24**: 24–8

Wilson R T (1989a) Livestock production in central Mali: economic characters and productivity indices for Sudanese Fulani cattle in the agro-pastoral system. *Tropical Agriculture* **66**: 49–53

Wilson R T (1989b) Reproductive performance of African indigenous small ruminants under various management systems: a review. *Animal Reproduction Science* **20**: 265–86

Wischmeier W H, Smith D D (1958) Rainfall energy and its relationship to soil loss. *Transactions, American Geophysical Union* **39**: 285–91

Wischmeier W H, Smith D D (1978) *Predicting rainfall erosion losses – a guide to conservation planning*. Agricultural Handbook 537, SEA. Washington DC, USDA

Wischmeier W H, Johnson C B, Cross B V (1971) A soil erodibility monograph for farmland and construction sites. *Journal Soil and Water Conservation* **26**: 189–93

Wong C C (1991) Shade tolerance of tropical forages: a review. In Shelton H M, Stür W W (eds) *Forages for plantation crops*. Canberra, ACIAR Proceedings 32: 64–9

Wong C C (1993) Growth and persistence of two *Paspalum* species to defoliation in shade. PhD thesis, University of Queensland

Wong C C, Wilson J R (1980) Effects of shading on the growth and nitrogen content of green panic and Siratro in pure and mixed swards defoliated at two frequencies. *Australian Journal of Agricultural Research* **31**: 269–85

Wong C C, Sharudin M A MohD, Rahim H (1985a) Shade tolerance potential of some tropical forages for integration with plantations. 2. Legumes. *MARDI Research Bulletin* **13**: 249–69

Wong C C, Rahim H, Sharudin M A MohD (1985b) Shade tolerance potential of some tropical forages for integration with plantations. 1. Grasses. *MARDI Research Bulletin* **13**: 225–47

Wong P W (1964) Evidence for presence of growth inhibitory substances in *Mikania cordata* (Barm. f) B.L. Robinson. *Journal of Rubber Research Institute of Malaysia* **18**: 231–42

Wood P J (1990) The scope and potential of agroforestry. *Outlook on Agriculture* **19**: 141–6

Woodmansee R G, Vallis I, Mott J J (1981) Grassland nitrogen. In Clark F C, Rosswall T (eds) *Terrestrial nitrogen cycles: Processes, ecosystem strategies and management impacts*. Stockholm, Natural Sciences Research Council: 443–62

Wooley J N, Rodriguez W (1987) Cultivar × cropping system interactions in relay and row intercropping of bush beans with different maize plant types. *Experimental Agriculture* **23**: 181–92

World Commission on Environment and Development (1987) *Our common future*. Oxford, Oxford University Press

Wu Y J, Rin U C (1976) [Study on *Paulownia* tree plantation mixed with forage culture and fertilization]. Bulletin, *Taiwan Forestry Research Institute* 294

Yaacob O, Blair G J (1980) Mineralization of ^{15}N-labelled legume residues in soils with different nitrogen contents and its uptake by Rhodes grass. *Plant and Soil* **57**: 237–48

Yaacob O, Blair G J (1981) Effect of legume cropping and organic matter accumulation on the infiltration rate and structural stability of a granite soil under a simulated tropical environment. *Plant and Soil* **60**: 11–20

Yamoah C (1991) Choosing suitable intercrops prior to pruning *Sesbania* hedgerows in an alley configuration. *Agroforestry Systems* **13**: 87–94

Yamoah C F, Agboola A A, Mulongoy K (1986) Decomposition, nitrogen release and weed control by prunings of selected alley cropping shrubs. *Agroforestry Systems* **4**: 239–46

Yayock J Y (1981) Crops and cropping patterns of the Savanna region of Nigeria: the Kaduna situation. In *Proceedings of the international workshop on intercropping, Hyderabad, India, 10–13 January 1979*. Patancheru, Hyderabad, India, ICRISAT: 69–77

Yazman J A, McDowell R E, Cestero H, Arroyo Aguilú J A, Rivera Anaya J D, Soldevila M, Román Garcia F (1982) Efficiency of utilization of tropical grass pastures by lactating cows with or without a supplement. *Journal of Agriculture of the University of Puerto Rico* **66**: 200–22

Young A (1987) Soil productivity, soil conservation and land evaluation. *Agroforestry Systems* **5**: 277–91

Young A (1988) Agroforestry and its potential to contribute to land development in the tropics. *Journal of Biogeography* **15**: 19–30

Young A (1989) *Agroforestry for soil conservation*. Wallingford, UK, CAB International

Young R A, Wiersma J L (1973) The role of rainfall impact in soil detachment and transport. *Water Resources Research* **9**: 1629–36

Yusran M A, Yudi P A (1991) Draught animal work practices in two agro-ecosystems and two land ownership classes (of farmers) in east Java. *Draught Animal Bulletin* (1): 1–8

Zandstra H G (1979) Cropping systems research for the Asian rice farmer. *Agricultural Systems* **4**: 135–53

Zinke P J (1962) The pattern of influence of individual forest trees on soil properties. *Ecology* **43**: 130–3

INDEX OF SCIENTIFIC NAMES

Acacia, 224, 265
  aneura, 13, 272
  angustissima, 260
  cunninghamii, 251
  fimbriata, 251
  harpophylla, 47, 118, 135, 226
  mangium, 66, 261
  nilotica, 226
  tortillis, 196–7, 226
  villosa, 272
Acanthomia, 137
Acetobacter nitrocaptans, 60
Acioa barteri, 199–200, 235
Acroceras macrum, 90
Adonsonia digitata, 196–7
Aeschynomene americana, 90
Agathis robusta, 194
Albizia
  chinensis, 231, 260, 272
  fulcataria, 66, 237, 261
  lebbek, 261–2, 272
  saman, 262
Alchornia cordifolia, 199–200
Alysicarpus vaginalis, 85, 188, 293, 304, 310, 321
Amnemus quadrituberculatus, 309
Anaebena azollae, 58–9
Anacardium occidentale, 221
Andropogon gayanus, 46, 146, 289–90, 348
Aracauria cunninghamii, 194
Arachis, 65, 190
  hypogea, 325, 328
  pintoi, 45–6, 53, 70, 109, 189, 222, 231, 309
Aristida armata, 13–14

Arundinaria
  ciliata, 114
  fusilla, 114
Asystasia
  intrusa, 188, 210, 218
  gangetica, 222
Avena
  fatua, 134
  ludoviciana, 134
Axonopus
  affinis, 213, 233
  compressus, 174, 185, 186, 187, 210
Azolla, 58–9
  caroliniana, 59
  microphylla, 59
  pinnata, 59
Azospirillum, 32, 60
  amazonense, 60
  brasiliense, 60
  halopraeferans, 60
  lipoferum, 60

Bacillus azotofixans, 60
Bertholletia excelsa, 222
Borreria latifolia, 210
Bothriochloa pertusa, 97–8
Brachiaria
  brizantha, 186, 189, 201, 211
  decumbens, 46, 53, 60, 91, 109, 173, 179, 185, 186, 187, 189, 190, 210, 212, 217, 220, 223, 227, 278, 281, 303–4, 314
  dictyoneura, 19–20, 46, 146
  humidicola, 24, 35, 46, 60, 173, 185, 186, 187, 189, 190, 210, 223

lata, 222
miliiformis, 114, 182, 186, 189, 201, 212, 218, 263
mutica, 90, 186, 210, 223
ruziziensis, 1, 181, 306–7
Bradyrhizobium, 4, 53, 56
Brassica
 campestris, 205
 oleracea, 205
Bromus unioloides, 134, 287

Cajanus cajan, 11, 81, 84, 89, 108, 239–40, 260–1, 330–4, 343
Calliandra calothyrsus, 28, 234, 239, 251–2, 261–3, 266, 268–70, 271–4
Calopogonium
 caeruleum, 186, 187, 188, 218, 230
 mucunoides, 186, 187, 188, 210
Campylobacter nitrofigilis, 60
Canavalia ensiformis, 328–9
Carissa ovata, 226
Cassia
 siamea, 199–200, 240, 247, 252, 270–1
 spectabilis, 237, 247
Casuarina, 247
 cunninghamii, 58–9
 equisetifolia, 58
Cenchrus ciliaris, 2, 13–14, 47, 85, 118, 309–10
Centrosema, 308
 acutifolium, 46, 146, 309, 348
 macrocarpum, 146
 pascuorum, 293, 304, 310, 321
 pubescens, 57, 61, 178, 186, 187, 188, 193, 201, 210, 217, 222, 223, 230, 303–4, 306–7, 309
Chamaecytisus palmensis, 261, 271–2
Chloris gayana, 51–2, 136, 224, 294, 309, 320, 328, 347
Chromolaena odorata, 210
Chrysopogon aciculatus, 114
Cicer arietinum, 205
Clidemia hirta, 210
Cocos nucifera, 216
Codariocalyx gyroides, 272
Coffea arabica, 221
Colletotrichum gloeosporioides, 139, 275, 309
Colocasia esculenta, 239
Cordia alliodora, 221
Crotalaria, 221, 309, 328, 345
Cyamopsis tetragonoloba, 324
Cymbopogon confertiflorus, 38, 221
Cynodon
 dactylon, 160
 nlemfuensis, 25, 27, 279, 288–9, 303–4, 320
 plectostachyus, 186, 278
Cyperus, 112
Cyrtococcum oxyphyllum, 210, 219

Dactyloctenium aegyptium, 114
Desmanthus virgatus, 62, 109, 272
Desmodium
 canum, 188, 194
 heterophyllum, 186, 187
 intortum, 65, 71, 90, 112, 186, 189, 193, 221, 224, 230, 282, 286–7, 309–10
 ovalifolium, 19–20, 46, 186, 189, 220, 221
 strigillosum, 274
 triflorum, 186, 188
 uncinatum, 71, 194, 309
Dicranopteris lincaris, 218
Digitaria
 abscendens, 114
 decumbens, 147, 160, 186
 didactyla, 213
 gibbosa, 85
 milanjiana, 137
 pentzii, 186
 setivalva, 186
Diplodocus, 229

Echinochloa
 polystachya, 90
 pyramidalis, 90
 stagnina, 90
Elaeis guineensis, 219
Enterolobium cyclocarpum, 262
Entolasia embricata, 90
Eragrostis
 curvula, 38
 viscosa, 214
Eremophila gilesii, 226
Erysiphae polygoni, 140
Erythrina, 260
 edulis, 261
 glauca, 261
 poeppigiana, 221, 250, 254, 261
Eucalyptus, 117, 172, 199, 205
 camaldulensis, 211, 244
 degupta, 223
 fibrosa, 224
 grandis, 192, 211, 213
 intermedia, 224

maculata, 224
melanophloia, 2
siderophloia, 224
tereticornis, 204–5, 224

Faidherbia albida, 196–8, 233, 272
Flemingia macrophylla (syn. *congesta*), 186, 199, 237, 241, 246–7, 250, 260, 272
Frankia, 58, 247

Gigaspora margarita, 66
Gliricidia sepium, 80, 125, 199, 218, 221, 231, 234, 237, 238–9, 242–4, 246–7, 250–1, 254, 260–2, 266–70, 271–4
Glomus fasciculatum, 66
Glycine max, 51, 61, 71, 88
Gmelina arborea, 199–200

Helminthosporum turcicum, 139
Herbaspirillum seropedicae, 60
Heteropogon contortus, 2, 97–8
Heteropsylla cubana, 140, 240, 251, 260, 271, 275
Hevea brasiliensis, 218
Hibiscus cannabinus, 237
Hordeum vulgare, 28
Hymenachae amplexicaulis, 90
Hyparrhenia rufa, 320

Imperata cylindrica, 27, 186, 187, 210, 224, 241, 281–2
Ischaemum aristatum, 186, 187, 189, 223

Lablab purpureus, 136, 228, 329, 345
Lantana camara, 210
Leersia hexandra, 90
Lens esculentum, 205
Leucaena, 218
  diversifolia, 252, 271–2
  leucocephala, 4, 11, 56, 57, 61, 66, 67, 70, 80, 108, 125, 140, 147, 186, 189, 193, 199, 211, 221, 224, 230, 231, 235–40, 242–4, 246–58, 260–75
  pallida, 252, 275
  pulverulenta, 260, 270
Lotononis bainesii, 5, 61, 65, 70, 194–5, 230, 309
Lotus

major, 309
pedunculatus, 90

Macadamia integrifolia, 222
Macroptilium
  atropurpureum, 2, 51, 65, 71, 88, 120, 136, 138, 177, 186, 194, 230, 285, 295, 309–10
  lathyroides, 56, 70–1, 90
Macrotyloma
  africanum, 62
  axillaris, 224
Manihot esculenta, 325
Maruca testalis, 138
Medicago, 278
  sativa, 134, 285–6, 294–5, 312
  scutellata, 295
Melastoma malabathricum, 210
Melinis minutiflora, 136, 223, 306–7, 347
Meloidogyne, 136–7, 293
Mikania
  cordata, 188, 199, 210
  micrantha, 210, 218
Millsonia anomala, 32, 297
Mimosa pudica, 186, 187, 210
Myoporum desertii, 226

Neonotonia wightii, 57, 65, 120, 139, 145, 230, 306–7, 309
Nephrolepis biserrata, 218
Nezara viridula, 137
Nitrobacter, 49
Nitrosomonas, 49

Ootheca bennigseni, 137
Oryctes rhinoceros, 203
Ottochloa nodosa, 185, 210, 218

Panicum
  coloratum, 319–20
  maximum, 38, 126, 136, 178, 179, 186, 187, 189, 201, 209, 212, 220, 221, 278, 306–7, 314
  maximum var. *trichoglume*, 17–9, 39, 71, 136, 145, 177, 191, 192, 223, 224, 310
  repens, 113, 320
Paspalum, 109
  commersonii, 136
  conjugatum, 174, 185, 186, 187, 210, 223

malacophyllum, 179, 190, 210
notatum, 160, 192, 213, 303–4
pectinatum, 2
plicatulum, 91, 136
wettsteinii, 179, 190, 210
Pennisetum
  clandestinum, 190
  glaucum, 254–8
  purpureum, 147, 186, 224, 279
Phaseolus vulgaris, 138
Pinus
  caribaea var. *hondurensis*, 194, 223–4
  elliottii, 194
  kesiya, 211
Prosopis, 273
  cineraria, 226, 261
  juliflora, 261
  tamarugo, 261
Psidium guajava, 210
Psophocarpus palustris, 33–4, 303–5
Pseudomonas, 60
  phaseolicola, 138
Pueraria phaseoloides, 25, 27, 46, 70, 183, 186, 187, 188, 201, 205, 210, 218, 222, 223, 230, 281, 303–4, 306–7, 309
Pyrenophera tritici-repentis, 139

Radopholus similis, 136
Rhizobium, 4, 56, 67, 309–10

Sehima nervosum, 78
Sesbania, 56, 261–2
  grandiflora, 261–2, 269–70, 272
  sesban, 90, 227, 228, 240, 250–2, 254, 261–2, 265, 268, 272, 276
Setaria sphacelata var. *sericea*, 65, 81, 112, 136, 186, 213, 224, 305–7, 347
Sida
  acuta, 210
  rhombifolia, 210
Sorghum, 135, 313
  plumosum, 78
  sudanense, 136
Sphaerostephanos unitus, 210
Stegosaurus, 229

Stenotaphrum secundatum, 186, 187, 189, 190
Stizolobium deeringianum, 278, 303–4
Stylosanthes, 56, 62, 65, 308
  capitata 45–6, 309, 348
  guianensis, 46, 62, 65, 129, 139, 162, 186, 193, 221, 283, 303–4, 306–7, 309, 320, 337–9, 341–2, 348
  guianensis var. *guianensis*, 46, 62, 65, 129, 139, 162, 186, 193, 221, 283, 303–4, 306–7, 309, 320, 337–9, 341–2, 348
  guianensis var. *intermedia*, 61,231, 308–9
  hamata, 1, 45–6, 58, 61, 85, 186, 231, 283, 293, 295, 300–4, 309–10, 312, 321, 323, 327, 334–5, 337, 343
  humilis, 58, 61, 65, 85, 286
  scabra, 5, 66, 117–8, 231, 273, 307, 309–10
Swietenia macrophylla, 211

Taenothrips sjostedi, 138
Terminalia oblongata, 226
Themeda, 78, 224
Thyridolepsis mitchelliana, 13–14
Trachypogon vestitus, 2
Trifolium, 345
  alexandrinum, 205
  subterraneum, 278
Tripsacum laxum, 38, 221
Triticum aestivum, 205

Urochloa mosambicensis, 45–6

Vigna
  mungo, 324, 327
  pilosa, 90
  radiata, 29, 84, 324, 327, 346
  umbellata, 68
  unguiculata, 235, 324–6
  vexillata, 90

Zea mays, 68
Zizyphus nummularia, 226
Zornia diphylla, 186

SUBJECT INDEX

adaptation
  climatic, 271–4, 309
  edaphic, 274, 309
  species, 3–4, 271–5, 308–10
agroforestry, 166–71
  classification, 167–8
  rationale, 169–71
allelopathy, 203–5
alley farming, 229–76
  climate, 252–8
  experience, 233–41
  negative effects, 238–41
  outputs, 229–33, 241–4
  practice, 265–75
  soil fertility, 245–52
aluminium, 27–31, 65, 72
animal production, 141–57, 328–30
  beef, 2, 114
  buffalo, 114
  draught, 114, 152–5
  goats, 147
  liveweight gain, 2–3, 17–18, 19–20, 146–8, 206–9, 216–9, 220–1, 223, 319–21
  milk, 144–6, 295–6
  objectives, 144, 150–7
  outputs, 144, 150–7
  reproduction, 2, 148–50
  sheep, 147–8, 276
  supplementation, 242–4, 262–3, 328
antinutritive factors, 263–5
assimilate distribution, 176, 210, 279–80

banana, 228
biological nitrogen fixation, see soil nitrogen

botanical composition, 5, 19–20, 124, 134–6, 293–4, 327–8
buffalo, 114
burning, 38, 46, 90–1, 226, 300

calcium, 27, 279
cashew, 221
cassava, 81–4, 244, 314–17, 325–7, 330–5, 343–4
cattle, 114
chemical residues, 21
chloride, 46
coconut, 172, 216–18
coffee, 221
commensalism, 242
competition, 68, 128, 213–15, 225–6, 242
complementarity, 128
conservation
  practice, 105–6
  soil, see erosion, soil
coumarin, 265
cowpea, 138, 325–7, 346
crop
  density, 213–15, 341–4
  management, 213–28
  practice, 43–7, 340–9
  protection, 133–40, 293
  relay, 345–7
  residues, 44–6, 114–17
  sowing time, 343–4
  species and cultivar, 330–4, 340–1, 343–4
  yield, 7–8, 13–14, 232–3, 251–8
cyanogenetic glycosides, 265

defoliation, 69–71, 209–10, 249, 267–71, 274, 279, 345
deforestation, 7, 117–19, 169, 225–7
degradation, land, 5–16
deterioration, 16–20
  botanical composition, 19–20, 134–6, 293–4, 327–8
  pasture age, 16–9
disease, 138–9, 293
diversification, 352
  income, 156, 158–165
  positive indicators, 163–5
draught, 114, 128, 152–5
dung, 120, 150–2

earthworms, 32–4, 300–1
economics, 242–4, 294–6, 329, 340
efficiency
  moisture use, 117–19
  resource use, 111–32, 141, 169–70, 324–7, 352
energy transfer, 114–17
environment
  resources, 114–17
  variability, 160–1
erosion, soil, 8–16, 92–110, 245, 275, 292–3, 350
  approaches to conservation, 11, 106–10
  consequences, 8–10
  cover, 95–9
  erodibility, 14–16
  extent, 10–11
  mechanical structures, 106–7
  processes, 92–100
  rate, 10–11
  slope, 93, 104, 227, 292–3
  splash, 92–4
  tillage, 107–8
  universal soil loss equation, 100–6
  vegetation, barriers, 108–10
  vulnerability, 14–16, 103–4
  wind, 99–100
excreta, 120, 150–2, 276

farm
  size, 163–4
  systems, 21–2, 111–12, 167–9, 355
farmer
  income, 158–65, 170–1, 242–4, 270–1, 294–6, 329
  involvement in research, 353
  needs, 156–7, 159
  women, 276, 355

fertilizer needs, 64, 212, 281, 283, 314–19, 344–5
fire, see burning
flexibility, production, 158–65
forage allowance, 5, 208–9
frost, 224

goats, 143, 211, 276
grazing management, 5–6, 203, 205–12, 279–80, 319–21, 329, 345

hedgerow, see alley farming

infiltration, 85–8, 289–90, 299
institutional policy, 357
integration
  feed resources, 2
  land resources, 112–14, 324
interception, radiation, 171–80, 330–4
intercropping, 323–4
  density, 341–4
  disease, 139–40
  energy capture, 330–4
  light, 330–4
  management, 313–15
  moisture, 334–7
  nutrients, 337–40
  pests, 137–8
  sowing time, 342–4
  variety effects, 330–4, 340–1, 343–4
  weeds, 135
interference, trees and pasture, 195–205

labour, 127–8, 270–71, 276, 328
land
  degradation, 5–16
  equivalent ratio, 324–5, 343
  resource use, 112–14
  tenure, 275
leaching, 47
leaf area index, 254, 325–6
learning, 21, 164–5, 355–7
legumes, herbage, 19–20
  adaptation, 308–10
  establishment, 310–15
  fertilizer needs, 314–19
  grazing management, 319–21
  nodules, see soil nitrogen, biological fixation
legumes, shrub, 229–76
  density, 265–7

Subject Index 413

establishment, 265–7
management, 265–71
nutritive value, 258–65
species, 271–5
legume symbiosis, 56–72
ley farming, 277–321
  crop protection, 134–5
  duration, 283–96
  economics, 294–6
  establishment, 296–315
  fertilizer needs, 314–19
  nitrogen, 277–88, 296, 300–2, 304–10, 312
  organic matter, 288–92
  pests, 135–7
  soil loss, 292
  soil structure, 278–9
  utilization, 279–80
  weeds, 293–4, 313
light
  annual crops, 330–4
  plantations, 171–95, 254–8
lime, 27, 279
litter, 25–6, 38, 45–6

magnesium, 27
maize, 68, 122, 129, 237–8, 242–4, 252, 254, 279–83, 286–8, 298–300, 303–7, 329, 334–5, 340
management skills, 165–6, 171
manganese, 72
microbial biomass, 32, 50, 55, 300–1
micronutrients, 65
millet, 254–8, 325, 347
mimosine, 265, 325
moisture
  soil, 39, 85–8, 252–8, 289, 309, 312, 334–7
  use, 117–19
mulch
  live, 301–8
  surface, 299–301
mungbean, 85, 346
mutualism, 242

nematodes, 135–7
nitrogen, 23
  associative fixation, 60
  biological fixation, *see* soil nitrogen
  cycling, 53–4
  fertilizer, 281, 283, 287–8, 305, 337–9
  soil, *see* soil nitrogen
nutrient

cycling, 17–18, 53–4, 119–27, 245–6, 351
flow, 119–27, 326–7
loss, 25, 120–1
transfer, 121–4, 150–2, 276, 279–80, 319–20
nutrition, family, 155–6
nutritive value, 5, 190–1, 231, 258–65

oil palm, 172, 219–21
organic matter, 23–46, 72, 87, 288–92
  accretion, 34–8, 246–7, 303–4
  biological activity, 31–4, 50, 55, 300–1
  components, 24
  crop production, 24
  decomposition, depletion, 24, 38–47
  humic substances, 26
  ion exchange capacity, 26–31
  light fraction, 42–3, 55
  nutrients, 24–31, 50
  priming effect, 51–3

pasture
  adaptation, 3–4, 90, 180–90, 308–10
  botanical composition, 19–20
  establishment, 310–5, 347
  management, 205–10
  nutritive value, 17–18
  persistence, 3
peanut (groundnut), 325
pests, 19, 135–8, 293
phosphorus, 5, 64–5, 66, 281, 312, 316–17, 344–5
photosynthesis, 5, 143, 175, 230
pigeon pea, 81–4, 252–3, 330–5, 343–4
pollution, 129–32, 141
  atmospheric, 10, 24, 34, 55, 72, 129–30
  soil, 72, 130–1
  stream, 10, 131–2
potassium, 27, 122–9
predation, 242
production
  animal, 141–57
  complementary, 162–3
  pasture, 35–8, 173–80
  seed, 1, 4, 176
  supplementary, 162–3
protein
  bypass, 262–3
  legume, 260–3

radiation, 171–80, 254–8, 330–4
relay cropping, 345–7
resilience, 21–2
rhizosphere, 31
rice, 12, 114, 280–1, 337, 341–3, 348
risk, 159–61
root
  density, 334–5
  growth, 29, 338
  spatial arrangement, 253–4, 336
roselle (kenaf), 314–17, 343
rubber, 172, 218–19
runoff, 94–99, 119
  antecedent moisture, 94–5
  cover, 95–9
  intensity of rainfall, 95–7

salinity, 117–19, 227
saponins, 265
seed, 1, 4, 176
senescence, 36
shrub legumes, see legumes, shrub
social benefit, 21, 111–12
sodium, 46, 86
soil erosion, see erosion
soil fertility, 4–5, 23–110, 170
  acidity, 4, 27–31, 72
  biological, 31–4, 300–1
  chemical, 24–31
  physical, 73–91
soil nitrogen, 47–72
  availability, 23, 250–2, 285
  biological fixation, 4, 50, 53, 55–72, 246–9, 284–6, 287–8, 327, 339, 344–5
  gains and losses, 50–5, 283–7
  grazing effects, 53–4, 279–80
  nitrate, 48–51, 66, 117, 286
  rhizobial specificity and ecology, 56–7, 61–2
  shade effects, 68, 190, 191–5
  transformations, 48–51
soil organic matter, see organic matter
soil structure, 73–91, 292–3
  aeration, 39–40, 88–90
  aggregation, 74–7, 85–8
  architecture, 74–6
  bulk density, 81–4
  compaction, 79, 81–4, 297–8
  crusts, 77–81
  density, 40–3

infiltration, 46, 78–81, 85–9
organic binding agents, 76–7
particle size, 40–3, 74–7
physical properties, 81–91
porosity, 81–4, 85–8
slaking of aggregates, 77
strength, 84, 90
temperature, 39, 90–1, 298–300
soil surface management, 45, 73, 296–308
sorghum, 239, 252–3, 298–300, 304, 340
soybean, 332–4, 339
stability, 21–22
stocking rate, 2, 6, 17, 114, 206–9, 211, 216–18, 220–21, 223, 224, 320
sulphur, 24–5, 282, 316
sustainability, 20–2
systems, 21–2, 111–12, 167–9, 355

tannins, 231, 251, 260–5
tea, 221
temperature, soil, see soil temperature
tillage, 43–4, 46–7, 78–80, 91, 134, 154–5, 297–301, 335–6
timber, 222–7
toxicity, mineral, 4, 27–31
tree
  clearing, 7, 117–19, 169
  crops, 166–228
  management, 213–28
  nutrient relations, 196–202, 212
  pasture interactions, 171–205, 225–6
  water relations, 202–3
  yield, 211–12

undersowing crops, 313–5, 347–9
urine, 126–7

vesicular arbuscular mycorrhiza, 32, 66–7, 312

waterlogging, 88–90, 274
weeds, 124, 134–5, 226, 241, 293–4, 313, 327–8
wheat, 287–8, 295, 345
women
  involvement in decisions, 355
  tree ownership, 276